VOLUME ONE HUNDRED AND EIGHTY NINE

# INTERNATIONAL REVIEW OF NEUROBIOLOGY

## Lights on for the Nervous System: Exploring the Effects of Photobiomodulation - Part A

# INTERNATIONAL REVIEW OF NEUROBIOLOGY

VOLUME 189

## SERIES EDITOR


**PATRICIA JANAK**

*Janak Lab, Department of Neuroscience, Dunning Hall, John Hopkins University, Baltimore, MD, USA*

**PETER JENNER**

*Division of Pharmacology and Therapeutics, GKT School of Biomedical Sciences King's College, London, UK*

VOLUME ONE HUNDRED AND EIGHTY NINE

# INTERNATIONAL REVIEW OF NEUROBIOLOGY

## Lights on for the Nervous System: Exploring the Effects of Photobiomodulation - Part A

Edited by

**JOHN MITROFANIS**

*Université Grenoble Alpes, Fonds Clinatec, Grenoble, France; Department of Nueuroscience "Rita Levi Montalcini", University of Turin; Neuroscience Institute Cavalieri Ottolenghi, Orbassano, Turin, Italy; Institute of Ophthalmology, University College London, London, United Kingdom*

ELSEVIER

ACADEMIC PRESS

An imprint of Elsevier

Academic Press is an imprint of Elsevier
125 London Wall, London, EC2Y 5AS, United Kingdom
50 Hampshire Street, 5th Floor, Cambridge, MA 02139, United States

First edition 2026



ISBN: 978-0-443-47220-6
ISSN: 0074-7742

For information on all Academic Press publications
visit our website at https://www.elsevier.com/books-and-journals

*Publisher:* Zoe Kruze
*Acquisition Editor:* Mariana Kuhl
*Editorial Project Manager:* Palash Sharma
*Production Project Manager:* Abdulla Sait
*Cover Designer:* Muthukumaran Thangaraj

Printed in England by CPI Group (UK) Ltd
Croydon CR0 4YY
United Kingdom

Typeset by MPS Limited, India

# Contents

# Contributors

**Luca Bartesaghi**
Department of Nueuroscience "Rita Levi Montalcini", University of Turin; Neuroscience Institute Cavalieri Ottolenghi, Orbassano, Turin, Italy

**Hélène Blasco**
Inserm U930, Universite Francois-Rabelais, Laboratoire de Biochimie et Biologie Moleculaire, CHRU de Tours, Tours, France

**Alain Buisson**
Grenoble Institut Neurosciences, University Grenoble Alpes, Inserm, U1216, Grenoble, France

**Corrado Cali**
Université Grenoble Alpes, Fonds Clinatec, Grenoble, France; Department of Nueuroscience "Rita Levi Montalcini", University of Turin; Neuroscience Institute Cavalieri Ottolenghi, Orbassano, Turin, Italy

**Tigrane Cantat-Moltrecht**
Université Grenoble Alpes, Fonds de Dotation Clinatec, Grenoble, France

**Philippe Corcia**
Inserm U930, Universite Francois-Rabelais, Laboratoire de Biochimie et Biologie Moleculaire, CHRU de Tours, Tours, France

**Marjorie Dole**
Université Grenoble Alpes, Fonds de Dotation Clinatec, Grenoble, France

**Francisco Gonzalez-Lima**
Departments of Psychology, Psychiatry and Behavioral Sciences, Texas Consortium in Behavioral Neuroscience, The University of Texas at Austin, Austin, TX, United States

**Luke C. Gordon**
School of Biomedical Engineering, University of Sydney, Sydney, New South Wales, Australia

**Michael R. Hamblin**
Laser Research Centre, Faculty of Health Science, University of Johannesburg, Doornfontein, South Africa

**Catherine Hamilton**
Well Red Pty Ltd, Launceston, Tasmania, Australia

**Laurent Herault**
Université Grenoble Alpes, Fonds Clinatec, Grenoble, France

**Olivier Herault**
UMR Inserm U1069, Niche, Nutrition, Cancer et métabolisme Oxydatif (N2COx) UFR de Médecine, Tours, France

**Jaimie Hoh Kam**
Université Grenoble Alpes, Fonds de Dotation Clinatec, Grenoble, France

**Daniel M. Johnstone**
School of Biomedical Sciences & Pharmacy, University of Newcastle, Callaghan, New South Wales, Australia

**Ifigeneia Kalampouka**
University of Westminster, London, United Kingdom

**Pierre Magistretti**
King Abdullah University of Science and Technology, Thuwal, Saudi Arabia

**Jose Martinez-Hernandez**
Fonds de Dotation Clinatec, University Grenoble Alpes, Grenoble, France

**John Mitrofanis**
Université Grenoble Alpes, Fonds Clinatec, Grenoble, France; Department of Nueuroscience "Rita Levi Montalcini", University of Turin; Neuroscience Institute Cavalieri Ottolenghi, Orbassano, Turin, Italy; Institute of Ophthalmology, University College London, London, United Kingdom

**Frank Nicklason**
Geriatric Medicine, Royal Hobart Hospital, Hobart, Tasmania, Australia

**Jonathan Stone**
School of Medical Science, University of Sydney, Sydney, Australia

**Audrey Valverde**
Université Grenoble Alpes, Fonds Clinatec, Grenoble, France

CHAPTER ONE

# Non-invasive brain stimulation to prevent aging-related diseases: What does photobiomodulation offer compared with other methods

**Marjorie Dole***
Université Grenoble Alpes, Fonds de Dotation Clinatec, Grenoble, France
*Corresponding author. e-mail address: marjorie.dole@cea.fr

## Contents



## Abstract

The issue of the maintenance of cognitive and brain health across ageing is a striking concern in our ageing societies. Non-invasive brain stimulation methods have been proposed as tools to fight against various neurological pathologies and to this aim,

*International Review of Neurobiology*, Volume 189
ISSN 0074-7742, https://doi.org/10.1016/bs.irn.2026.03.001

could also be used against age-related brain changes. In this chapter, we examine the potential of transcranial photobiomodulation (tPBM), in comparison with other neurostimulation methods, to improve brain function and protect neurons.

In particular, it is hypothesised that the double action of tPBM – both on neuronal function and at a more systemic level, particularly on neuroinflammation, is a particular strength of tPBM relatively to the other techniques. The main challenge faced by the tPBM research community include the question of quantifying more precisely the energy quantity reaching the targeted regions; to this aim, computational modelling of light propagation can help defining more precisely the dosing parameters. Another challenge is related to the standardisation of the stimulation protocols, and is shared by all the neurostimulation methods examined in this review.

## 1. Introduction

In our ageing world, age-related cognitive diseases and dementias are a striking concern of many people around the world. Ageing is a prominent risk factor to develop neurodegenerative diseases such as Alzheimer's or Parkinson's diseases. Cognitive health is essential in daily functioning, therefore, finding a way to promote brain resilience and fight against the deleterious effects of age is crucial to help people keeping their independence. In the last decades, considerable effort has been done to develop strategies to preserve cognitive and brain health. Non-invasive brain stimulation, *eg* transmagnetic stimulation (TMS), transcranial electric stimulation, or newer methods, such as transcranial focused ultrasound and transcranial photobiomodulation, have been proposed as tools to fight against various neurological pathologies. In this context, transcranial photobiomodulation could represent an easy-to-access, and probably safe method to prevent cognitive age-related decline. In this chapter, I will describe the cerebral changes related to age, present the main neurostimulation methods, their mechanism of action and penetration into the brain, and then, finally, describe the data available concerning their usage in brain aging research.

## 2. The aging brain

### 2.1 Age-related brain changes

It is needless to say that as we age, our cognitive performances decline, even relatively early in life, from middle age onwards (Prince et al., 2024). This cognitive decline is not a linear and homogeneous trajectory, however, and depends on the cognitive domains assessed. For example, processing speed is one of the earliest domains affected (Salthouse, 1996), as well as working memory (Park & Reuter-Lorenz, 2009) or encoding of episodic memory. Other cognitive abilities such as vocabulary or language comprehension remain stable later in life (Salthouse, 2019). It is also known that a lot of individual factors influence positively or negatively cognitive trajectories. Indeed, whereas some individuals are severely affected by the effects of age on cognition, even in the absence of underlying pathologies, others show remarkable resilience, even when affected by brain atrophy and severe neuronal loss (Stern et al., 2019). From longitudinal studies, it is known that a lot of factors – genetic, environmental, social, educational, as well as health risk factors – alcohol consumption, smoking, lack of physical activities, diabetes, cardiovascular risk – affect cognitive trajectories and play a role in brain resilience. The complex interplay between all these factors in the development or acceleration of cognitive disabilities is not totally understood but is conceptualised in several notions to describe brain resilience, such as cognitive reserve (*ie*, preservation of cognitive abilities despite age-related brain alterations due to cognitive resources during adulthood; Stern et al., 2020; Stern et al., 2023); brain reserve (the neurobiological capital defined as the number of neurons, synapses, etc.), brain maintenance (preservation of brain structures despite age-related insults, due to genetics and lifestyle; Stern et al., 2020) or the brain age gap (the deviation between the estimated age of the brain and the real chronological age; Beck et al., 2022; Yi et al., 2025; Marseglia et al., 2025).

Biologically, cognitive decline is related to changes in brain function and morphology. Brain atrophies occur with growing age and both grey and white matter are concerned. The frontal cortex, middle temporal and parietal areas are the most affected cortical regions and these regional atrophies are correlated with performances in various cognitive tests such as memory, executive functions, language or visuo-spatial abilities (Yang et al., 2024). These atrophies are paralleled with increased volumes of cerebrospinal fluid and enlargement of the ventricles (MacDonald & Pike, 2021). This is due to decreased neurogenesis, mainly in brain regions involved in memory and executive functions, together with an increase in cell death. These alterations of brain volumes are often associated with the presence of microbleeds and white matter hyperintensities (MacDonald & Pike, 2021), which are considered as a manifestation of small vessel disease linked to neuroinflammation. White matter hyperintensities severity has been closely associated to cognitive decline and dementia (Wiseman et al., 2017; Wang et al., 2022).

These vascular changes come along with alterations of cerebral perfusion, particularly in grey matter. There are also alterations of the neurovascular coupling (Fabiani et al., 2014; D'Esposito et al., 2003) associated with impaired hemodynamic responses (Mukli et al., 2023). Indeed, number of studies show decreased activations in task-related fMRI studies, for example in parietal areas during a working memory task (Archer et al., 2018) or medial temporal lobe during episodic memory encoding or retrieval (Tromp et al., 2015). Quite remarkably however, there is also increased activity in other areas (mainly homologous contralateral or frontal areas), generally interpreted as compensation or dedifferentiation processes (Cabeza, 2002; Davis et al., 2008). For example, when executing a motor movement, the aged brain generally shows a more widespread pattern of activation. These functional changes affect also the communication between different parts of the brain resulting in changes in functional connectivity. For example, increased connectivity between the sensorimotor network and basal ganglia have been found (Seidler et al., 2015), as well as reduced connectivity in the default mode network or the central executive network (Damoiseaux et al., 2008; Zonneveld et al., 2019). More generally, older individuals exhibit increased between-network and decreased within-network functional connectivity, as well as increased integration and reduced local efficiency in those networks (Deery et al., 2023; Mukli et al., 2023).

At a more basic level, these changes are thought to arise from a variety of synaptic and cellular modifications. With aging, the neurotransmission in the cholinergic and dopaminergic system decreases. Concerning the dopaminergic system, decreased levels of dopamine, reduced expression of dopaminergic receptors D1 and D2 as well as dopamine transmitters in the cortex and striatum are found in aged subjects (Seidler et al., 2010; Tromp et al., 2015). Similar changes are found in the cholinergic system. Neurons are sensitive to age-induced oxidative stress, and alterations of the mitochondrial function appear, as well as impaired mitochondrial recycling (Picca et al., 2023). As mitochondria is the principal provider of cell energy via oxidative respiration, using the glucose and producing adenosine triphosphate (ATP), altered mitochondrial function results in an alteration of cell metabolism, particularly glucose metabolism which is a hallmark of neurodegenerative diseases but also healthy aging. In line with this, impaired glucose metabolism is found in various regions of the aging brain, particularly frontal and temporal regions (Deery et al., 2022).

## 2.2 How can brain stimulation help to fight against these dysfunctions?

In this context, finding strategies to help the brain increase resilience and boost compensatory mechanisms is a crucial issue. The first therapeutical options are pharmacological treatments that are commonly used for treating neurodegenerative diseases, but they can cause significant side effects and their efficacy remain limited. One possible reason is that biological and brain alterations often appear decades before the onset of cognitive disabilities and the diagnosis rendering the treatment 'too late to be effective'. This has prompted researchers and clinicians to work on prevention and focus their efforts on the preclinical stages of the diseases or in the development of lifestyle prevention programmes, such as the control of cardio-vascular risk factors, nutrition programs, or cognitive trainings (Beck et al., 2022; Ngandu et al., 2015). Anti-aging drugs have also been tested, such as the antidiabetics Metformin, Rapamycin, probiotics targeting gut microbiota, or anti-inflammatory drugs (Li et al., 2021).

Another promising option is provided by non-invasive brain stimulation. The general idea behind these methods is to modulate cortical excitability/activity and induce brain plasticity. The most popular methods are without a doubt transmagnetic stimulation (TMS) and transcranial electric stimulation (which can be subdivided in tDCS, transcranial Direct Current Stimulation and tACS, transcranial Alternative Current Stimulation). More recently, new techniques have emerged such as transcranial photobiomodulation and transcranial focused ultrasounds. Fig. 1 presents a schematic view of these four techniques.

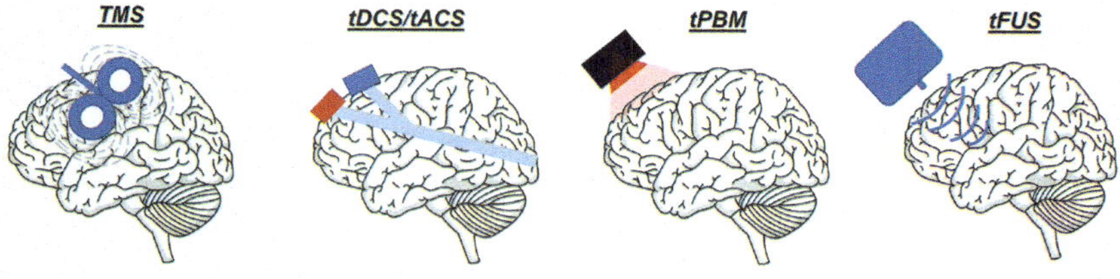


**Fig. 1** Schematic illustration of the four methods discussed in this Chapter. TMS: transmagnetic stimulation. tDCS: transcranial Direct Current Stimulation. tACS: transcranial Alternating Current Stimulation. tPBM: transcranial photobiomodulation. tFUS: transcranial focused ultrasound.

## 3. Description of the methods and mechanisms of action

### 3.1 Transmagnetic stimulation (TMS)

TMS relies on the application of a magnetic field at the surface of the scalp, that generates an electrical field that modulates intracortical excitability in the targeted region (Pascual-Leone et al., 1994; Valero-Cabré et al., 2017; Bhattacharya et al., 2022). A TMS pulse results from a rapidly changing high intensity current circulating in a coil, that produces a magnetic field transmitted to the scalp that propagates into the brain. The equipment consists in a coil containing a loop of copper wire to generate the magnetic field, a central unit to generate the current, and capacitors to accumulate high loads of electrical charges. Several types of coils have been designed, allowing different profiles of penetration into the brain: flat circular coils ensure high penetration power but lack spatial selectivity, whereas figure-of-eight coils, composed of two wires that transmit the current in opposite directions, ensure the maximal focus at the junction between the two, resulting in a higher spatial selectivity but weaker magnetic field. There are also H-shape coils to stimulate deeper brain regions.

Since its introduction in 1985, a variety of TMS protocols have been developed, producing different biological effects. The simplest one, single-pulse TMS, is widely used in clinics to assess cortical excitability for diagnostic purposes, mainly to investigate the cortico-spinal tract through the recording of motor evoked potentials, at rest or when active, as well as the recruitment curves (higher motor evoked potentials amplitude with increasing stimulation intensity). It can also be used to investigate silent periods, that is, the resting period in the electromyogram after a motor evoked potential has been generated reflecting cortical inhibition (Hupfeld et al., 2020). Paired-pulse TMS is used to investigate facilitatory or inhibitory cortical circuits, such as short-interval cortical inhibition or intracortical facilitation circuits (Ansari & Tremblay, 2019). During intracortical facilitation, a first weak stimulation is applied which activates inhibitory interneurons; it is followed, after a short interval, by a second stimulation. The stimulation of excitatory neurons produced by the second pulse is therefore reduced because of the previous activation of interneurons. This represents a short-latency inhibitory effect. During short-interval cortical inhibition, the interval inter-burst is increased, that results in activation of glutamatergic excitatory synapses, representing a facilitatory effect (Chen et al., 1998).

Besides these immediate effects, TMS is also used repetitively (repetitive TMS) to obtain long-lasting effects. High-frequency repetitive TMS

(>5 kHz) results in an increased cortical excitability, potentially through long-term potentiation, whereas low-frequency repetitive TMS (<1hz) results in decreased neural excitability through long-term depression (Bhattacharya et al., 2022; Valero-Cabré et al., 2017). Repetitive TMS is also widely used with a theta-burst stimulation protocol; three bursts at high frequency (50 Hz) are applied, with an interval inter-burst of 200 ms (*ie* 5 Hz, corresponding to the theta frequency). Theta burst stimulation can also be applied in an intermittent manner, repeating theta bursts for 2s-long trains followed by an 8 s pause. Intermittent theta-burst stimulation in the left dorsolateral prefrontal cortex has proven its efficacy in resistant depression, with improvements in depressive symptoms (Kishi et al., 2024) and increasing functional connectivity between the left dorsolateral prefrontal cortex and insula (Struckmann et al., 2022).

From a mechanistic perspective, theta burst induced plasticity is thought to go through a long-term potentiation/depression like effect by acting on the glutamatergic system. Indeed, repetitive TMS increases glutamate transmission, modulates expression of GLUR1 subunits of glutamatergic AMPA and NMDA receptors and increases calcium ions. As evidence, a blockage of NMDA receptors suppresses repetitive TMS effects (Huang et al., 2007). Furthermore, plasticity of inhibitory synapses has also been showed, with a reduction of GABAergic synaptic transmission in pyramidal neurons (Lenz et al., 2016).

Regarding safety aspects, TMS is well tolerated, with over 40 years' of use. Recommendations and standard protocols are well-defined and regularly updated by the International Federation of Clinical Neurophysiology (Rossi et al., 2009; Rossi et al., 2021). The major safety issue concerns the risks of seizure, which are considered low if standard protocols are used (Rossi et al., 2009). Minor side effects are related to tissue heating, hearing safety concerns limited by the use of earplugs, headaches, minor pains, discomfort, or anxiety linked to the procedure.

The major limitation of TMS is the focalisation specificity that has been improved by the development of neuronavigation systems and computational modelling of the TMS-induced electric field to improve coil placement and orientation (Rossi et al., 2021; Thielscher et al., 2015). Another limitation is the reduced propagation toward deep regions, even if recent developments render it possible for deeper stimulation, at the cost of a reduced focality and possible coactivation of more superficial regions (Gomez-Tames et al., 2020).

## 3.2 Transcranial electrical stimulation

Transcranial electric stimulation is the second family of methods used to stimulate non-invasively the brain. It relies on the application of a weak current at the surface of the scalp, by the mean of two or more electrodes. This current can be either continuous in the case of transcranial Direct Current Stimulation (tDCS) or alternating with a sinusoidal shape in the case of transcranial Alternating Current Stimulation (tACS). A standard montage typically includes at least two electrodes (one anode, one cathode) but more sophisticated montages exist. Electrodes delivering the current are fixed on the head with a headgear, and are immersed in an electrolyte solution containing ion charges (Bikson et al., 2016). In two-electrodes montages, the current enters the head via the anode and exits via the cathode. As for repetitive transmagnetic stimulation, tDCS effects on synaptic plasticity would rely on long-term potentiation/depression through modulation of the excitatory glutamatergic and inhibitory GABAergic synapses (Cavaleiro et al., 2020). Anodal tDCS is thought to increase neural excitability, whereas cathodal tDCS reduces it through hyperpolisation of the neuronal membrane potential (Battacharya et al., 2020). Blockage of NMDA receptors abolish tDCS-induced responses (Nitsche et al., 2003). Both transcranial electric stimulation methods use a limited current amplitude and do not trigger directly action potentials. A typical session implies a 1–2 mA current and the session duration is several seconds to 40 min.

tACS shares the same principles as tDCS, but the applied current oscillates sinusoidally with a given frequency instead of being constant. This oscillation is thought to entrain cortical oscillations. As neural oscillations reflect the fine balance between excitatory and inhibitory circuits of synchronised neuron populations, supporting integration between distant regions, stimulating the brain at these frequencies would facilitate integration processes (Wischnewski et al., 2023). At the cellular level, tACS is showed to modulate the local spike timing, causing a local neural entrainment (Krause et al., 2019). Stimulating the brain in the theta frequency, for example, increases theta power spectrum; these effects would also affect cognitive performances (Aktürk et al., 2022). tACS would modulate functional connectivity, but the characterisation of its effects of global networks still warrants more research to better understand its consequences (Carrasco-GÓmez et al., 2025; Kaiser et al., 2025).

More recently, a new promising method have emerged, temporal interference stimulation. Also called non-invasive deep brain stimulation, this technique has been developed to stimulate deep regions with good spatial resolution, initially to target subcortical structures in movement disorders (Wessel et al., 2023). It consists in two pairs of electrodes, each delivering alternating high frequencies, with a slight frequency shift between the two. In the targeted area, these two alternating frequencies superpose, stimulating the region with a frequency corresponding to the difference between the two (for example, for a 2000 Hz and 2010 Hz, the resulting frequency is 10 Hz). This method may provide interesting alternative to invasive deep brain stimulation, and has been successfully tested in healthy volunteers (Wessel et al., 2023), in a pilot study targeting Parkinson's Diseases (Yang et al., 2025), and several clinical trials are currently ongoing for other applications (NCT06467422, NCT06931483, NCT06851156). As an emerging method, a refinement of the protocols, fields of application, and safety assessments, are still needed.

Apart from temporal interference stimulation, the limitations of transcranial electric stimulation are roughly identical as TMS: the focus of the stimulation and the difficulties in targeting deep regions, inherent of the application of a weak current mainly absorbed by the scalp and surrounding tissues. The biological underlying mechanisms are not entirely clear and subject to controversy. There is a high heterogeneity in the protocols used (session number and duration, frequency, intensities, etc) that compromise results consistency. To overcome these limitations, effort has been done recently to optimise the protocols through the use of individual morphology (anatomical MRI) and electric field modelling, which will be of particular interest for clinical translation.

Safety issues are almost inexistant, at least for studies respecting standardised protocols. In human clinical studies (>33.000 sessions reported in Bikson et al., 2016), no serious adverse events have been observed. Adverse events, if any, that have been reported are mild skin erythema or burning sensation at the electrode contact, headaches, fatigue, and are generally resolutive (Antal et al., 2017). Systematic assessment of adverse events on biological parameters in the long-term is still lacking, but Food and Drug Administration considers transcranial electric stimulation as low-risk (Bikson et al., 2016).

## 3.3 Transcranial photobiomodulation

While the previous methods act directly on neural activity by changing neural excitability or synaptic plasticity, transcranial photobiomodulation appears to act more indirectly by improving energy production and inducing neuroprotection mechanisms. The underlying mechanism of action is therefore completely different; light is thought to activate mitochondrial respiration in the neural cells through the activation of a chromophore located in the mitochondria, cytochrome-C-oxidase, that would catalyse enzymatic reaction, ending in (1) a liberation of nitric oxide, a vasodilator resulting in the increase of regional cerebral blood flow; (2) increased ATP production, therefore, metabolic energy in the cell (3) expression of various neuroprotective and anti-inflammatory factors. In animal studies, near infrared light has been showed to change neuronal morphology (Cardoso et al., 2022), improve cell surviving in MPTP-induced models of Parkinson's Disease (Moro et al., 2013; Shaw et al., 2012), enhance the expression of synaptic proteins (Hosseini et al., 2022), reduce the production of pro-inflammatory cytokines (Cardoso et al., 2021; Hosseini et al., 2022) and glial activation (Xie et al., 2023) and reduce stress-induced reactive oxygen species and nitric oxide production in the brain (Meynaghizadeh-Zargar et al., 2020). Normalisation of mitochondrial membrane potential and ATP production is shown in a hypoxia model of mice (Pchelin et al., 2023). The effect of photobiomodulation on the neural networks are not entirely understood, but light, potentially through the liberation of nitric oxide and alteration of regional cerebral blood flow, could increase oxy-haemoglobin/deoxyhaemoglobin ratio in the stimulated region (Saucedo et al., 2021), potentially leading to alteration of the neurovascular coupling (Shahdadian et al., 2024) and change in functional connectivity (Yang et al., 2025; Dmochowski et al., 2020). As a consequence, alterations of cortical oscillations have been consistently showed (Wang et al., 2021; Truong et al., 2023; Wang et al., 2022). These encouraging results drive a hope for designing clinical studies aiming at testing transcranial photobiomodulation in a variety of clinical populations, such as neurodegenerative diseases (Blivet et al., 2022), traumatic brain injuries (Figueiro Longo et al., 2020) or other applications.

The advantages of transcranial photobiomodulation are its simplicity of use, making it possible to develop devices accessible at home. At present, safety issues are almost absent, even after long-term use (>twelve months; Boyer et al., 2024), and no major side effects have been reported. Systematic and large-scale safety data are still lacking, however. In a recent clinical study, (Coelho et al., 2025) examine the dose-dependent safety of a weekly treatment with 808 nm light using three different energy densities ($60\,J/cm^2$, $100\,J/cm^2$, $180\,J/cm^2$). The investigation of the frequency of side effect using the SAFTEE-SI scale, result in no change after treatment on the overall scores nor on systolic and diastolic blood pressure. When reported, side effects are rather weak. Recently a study has examined the effect of a single 1064-nm photobiomodulation session delivered with an energy density of 80 and 180 J/cm$^2$ on blood parameters, structural MRI, and subjective questionnaires (Li et al., 2024). They report no deleterious changes in any of these parameters. However, the single-session design in this study makes impossible drawing conclusions about the long-term safety of photobiomodulation.

An important issue of transcranial photobiomodulation is the question of light propagation throughout the head tissues, rendering difficult to establish precise dosing and limiting the ability of stimulating deep regions of the brain (Dole et al., 2024; Cassano et al., 2019). It is generally thought that light can reach superficial layers of the cortex, 2–4 cm under the scalp surface. Another problem is the high heterogeneity in the parameters used in the different studies, rendering it difficult to establish a comprehensive map of the brain changes elicited by photobiomodulation (Dole et al., 2023). As well, the high prevalence of case studies and limited sample sizes in the current literature limit the generability of the findings and renders necessary the development of carefully controlled trials in large samples of patients.

## 3.4 Transcranial focused ultrasound

Transcranial Focused Ultrasound is a new technique that uses low-intensity ultrasonic waves applied on the scalp by the mean of transducers to induce neuromodulation. This method is therefore distinct to diagnostic ultrasound. Due to its recent development, clinical data are relatively scarce and the mechanisms of action are not entirely understood (Darmani et al., 2022; Blackmore et al., 2019). Contrary to other ultrasound applications (such as blood-brain barrier opening or tumour ablation), in neuromodulation, ultrasounds are applied with low intensity. It consists in an acoustic wave with a frequency >20 kHz, generating a mechanical pressure wave front that propagates through the skull to the brain. A focal transducer or phase-correcting lenses are generally used to reduce the focus toward the targeted region. Indeed, ultrasounds propagate well in the soft tissues and are minimally affected by brain geometry, but the bone tends to defocus the waves. Nevertheless, the strength of transcranial focused ultrasound is a good spatial resolution (≈3 mm at 0.5 MHz; Blackmore et al., 2019). To avoid detrimental thermal effects, ultrasounds are generally applied as bursts; as a consequence, dosage can be defined in different time-scales, at the level of the pulse or the level of the burst, complexifying the study of parameters.

In animal studies, both excitatory or inhibitory effects are found, depending on the pulse regime. In the motor cortex, ultrasounds increase motor evoked potential thresholds, in a frequency dependent manner (Tufail et al., 2010). Studies investigating the role of other stimulation parameters have drawn inconclusive results. Perhaps even more than for other methods, the complex interplay between multiple stimulation parameters renders difficult the systematic study of each and their interaction. Interpretations are complexified by the fact that mechanisms of action have not been fully elucidated. Three hypotheses are proposed (Blackmore et al., 2019; Darmani et al., 2022): (1) ultrasound wave exerts a mechanical pressure on cell membranes, changing membrane capacitance and resulting in the generation of nerve impulse or activating mechanosensitive ion channels. Ultrasound effects have for example been showed to mediated by the channel Piezo1 (Zhu et al., 2023). (2) mechanism of cavitation and formation of bubbles into the lipid membrane which changes the structure and potential of lipidic membranes (3) thermal effects, due to the ultrasound absorption. Ultrasound stimulation induces heating, which could lead to thermal neurostimulation, as temperature increases are found to have an effect of neural activity.

Given the novelty of the methods, clinical research on transcranial focused ultrasound is still in its infancy. Applied to the primary somatosensory cortex, it results in somatosensory sensations in the hand and associated evoked potentials (Lee et al., 2015). Similarly, phosphenes are also induced by stimulation in the visual cortex (Lee et al., 2016). Modulation of cortical oscillatory activity have been reported (Samuel et al., 2022). It could also inhibit the amplitude of TMS-induced motor evoked potentials and attenuate intracortical facilitation, suggesting neuronal inhibition (Legon et al., 2018). Theta-burst stimulation protocols have also been used (Samuel et al., 2022; Xia et al., 2024) and suggest its ability to change motor cortex excitability and local connectivity.

Clinically, the most important safety issues are mainly related to thermal and mechanical bioeffects. Mechanical bioeffects are linked to cavitation, which can result in bleedings, cell damage or opening of the blood brain barrier, rendering mandatory the careful following of recommendations, particularly in clinical populations, and application of the stimulation by qualified operators (Beisteiner et al., 2024; Murphy et al., 2025). However, the use of low-intensity stimulation, renders it sure for use in animal and human studies and at that date, no serious adverse events have been noted.

A recent review of 35 human studies reveals no serious adverse events; mild adverse events include anxiety, headaches, heating, neck pain or mood deterioration (Sarica et al., 2022).

The limitations of transcranial ultrasounds are linked to the novelty of the method: high heterogeneity in the results and protocols; the mechanisms of action are still unclear. Another limitation is related to auditory costimulation, even though not audible by human, but that can influence some results, particularly in animal studies (Darmani et al., 2022).

## 4. Reach the target: How does stimulation propagate into the brain

### 4.1 Modelling electric fields in TMS and electric stimulation

The idea of optimising target localisation and stimulation parameters using computational modelling has recently taken importance for all neuro-stimulation techniques. In the case of trans-magnetic/electric simulation, computational simulation can take into account individual differences in brain anatomy influencing the distribution of electric fields. It allows to precisely determine the peak intensity, its anatomical location and the field decay, as function of the conductive capacities of the tissues and the localization of the coils/electrodes, intensity, or other stimulation parameters (Laakso et al., 2018; Thielscher et al., 2015). Fig. 2 illustrates examples of computational modelling for tDCS and TMS stimulation.

Concerning TMS-induced electric field modelling, the mostly common method is the Finite Element Method, but others have also been used, such as the Boundary Element Method and Finite Differences Method (Gomez et al., 2020). Specifically, the procedure requires a model of the subject's head (typically issued from MRI) segmented into its different tissues (scalp, skull, grey/white matter, cerebrospinal fluid, air). Each tissue has its own conductivity parameters, which are entered in the model. A model of the TMS coil with its parameters and localization relatively to the subject's head is added to the model. The time derivative of the magnetic vector potential A, (δA/δt), generated by the coil is calculated using the Biot-Savart law, then, the induced field is calculated as:

$$E = -\nabla\varphi - (\delta A/\delta t)$$

where $\varphi$ is the electric potential.

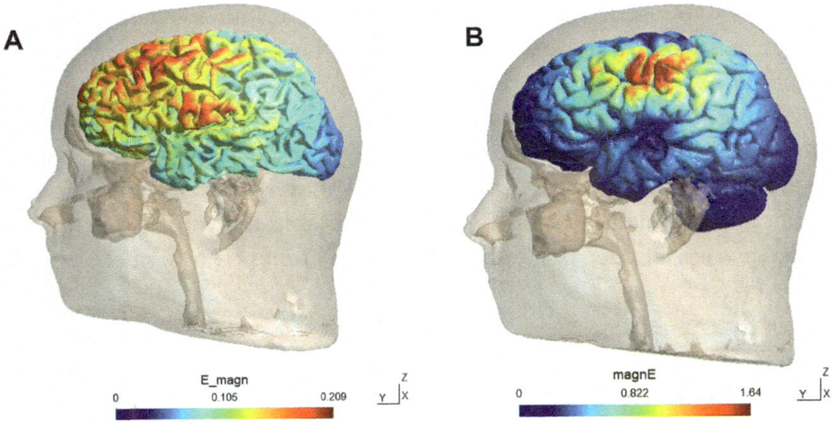


**Fig. 2** (A) Example of a simulation result obtained using SimNIBS example set, showing the magnitude of the electric field (V/m) after a tDCS stimulation with the anode located at C3. (B) Example of a simulation result showing the magnitude of the electric field (V/m) after a TMS stimulation located at C3.

In the case of tDCS, the Laplace equation is resolved (Thielscher et al., 2015):

$$E = -\nabla\varphi$$

Several open-source softwares have been developed allowing researchers and clinicians to simulate electric fields using various coils and electrodes montages. Among those, the most used are SimNIBS (Thielscher et al., 2015) or ROAST (Huang et al., 2019). They allow optimising the placement of the coils/electrodes and the stimulation parameters to obtain the maximal peak intensity at the target. Another application can be to investigate electric field distribution under different pathological conditions or in the elderliness. For example, Alawi et al., (2023) use MRI images from participants aged 4–84 years and show that age has a substantial negative effect on electric fields induced by TMS. Recently, advanced modelling techniques have been developed to solve the computation in real-time, allowing for optimisation of parameters during clinical or experimental settings.

## 4.2 Modelling light propagation in photobiomodulation

The question of dosage is a crucial and yet unclear issue in transcranial photobiomodulation. The modelling of light propagation across brain tissues relies mainly on the Monte Carlo simulation which is a stochastic solver for the radiative transfer equation. In Monte Carlo simulation, a packet of photons is launched from one (or more) light sources, and travels from interaction site to interaction site. At each interaction site, a fraction of the photon packet is absorbed according the properties of the given tissue (determined by the absorption coefficient $\mu_a$), and the weight of the photon packet is updated accordingly. Then the photons travel to the next interaction site, with a scattering angle dependent on the scattering coefficient $\mu_s$ and the anisotropy factor $g$ of the tissue. The coordinates and the weight of the photons are updated, and so on, until all photons are absorbed (Prahl, 1989).

Practically, the procedure starts from a segmented MRI image. Each element (voxel or mesh element) is labelled (white/grey matter, blood, cerebrospinal fluid, skull, air or scalp) and given the known optical properties of each tissue type ($\mu_a$, $\mu_s$, $g$ and refractive index $n$) the Monte Carlo computation is calculated (Fig. 3). Several studies have used this method to quantify the energy reaching grey matter from an external source (Cassano et al., 2019; Dole et al., 2024). Simulation softwares include MCX suite and its derivative (Fang, 2010), FullMonte (Cassidy et al., 2018), or ValoMC (Leino et al., 2019). Using a voxel-based version of this technique, Cassano et al. (2019) have investigated the deposition of energy in the dorsolateral and ventromedial prefrontal regions for a light source positioned over F3-F4 and Fp1-Fp2 regions. They show an energy deposition from $\approx$ 1.27E-4 J/ $cm^3$ to $\approx$ 2.34E-4 J/$cm^3$ per administered joule in both prefrontal regions. The authors concluded that, to simulate effectively the underlying regions, duration of session should be set at 30 min for output power 30 mW/$cm^2$ at the scalp level. Extending this work, we have provided a parcellation of the whole brain and simulated a multisource helmet with 670 nm and 810 nm wavelengths (Dole et al., 2024). Here, the results show that the main location of energy deposition is the frontal cortex, with better penetration for 810 nm wavelength. Using a single source located over frontal regions, Yuan et al. (2020) use mesh-based simulation from infancy to elderliness, and find that energy deposition decreases slightly with increasing age. This confirms the importance of developing individual modelling to individualise photobiomodulation treatments to improve its efficacy.

## 4.3 Modelling ultrasound propagation in transcranial ultrasounds

Similarly, computational modelling of transcranial ultrasounds propagation allows researchers taking into account the attenuation, scattering and defocusing of the ultrasound front wave, mainly created by the skull that

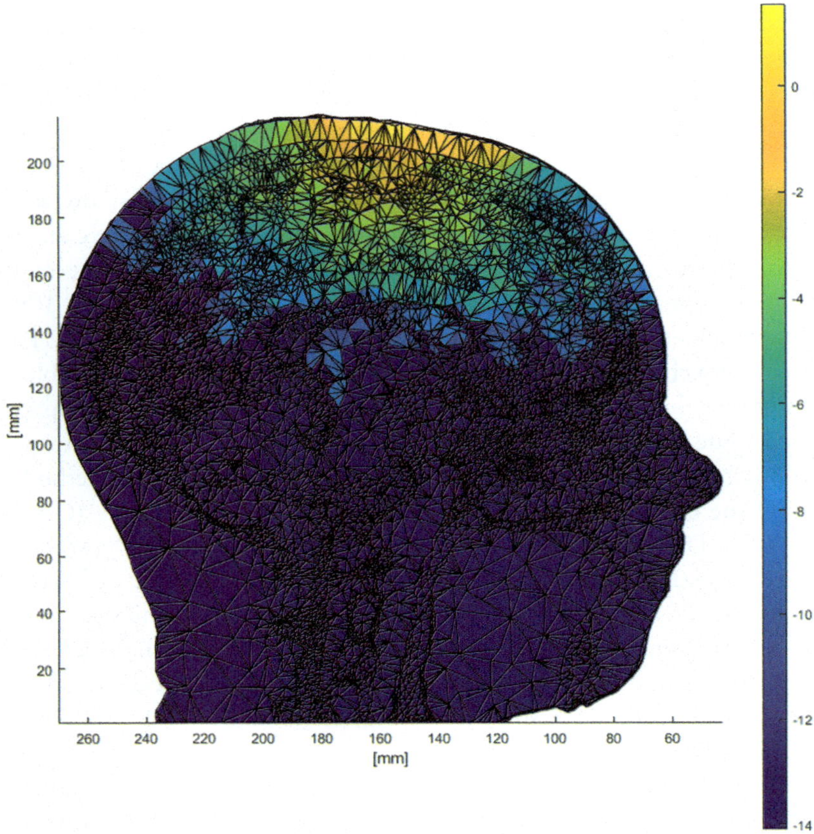


**Fig. 3** Results of a mesh-based simulation of light propagation using a single 810-nm light source located over the head. Energy density in log scale (J/mm$^3$).

shifts the focus away from the target. This is done using a model of a skull obtained from a CT scan of the subjects (Mueller et al., 2017). In this model, acoustic properties of the different tissues, such as sound speed in each tissue (m/s), tissue density (kg/m$^3$) or attenuation coefficient (Np/MHz/m) are calculated using CT Hounsfield Units and the porosity (Chen et al., 2023). Then, different simulation methods can be used, k-space (Chen et al., 2023; Mueller et al., 2017), Finite Element Method or Finite-Difference Time Domain (Yoon et al., 2018). Studies using these methods show a ratio of transmitted intensity of about 25 % (Chen et al., 2023) and confirm that deep regions can be targeted (Samoudi et al., 2019). Parameters tested can be transducer positioning, ultrasound frequency, intensity, aperture size, pulse frequency or pulse duration. Chen et al. (2023), for example, using this method, have found a transmission efficacy of about 20–30 % at 150 kHz, decreasing exponentially with increased frequency, and that higher frequencies produce narrowed bandwidth. Recently, Park et al. (2023) have proposed accelerated acoustic simulation using deep learning coupled to a navigation system, predicting the acoustic field in real time with good accuracy.

## 5. Prevent aged-related diseases using non invasive brain stimulation: State of the art

Non-invasive brain stimulation has been used in many pathological conditions to stimulate brain function and induce synaptic changes. It could also be used in a more preventive way in aging in an attempt to offset the various age-related changes. In the following section, we will focus our discussion on the findings of each method in the context of healthy aging, and end with a discussion about the specific advantages/limitations of transcranial photobiomodulation relatively to the other methods.

### 5.1 Age-related studies using (repetitive) TMS

Preclinical and clinical studies investigating various (repetitive) TMS protocols in healthy aging reveal interesting benefits. In a D-galactose induced animal model of ageing, increased performances in a spatial memory test and partial preservation of the hippocampal neurons against the toxin are found using a theta-gamma frequency stimulation (Wang et al., 2023). In human, TMS-EEG recordings or TMS together with motor evoked potentials are widely used to investigate age-related changes in cortical excitability, plasticity, and connectivity (Kallioniemi et al., 2022; Guerra et al., 2021). Generally, these studies indicate an altered excitability of the cortico-spinal tract and reduced cortical excitability in the elderly (Oliviero et al., 2006), presumably via an increased inhibition due to GABA system potentiation (Opie et al., 2018). Long-range connectivity shows no difference between young and aged subjects at rest, but a decreased inhibition of the ipsilateral primary motor cortex by the contralateral one during a motor task (Talelli et al., 2008).

TMS is also used to induce changes in neural excitability, modify brain function and induce cognitive improvements. To this end, repetitive TMS protocols are used for their action on synaptic plasticity, including long-term potentiation/depression mechanisms. For example, an intermittent theta-burst stimulation protocol over the left inferior frontal gyrus in 24 healthy older adults (single-session) fails to induce improvement in a memory task, but increases functional connectivity between inferior frontal gyrus and cerebellum-occipital cortex (Vidal-Piñeiro et al., 2014). A repetitive paired-pulse paradigm increases visuo-motor skill acquisition and excitability in the primary motor cortex, in both young and aged healthy participants (Hand et al., 2023). For therapeutic purpose, repetitive TMS has also been applied to treat geriatric depression: over the left or bilateral dorsolateral prefrontal cortex (hypoactive in depression), TMS at 10 Hz or with a theta burst stimulation protocol reduces depressive symptoms (Cappon et al., 2022). In aged subjects with mild cognitive impairments or Alzheimer's Disease, a recent meta-analysis also concludes on positive effects of repetitive TMS on memory, language, or more general cognitive abilities. Repetitive TMS is generally applied over the dorsolateral prefrontal cortex, inferior frontal gyrus, or temporal regions, and high-frequency TMS (>5 Hz) on the left dorsolateral prefrontal cortex and low-frequency TMS (<1 kHz) on the right dorsolateral prefrontal cortex can increase memory function whereas high frequency TMS over the right inferior frontal gyrus improves executive functions (Chou et al., 2020). The effects of 5–30 sessions could persist until 12 weeks and the reported effect sizes concerning memory are moderate to large (Koch et al., 2024). In neural networks, 4-weeks 10 Hz repetitive TMS targeting the left dorsolateral prefrontal cortex increases functional connectivity in the fronto-parietal network in patients with mild cognitive impairments (Esposito et al., 2022). Furthermore, a promising approach is currently developing through the emergence of personalised stimulation and multi-site targeting. In (Taylor et al., 2025), for example, resting-state functional connectivity maps were taken before intervention, to help defining the optimal target sites. This ensures better localisation of the target and better efficiency.

## 5.2 Age-related studies with transcranial electric stimulation

There are very few studies on the effect of tDCS/tACS on animal models of aging. One study (Sun et al.,2024) investigating the effects of anodal tDCS and 10 Hz tACS in a model of Alzheimer's Disease and healthy aged mice, shows that tDCS and tACS change EEG spectral powers in both groups, mainly in alpha, theta and gamma frequency bands. In human, transcranial electric stimulation is mainly used to improve cognitive functions, particularly memory abilities. One systematic review (Bjekić et al., 2023) compares the beneficial effects of different methods to improve associative memory. It shows that the efficiency depends on the localisation of the stimulation (targeting the precuneus results in better improvement than the left dorsolateral prefrontal cortex) and the stimulation method (tDCS would be superior than tACS). However, the overall picture does not look so clear, and mixed results are found in other memory domains (Goldthorpe et al., 2020). In particular, meta-analyses or systematic reviews generally highlight the high degree of heterogeneity between the methods, rendering it difficult for a comparative analysis between the studies (Goldthorpe et al., 2020; Bjekić et al., 2023).

tDCS has been also used in a variety of other cognitive or motor domains. For example, anodal tDCS applied over the left primary motor cortex increases cortical excitability, with prolonged effects for higher intensities (Farnad et al., 2021). However it has no additional effect on the learning of finger tapping sequences complementary to a motor imaging practice (Metais et al., 2022). Anodal tDCS applied to the left dorsolateral prefrontal cortex is found to increase verbal episodic memory (Sandrini et al., 2019). In patients with mild cognitive impairments, encouraging results are found in word retrieval or recognition tasks (Meinzer et al., 2015). In 201 patients with cognitive impairments due to Alzheimer's Disease, 12 sessions of anodal tDCS on temporal areas combined with a cognitive training show positive effects at 4 weeks in various memory tasks (Lu et al., 2019). These benefits on cognitive abilities are paralleled by enhanced cerebral activation in the stimulated areas, and improvement of functional connectivity (Meinzer et al., 2013; 2015; Iordan et al., 2022). In general, the effects are potentialised with repeating the sessions (Dumel et al., 2016), and the effect sizes for offline effects would be higher than for online effects (Hsu et al., 2015).

Relatively to tDCS, few studies have investigated the effects of tACS on cognitive abilities of the elders. Researchers typically use tACS to entrain cortical oscillations in the brain, improving communication within and between brain networks. For example, high density tACS applied in the beta-frequency range (20 Hz) over the left motor cortex during a hand grip task reduces movement-related beta desynchronization and delays resting beta power (Morales-Fajardo et al., 2024). Theta-gamma tACS is also shown to increase left primary motor cortex excitability in both young and older subjects, and improve motor performances to a greater extend in young than aged participants (Gamage et al., 2025). Improvements have been found in various cognitive domains including visual attention, working memory, long-term memory, executive control, fluid intelligence or motor learning (Grover et al., 2023; Reinhart & Nguyen, 2019).

Notwithstanding these positive results, the effects of tDCS and tACS are still matter of debate, particularly concerning the transition from basic research to clinics. Indeed, a number of studies fail to show such beneficial effects (Huo et al., 2018; Klees-Themens & ThÉoret, 2023). As already stated, one reason for this inconstancy probably relies in a lack of standardisation of the protocols. There is also probably a high interindividual variability in the responses, which should be improved by optimisation of the protocols thanks to computational modelling of E-field and use of individual morphological imaging.

## 5.3 Age-related studies using transcranial photobiomodulation

In animal models of aging, photobiomodulation is used for its neuroprotective effects and its ability to stimulate neural metabolism. The preclinical studies have showed that 810-nm light with 15J of total energy for 58 consecutive days reduces the expression of cortical of interleukin 5 and increased the expression of TNF-alpha, interleukin-6 and interleukin-10 suggesting a possible anti-inflammatory effect; it also improves spatial memory (Cardoso et al., 2021). Increased cytochrome C oxidase activity in striatum, thalamus, hippocampus, primary auditory and visual areas is found in a following study (Cardoso et al., 2022). Recently, Hosseini et al. (2022) show in aged mice, that 810-nm light applied during 8 weeks with an energy density of 8 and 16 J/cm$^2$ (at the scalp surface) increases synaptic structural associated proteins (GAP-43 and synaptophysin) expression in the hippocampus and reduced interleukin-6 and TNF alpha. Red and near infrared light would also improve spatial and episodic-like memory at 8 J/cm$^2$ (at the cortical surface), increase mitochondrial membrane potential and ATP production, and reduce reactive oxygen species production (Salehpour et al., 2017). Long-term (8 months) photobiomodulation has also been showed to reduce astrogliosis, with a decreased number of astrocytes and microglial cell, in the striatum of aged rats (El Massri et al. 2018); similar results have been found on MPTP (1-methyl-4-phenyl-1,2,3,6-tetrahydropyridine) monkey models of Parkinson's Disease (El Massri et al. 2016).

Clinical studies involving healthy aged participants mainly use single session (Qu et al., 2022; Zomorrodi et al., 2019; Yang et al., 2025).

Working memory has been the most investigated cognitive ability so far, and generally shows increased accuracy and reduced reaction times after photobiomodulation, particularly under high working memory load (Yang et al., 2025; Qu et al., 2022; Chan et al., 2019). Targeting executive functions (Eriksen flanker and verbal fluency) also suggest improvements after one 7.5 min session with a 633 and 870 nm light. However, some methodological considerations (absence of adequate control, small sample sizes) render necessary additional randomised and controlled studies before drawing firm conclusions about efficacy of photobiomodulation on cognition during aging. Moreover, further investigations of additional cognitive domains are still lacking. In the brain, a recent pilot study using $^{31}P$ magnetization transfer MR-spectroscopy reveals for the first time using neuroimaging, increased ATP levels after 7 days with a 670 nm light in the occipital cortex of healthy aged participants (Fear et al. 2023). Using broadband near infrared spectroscopy at rest, (Saucedo et al., 2021) suggest an increased oxyhaemoglobin/deoxyhaemoglobin ratio during and a few minutes after stimulation, and the effect seems more important in the older participants. This suggests an improvement in the cerebral oxygenation of the brain after transcranial photobiomodulation, and this is associated with increased functional connectivity revealed by near infrared spectroscopy in the fronto-parietal network (Yang et al., 2025). Effects on functional connectivity and global and local network efficacy have also been suggested using EEG (Zomorrodi et al., 2019). In aged patients with mild cognitive impairments, a recent doubled blinded, controlled study with 24 sessions of 670/870 nm light (de Oliveira et al., 2024) shows improvement in the global MoCA (Montreal Cognitive Assessment) scores, memory subscores, and an increased level of brain derived neurotrophic factor related to neuroplasticity. No effect was found in the levels of neuron specific enolase and S100β, two markers of neuroinflammation and neurodegenerescence.

Taken together, studies examining the effects of transcranial photobiomodulation on brain and cognition in the elderly suggest that it has the potential to influence brain activity and connectivity, cerebral oxygenation, and maybe improve some cognitive abilities. However, as for other methods, the field is lacking standardised practices. The use of relatively small sample sizes and lack of adequate controls render difficult to draw conclusive interpretation at present. Preclinical studies suggest beneficial effects on neuroinflammation; however this is still poorly investigated in human. It is unclear how long the effects last, what regions can/cannot be targeted, and whether this can translate in a long-term beneficial effect in clinics.

## 6. Conclusion: How does photobiomodulation compare to the other non invasive brain stimulation techniques

Taken together, all methods that have been examined in this review exhibit encouraging results on brain function and cognition. In the following section, an overall picture of the strengths and weaknesses of photobiomodulation as a neurostimulation technique, relatively to the other methods will be explored and some conclusions will be drawn.

Starting with the strengths, a key advantage of photobiomodulation is that the devices are easily accessible for home-using, as they mainly rely on LEDs that can be usable with or without the help of a carer. This is obviously a great advantage and of relevant importance in terms of treatment observance in the long-term. In comparison, TMS and transcranial focused ultrasounds in particular, have to be administered in clinics (at least for the moment), requiring several venues per weeks in the hospital, which can in the long-term compromise treatment adherence. A second advantage of photobiomodulation is the near absence of report of serious adverse events, being it in preclinical or clinical studies. This observation should nevertheless be treated with some caution as we do not, at present, have enough information on the long-term use in human subjects, with multicentric studies and large sample size. In comparison, TMS has been used for more than forty years in thousands of patients/participants, and its potential seizure risks, for example, are well documented. Moreover, the development of guidelines by the International Federation of Clinical Neurophysiology have undoubtedly contributed to the development of standardised protocols that can be safely used in various clinical populations. The photobiomodulation community would without doubt benefit from such guidelines, and this would also bring positive outcomes concerning the standardisation of the protocols.

A third advantage is its ability to stimulate easily different regions of a same network due to the development of LEDs helmets, rather than stimulating one given region. It is being ever more acknowledged in the neuroscience community that specific cognitive functions are subserved by networks rather than by single regions. In this context, transcranial photobiomodulation is probably more flexible in targeting entire networks and influencing functional connectivity.

Finally, a major strength of photobiomodulation relies on the ability of acting on cellular energy metabolism and neuroinflammation. Alteration of mitochondrial function, neural and astrocytes metabolism and dysregulation of neuroinflammation processes are key factors in the development of neurodegenerative diseases. This makes of photobiomodulation a powerful tool to target not only neural activity (like other non-invasive brain stimulation methods do), but also, act on cerebral oxygenation and neuroinflammation. This offers the opportunity to act on a more preventive way in comparison with the other techniques. Of course, this suggestion and the precise mechanisms of action are still to be defined precisely and more research should be dedicated to this topic.

With regards to the limitations, perhaps the most pivotal aspect is the issue of the dosage and how deep light delivered transcranially can penetrate into the brain. It is generally agreed that deeper brain regions, such as thalamus and brainstem, cannot be targeted (at least, directly); even at the cortical level, the particular regions that can be targeted effectively is far from clear. Direct measurements are nearly impossible, at least in the living human brain, and whereas computational modelling can provide some insights, this modelling has its own limitations: it relies on MRI images that do not consider the microvasculature in the brain, and on optical coefficients that have not been directly measured in the living human brain. Further, as light has a strong scattering across biological tissues, it is difficult to focus precisely the transcranial light, thereby preventing illumination of surrounding regions. This concern has also been raised for TMS and transcranial electric stimulation techniques; transcranial focused ultrasounds and temporal interference stimulation show deeper penetration power in the brain and better focalisation. This issue is linked to the problem concerning the heterogeneity in the parameters; this is a recurrent problem for all non-invasive neurostimulation methods. This makes an effective and worthwhile comparative analysis between the studies very difficult. Over the last decade, efforts have been made to optimise the protocols using the use of individual anatomy and computational modelling, and this should be also developed for photobiomodulation.

To conclude, transcranial photobiomodulation is a promising treatment in the strategy of age-related diseases prevention, as it could act as a method to improve neural metabolism and activity directly, as well as on a systemic level, promoting anti-inflammatory processes and improving cerebral oxygenation. However, there must be focused effort to standardise the protocols in clinical research in order for further development to be made.

## References

Aktürk, T., de Graaf, T. A., Güntekin, B., Hanoğlu, L., & Sack, A. T. (2022). Enhancing memory capacity by experimentally slowing theta frequency oscillations using combined EEG-tACS. *Scientific Reports, 12*(1), 14199. https://doi.org/10.1038/s41598-022-18665-z.

Alawi, M., Lee, P. F., Deng, Z.-D., Goh, Y. K., & Croarkin, P. E. (2023). Modelling the differential effects of age on transcranial magnetic stimulation induced electric fields. *Journal of Neural Engineering, 20*(2), https://doi.org/10.1088/1741-2552/ac9a76.

Ansari, Y., & Tremblay, F. (2019). Short-latency afferent-induced facilitation and inhibition as predictors of thermally induced variations in corticomotor excitability. *Experimental Brain Research, 237*(6), 1445–1455. https://doi.org/10.1007/s00221-019-05522-1.

Antal, A., Alekseichuk, I., Bikson, M., Brockmöller, J., Brunoni, A. R., Chen, R., ... Paulus, W. (2017). Low intensity transcranial electric stimulation: Safety, ethical, legal regulatory and application guidelines. *Clinical Neurophysiology: Official Journal of the International Federation of Clinical Neurophysiology, 128*(9), 1774–1809. https://doi.org/10.1016/j.clinph.2017.06.001.

Archer, J. A., Lee, A., Qiu, A., & Chen, S.-H. A. (2018). Working memory, age and education: A lifespan fMRI study. *PLoS One, 13*(3), e0194878. https://doi.org/10.1371/journal.pone.0194878.

Beck, D., de Lange, A.-M. G., Pedersen, M. L., Alnaes, D., Maximov, I. I., Voldsbekk, I., ... Westlye, L. T. (2022). Cardiometabolic risk factors associated with brain age and accelerate brain ageing. *Human Brain Mapping, 43*(2), 700–720. https://doi.org/10.1002/hbm.25680.

Beisteiner, R., Lozano, A., Di Lazzaro, V., George, M. S., & Hallett, M. (2024). Clinical recommendations for non-invasive ultrasound neuromodulation. *Brain Stimulation, 17*(4), 890–895. https://doi.org/10.1016/j.brs.2024.07.013.

Bhattacharya, A., Mrudula, K., Sreepada, S. S., Sathyaprabha, T. N., Pal, P. K., Chen, R., & Udupa, K. (2022). An overview of noninvasive brain stimulation: Basic principles and clinical applications. *The Canadian Journal of Neurological Sciences, 49*(4), 479–492. https://doi.org/10.1017/cjn.2021.158.

Bikson, M., Grossman, P., Thomas, C., Zannou, A. L., Jiang, J., Adnan, T., ... Woods, A. J. (2016). Safety of transcranial direct current stimulation: Evidence based update 2016. *Brain Stimulation, 9*(5), 641–661. https://doi.org/10.1016/j.brs.2016.06.004.

Bjekić, J., Manojlović, M., & Filipović, S. R. (2023). Transcranial electrical stimulation for associative memory enhancement: State-of-the-art from basic to clinical research. *Life (Basel, Switzerland), 13*(5), 1125. https://doi.org/10.3390/life13051125.

Blackmore, J., Shrivastava, S., Sallet, J., Butler, C. R., & Cleveland, R. O. (2019). Ultrasound neuromodulation: A review of results, mechanisms and safety. *Ultrasound in Medicine & Biology, 45*(7), 1509–1536. https://doi.org/10.1016/j.ultrasmedbio.2018.12.015.

Blivet, G., Relano-Gines, A., Wachtel, M., & Touchon, J. (2022). A randomized, double-blind, and sham-controlled trial of an innovative brain-gut photobiomodulation therapy: Safety and patient compliance. *Journal of Alzheimer's Disease, 90*(2), 811–822. https://doi.org/10.3233/JAD-220467.

Boyer, D., Hu, A., Warrow, D., Xavier, S., Gonzalez, V., Lad, E., ... Tedford, C. E. (2024). LIGHTSITE III: 13-month efficacy and safety evaluation of multiwavelength photobiomodulation in nonexudative (Dry) age-related macular degeneration using the lumithera valeda light delivery system. *Retina (Philadelphia, Pa.), 44*(3), 487–497. https://doi.org/10.1097/IAE.0000000000003980.

Cabeza, R. (2002). Hemispheric asymmetry reduction in older adults: The HAROLD model. *Psychology and Aging, 17*(1), 85–100. https://doi.org/10.1037//0882-7974.17.1.85.

Cappon, D., den Boer, T., Jordan, C., Yu, W., Metzger, E., & Pascual-Leone, A. (2022). Transcranial magnetic stimulation (TMS) for geriatric depression. *Ageing Research Reviews, 74*, 101531. https://doi.org/10.1016/j.arr.2021.101531.

Cardoso, F. D. S., Serra, F. T., Coimbra, N. C., Gonzalez-Lima, F., & Gomes da Silva, S. (2022). Transcranial photobiomodulation changes neuronal morphology in the cerebral cortex of rats. *Neuroscience Letters, 781*, 136681. https://doi.org/10.1016/j.neulet.2022.136681.

Cardoso, F. dos S., de Souza Oliveira Tavares, C., Araujo, B. H. S., Mansur, F., Lopes-Martins, R.Á. B., & Gomes da Silva, S. (2021). Improved spatial memory and neuroinflammatory profile changes in aged rats submitted to photobiomodulation therapy. *Cellular and Molecular Neurobiology, 42*(6), 1875–1886. https://doi.org/10.1007/s10571-021-01069-4.

Carrasco-Gómez, M., García-Colomo, A., Cabrera-Álvarez, J., Del Cerro-León, A., GÓmez-Ariza, C. J., Santos, A., & Maestú, F. (2025). Individual alpha frequency tACS reduces static functional connectivity across the default mode network. *Frontiers in Human Neuroscience, 19*, 1534321. https://doi.org/10.3389/fnhum.2025.1534321.

Cassano, P., Tran, A. P., Katnani, H., Bleier, B. S., Hamblin, M. R., Yuan, Y., & Fang, Q. (2019). Selective photobiomodulation for emotion regulation: Model-based dosimetry study. *Neurophotonics, 6*(1), 015004. https://doi.org/10.1117/1.NPh.6.1.015004.

Cassidy, J., Nouri, A., Betz, V., & Lilge, L. (2018). High-performance, robustly verified Monte Carlo simulation with FullMonte. *Journal of Biomedical Optics, 23*(8), 1–11. https://doi.org/10.1117/1.JBO.23.8.085001.

Cavaleiro, C., Martins, J., Gonçalves, J., & Castelo-Branco, M. (2020). Memory and cognition-related neuroplasticity enhancement by transcranial direct current stimulation in rodents: A systematic review. *Neural Plasticity, 2020*, 4795267. https://doi.org/10.1155/2020/4795267.

Chan, A. S., Lee, T. L., Yeung, M. K., & Hamblin, M. R. (2019). Photobiomodulation improves the frontal cognitive function of older adults. *International Journal of Geriatric Psychiatry, 34*(2), 369–377. https://doi.org/10.1002/gps.5039.

Chen, M., Peng, C., Wu, H., Huang, C.-C., Kim, T., Traylor, Z., ... Jiang, X. (2023). Numerical and experimental evaluation of low-intensity transcranial focused ultrasound wave propagation using human skulls for brain neuromodulation. *Medical Physics, 50*(1), 38–49. https://doi.org/10.1002/mp.16090.

Chen, R., Tam, A., Bütefisch, C., Corwell, B., Ziemann, U., Rothwell, J. C., & Cohen, L. G. (1998). Intracortical inhibition and facilitation in different representations of the human motor cortex. *Journal of Neurophysiology, 80*(6), 2870–2881. https://doi.org/10.1152/jn.1998.80.6.2870.

Chou, Y.-H., Ton That, V., & Sundman, M. (2020). A systematic review and meta-analysis of rTMS effects on cognitive enhancement in mild cognitive impairment and Alzheimer's disease. *Neurobiology of Aging, 86*, 1–10. https://doi.org/10.1016/j.neurobiolaging.2019.08.020.

Coelho, D. R. A., Fernando Vieira, W., Hurtado Puerto, A. M., B Gersten, M., Anne Collins, K., Peterson, A., ... Cassano, P. (2025). Dose-dependent tolerability and safety of transcranial photobiomodulation: A randomized controlled trial. *Lasers in Medical Science, 40*(1), 248. https://doi.org/10.1007/s10103-025-04501-z.

Damoiseaux, J. S., Beckmann, C. F., Arigita, E. J. S., Barkhof, F., Scheltens, P. H., Stam, C. J., ... Rombouts, S. A. R. B. (2008). Reduced resting-state brain activity in the "default network" in normal aging. *Cerebral Cortex, 18*(8), 1856–1864. https://doi.org/10.1093/cercor/bhm207.

Darmani, G., Bergmann, T. O., Butts Pauly, K., Caskey, C. F., de Lecea, L., Fomenko, A., ... Chen, R. (2022). Non-invasive transcranial ultrasound stimulation for neuromodulation. *Clinical Neurophysiology, 135*, 51–73. https://doi.org/10.1016/j.clinph.2021.12.010.

Davis, S. W., Dennis, N. A., Daselaar, S. M., Fleck, M. S., & Cabeza, R. (2008). Que PASA? The posterior-anterior shift in aging. *Cerebral Cortex, 18*(5), 1201–1209. https://doi.org/10.1093/cercor/bhm155.

de Oliveira, B. H., Lins, E. F., Kunde, N. F., Salgado, A. S. I., Martins, L. M., Bobinski, F., ... Martins, D. F. (2024). Transcranial photobiomodulation increases cognition and serum BDNF levels in adults over 50 years: A randomized, double-blind, placebo-controlled trial. *Journal of Photochemistry and Photobiology. B, Biology, 260*, 113041. https://doi.org/10.1016/j.jphotobiol.2024.113041.

Deery, H. A., Di Paolo, R., Moran, C., Egan, G. F., & Jamadar, S. D. (2022). Lower brain glucose metabolism in normal ageing is predominantly frontal and temporal: A systematic review and pooled effect size and activation likelihood estimates meta-analyses. *Human Brain Mapping, 44*(3), 1251–1277. https://doi.org/10.1002/hbm.26119.

Deery, H. A., Di Paolo, R., Moran, C., Egan, G. F., & Jamadar, S. D. (2023). The older adult brain is less modular, more integrated, and less efficient at rest: A systematic review of large-scale resting-state functional brain networks in aging. *Psychophysiology, 60*(1), e14159. https://doi.org/10.1111/psyp.14159.

D'Esposito, M., Deouell, L. Y., & Gazzaley, A. (2003). Alterations in the BOLD fMRI signal with ageing and disease: A challenge for neuroimaging. *Nature Reviews. Neuroscience, 4*(11), 863–872. https://doi.org/10.1038/nrn1246.

Dmochowski, G. M., Shereen, A. D., Berisha, D., & Dmochowski, J. P. (2020). Near-infrared light increases functional connectivity with a non-thermal mechanism. *Cerebral Cortex Communications, 1*(1), tgaa004. https://doi.org/10.1093/texcom/tgaa004.

Dole, M., Auboiroux, V., Langar, L., & Mitrofanis, J. (2023). A systematic review of the effects of transcranial photobiomodulation on brain activity in humans. *Reviews in the Neurosciences, 34*(6), 671–693. https://doi.org/10.1515/revneuro-2023-0003.

Dole, M., Bleuet, P., Auboiroux, V., Billères, M., & Mitrofanis, J. (2024). Monte Carlo simulations of a multisource transcranial photobiomodulation helmet device: Application to young and aged brains. *Advanced Technology in Neuroscience, 1*(2), 261. https://doi.org/10.4103/ATN.ATN-D-24-00022.

Dumel, G., Bourassa, M.-E., Desjardins, M., Voarino, N., Charlebois-Plante, C., Doyon, J., & De Beaumont, L. (2016). Multisession anodal tDCS protocol improves motor system function in an aging population. *Neural Plasticity, 2016*, 5961362. https://doi.org/10.1155/2016/5961362.

El Massri, N., Moro, C., Torres, N., Farlot, F., Agay, D., Chabrol, C., ... Mitrofanis, J. (2016). Near-infrared light treatment reduces astrogliosis in MPTP-treated monkeys. *Experimental Brain Research, 234*(11), 3225–3232.

El Massri, N., Weinrich, T. W., Hoh Kam, J., Jeffert, G., & Mitrofanis, J. (2018). Photobiomodulation reduces gliosis in the basal ganglia of aged mice. *Neurobiology of Aging, 66*, 131–137.

Esposito, S., Trojsi, F., Cirillo, G., de Stefano, M., Di Nardo, F., Siciliano, M., ... Tedeschi, G. (2022). Repetitive transcranial magnetic stimulation (rTMS) of dorsolateral prefrontal cortex may influence semantic fluency and functional connectivity in fronto-parietal network in mild cognitive impairment (MCI). *Biomedicines, 10*(5), 994. https://doi.org/10.3390/biomedicines10050994.

Fabiani, M., Gordon, B. A., Maclin, E. L., Pearson, M. A., Brumback-Peltz, C. R., Low, K. A., ... Gratton, G. (2014). Neurovascular coupling in normal aging: A combined optical, ERP and fMRI study. *Neuroimage, 85*(1), 592–607. https://doi.org/10.1016/j.neuroimage.2013.04.113.

Fang, Q. (2010). Mesh-based Monte Carlo method using fast ray-tracing in Plücker coordinates. *Biomedical Optics Express, 1*(1), 165–175. https://doi.org/10.1364/BOE.1.000165.

Farnad, L., Ghasemian-Shirvan, E., Mosayebi-Samani, M., Kuo, M.-F., & Nitsche, M. A. (2021). Exploring and optimizing the neuroplastic effects of anodal transcranial direct current stimulation over the primary motor cortex of older humans. *Brain Stimulation, 14*(3), 622–634. https://doi.org/10.1016/j.brs.2021.03.013.

Fear, E. J., Torkelsen, F. H., Zamboni, E., Chen, K.-J., Scott, M., Jeffery, G., ... Kennerley, A. J. (2023). Use of 31 P magnetisation transfer magnetic resonance spectroscopy to measure ATP changes after 670 nm transcranial photobiomodulation in older adults. *Aging Cell, 22*(11), e14005. https://doi.org/10.1111/acel.14005.
Figueiro Longo, M. G., Tan, C. O., Chan, S.-T., Welt, J., Avesta, A., Ratai, E., ... Gupta, R. (2020). Effect of transcranial low-level light therapy vs Sham therapy among patients with moderate traumatic brain injury: A randomized clinical trial. *JAMA Network Open, 3*(9), e2017337. https://doi.org/10.1001/jamanetworkopen.2020.17337.
Gamage, N. N., Liao, W.-Y., Hand, B. J., Atherton, P. J., Piasecki, M., Opie, G. M., & Semmler, J. G. (2025). Theta-gamma transcranial alternating current stimulation enhances ballistic motor performance in healthy young and older adults. *Neurobiology of Aging, 152*, 1–12. https://doi.org/10.1016/j.neurobiolaging.2025.04.010.
Goldthorpe, R. A., Rapley, J. M., & Violante, I. R. (2020). A systematic review of non-invasive brain stimulation applications to memory in healthy aging. *Frontiers in Neurology, 11*, 575075. https://doi.org/10.3389/fneur.2020.575075.
Gomez, L. J., Dannhauer, M., Koponen, L. M., & Peterchev, A. V. (2020). Conditions for numerically accurate TMS electric field simulation. *Brain Stimulation, 13*(1), 157–166. https://doi.org/10.1016/j.brs.2019.09.015.
Gomez-Tames, J., Hamasaka, A., Hirata, A., Laakso, I., Lu, M., & Ueno, S. (2020). Group-level analysis of induced electric field in deep brain regions by different TMS coils. *Physics in Medicine and Biology, 65*(2), 025007. https://doi.org/10.1088/1361-6560/ab5e4a.
Grover, S., Fayzullina, R., Bullard, B. M., Levina, V., & Reinhart, R. M. G. (2023). A meta-analysis suggests that tACS improves cognition in healthy, aging, and psychiatric populations. *Science Translational Medicine, 15*(697), eabo2044. https://doi.org/10.1126/scitranslmed.abo2044.
Guerra, A., Rocchi, L., Grego, A., Berardi, F., Luisi, C., & Ferreri, F. (2021). Contribution of TMS and TMS-EEG to the understanding of mechanisms underlying physiological brain aging. *Brain Sciences, 11*(3), 405. https://doi.org/10.3390/brainsci11030405.
Hand, B. J., Merkin, A., Opie, G. M., Ziemann, U., & Semmler, J. G. (2023). Repetitive paired-pulse TMS increases motor cortex excitability and visuomotor skill acquisition in young and older adults. *Cerebral Cortex, 33*(20), 10660–10675. https://doi.org/10.1093/cercor/bhad315.
Hosseini, L., Farazi, N., Erfani, M., Mahmoudi, J., Akbari, M., Hosseini, S. H., & Sadigh-Eteghad, S. (2022). Effect of transcranial near-infrared photobiomodulation on cognitive outcomes in D-galactose/AlCl3 induced brain aging in BALB/c mice. *Lasers in Medical Science, 37*(3), 1787–1798. https://doi.org/10.1007/s10103-021-03433-8.
Hsu, W.-Y., Ku, Y., Zanto, T. P., & Gazzaley, A. (2015). Effects of noninvasive brain stimulation on cognitive function in healthy aging and Alzheimer's disease: A systematic review and meta-analysis. *Neurobiology of Aging, 36*(8), 2348–2359. https://doi.org/10.1016/j.neurobiolaging.2015.04.016.
Huang, Y., Datta, A., Bikson, M., & Parra, L. C. (2019). Realistic volumetric-approach to simulate transcranial electric stimulation-ROAST-a fully automated open-source pipeline. *Journal of Neural Engineering, 16*(5), 056006. https://doi.org/10.1088/1741-2552/ab208d.
Huang, Y.-Z., Chen, R.-S., Rothwell, J. C., & Wen, H.-Y. (2007). The after-effect of human theta burst stimulation is NMDA receptor dependent. *Clinical Neurophysiology, 118*(5), 1028–1032. https://doi.org/10.1016/j.clinph.2007.01.021.
Huo, L., Zheng, Z., Li, J., Wan, W., Cui, X., Chen, S., ... Li, J. (2018). Long-term transcranial direct current stimulation does not improve executive function in healthy older adults. *Frontiers in Aging Neuroscience, 10*, 298. https://doi.org/10.3389/fnagi.2018.00298.

Hupfeld, K. E., Swanson, C. W., Fling, B. W., & Seidler, R. D. (2020). TMS-induced silent periods: A review of methods and call for consistency. *Journal of Neuroscience Methods, 346*, 108950. https://doi.org/10.1016/j.jneumeth.2020.108950.

Iordan, A. D., Ryan, S., Tyszkowski, T., Peltier, S. J., Rahman-Filipiak, A., & Hampstead, B. M. (2022). High-definition transcranial direct current stimulation enhances network segregation during spatial navigation in mild cognitive impairment. *Cerebral Cortex, 32*(22), 5230–5241. https://doi.org/10.1093/cercor/bhac010.

Kaiser, M., Wang, Y., Ten Oever, S., Duecker, F., Sack, A. T., & van de Ven, V. (2025). Simultaneous tACS-fMRI reveals state- and frequency-specific modulation of hippocampal-cortical functional connectivity. *Communications Psychology, 3*(1), 19. https://doi.org/10.1038/s44271-025-00202-z.

Kallioniemi, E., Saari, J., Ferreri, F., & Määttä, S. (2022). TMS-EEG responses across the lifespan: Measurement, methods for characterisation and identified responses. *Journal of Neuroscience Methods, 366*, 109430. https://doi.org/10.1016/j.jneumeth.2021.109430.

Kishi, T., Ikuta, T., Sakuma, K., Hatano, M., Matsuda, Y., Wilkening, J., ... Iwata, N. (2024). Theta burst stimulation for depression: A systematic review and network and pairwise meta-analysis. *Molecular Psychiatry, 29*(12), 3893–3899. https://doi.org/10.1038/s41380-024-02630-5.

Klees-Themens, G., & Théoret, H. (2023). The effects of transcranial direct current stimulation on corticospinal excitability: A systematic review of nonsignificant findings. *The European Journal of Neuroscience, 58*(4), 3074–3097. https://doi.org/10.1111/ejn.16073.

Koch, G., Altomare, D., Benussi, A., Bréchet, L., Casula, E. P., Dodich, A., ... Frisoni, G. B. (2024). The emerging field of non-invasive brain stimulation in Alzheimer's disease. *Brain: A Journal of Neurology, 147*(12), 4003–4016. https://doi.org/10.1093/brain/awae292.

Krause, M. R., Vieira, P. G., Csorba, B. A., Pilly, P. K., & Pack, C. C. (2019). Transcranial alternating current stimulation entrains single-neuron activity in the primate brain. *Proceedings of the National Academy of Sciences of the United States of America, 116*(12), 5747–5755. https://doi.org/10.1073/pnas.1815958116.

Laakso, I., Murakami, T., Hirata, A., & Ugawa, Y. (2018). Where and what TMS activates: Experiments and modeling. *Brain Stimulation, 11*(1), 166–174. https://doi.org/10.1016/j.brs.2017.09.011.

Lee, W., Kim, H., Jung, Y., Song, I.-U., Chung, Y. A., & Yoo, S.-S. (2015). Image-guided transcranial focused ultrasound stimulates human primary somatosensory cortex. *Scientific Reports, 5*, 8743. https://doi.org/10.1038/srep08743.

Lee, W., Kim, H.-C., Jung, Y., Chung, Y. A., Song, I.-U., Lee, J.-H., & Yoo, S.-S. (2016). Transcranial focused ultrasound stimulation of human primary visual cortex. *Scientific Reports, 6*(1), 34026. https://doi.org/10.1038/srep34026.

Legon, W., Bansal, P., Tyshynsky, R., Ai, L., & Mueller, J. K. (2018). Transcranial focused ultrasound neuromodulation of the human primary motor cortex. *Scientific Reports, 8*(1), 10007. https://doi.org/10.1038/s41598-018-28320-1.

Leino, A. A., Pulkkinen, A., & Tarvainen, T. (2019). ValoMC: A Monte Carlo software and MATLAB toolbox for simulating light transport in biological tissue. *OSA Continuum, 2*(3), 957–972. https://doi.org/10.1364/OSAC.2.000957.

Lenz, M., Galanis, C., Müller-Dahlhaus, F., Opitz, A., Wierenga, C. J., Szabó, G., ... Vlachos, A. (2016). Repetitive magnetic stimulation induces plasticity of inhibitory synapses. *Nature Communications*, 7(1), 10020. https://doi.org/10.1038/ncomms10020.

Li, Z., Zhang, Z., Ren, Y., Wang, Y., Fang, J., Yue, H., ... Guan, F. (2021). Aging and age-related diseases: From mechanisms to therapeutic strategies. *Biogerontology, 22*(2), 165–187. https://doi.org/10.1007/s10522-021-09910-5.

Li, Z., Zhao, Y., Hu, Y., Li, Y., Zhang, K., Gao, Z., ... Zhao, C. (2024). Transcranial low-level laser stimulation in the near-infrared-II region (1064 nm) for brain safety in healthy humans. *Brain Stimulation, 17*(6), 1307–1316. https://doi.org/10.1016/j.brs.2024.11.010.

Lu, H., Chan, S. S. M., Chan, W. C., Lin, C., Cheng, C. P. W., & Linda Chiu Wa, L. (2019). Randomized controlled trial of TDCS on cognition in 201 seniors with mild neurocognitive disorder. *Annals of Clinical and Translational Neurology, 6*(10), 1938–1948. https://doi.org/10.1002/acn3.50823.

MacDonald, M. E., & Pike, G. B. (2021). MRI of healthy brain aging: A review. *NMR in Biomedicine, 34*(9), e4564. https://doi.org/10.1002/nbm.4564.

Marseglia, A., Dartora, C., Samuelsson, J., Poulakis, K., Mohanty, R., Shams, S., ... Westman, E. (2025). Biological brain age and resilience in cognitively unimpaired 70-year-old individuals. *Alzheimer's & Dementia, 21*(2), e14435. https://doi.org/10.1002/alz.14435.

Meinzer, M., Lindenberg, R., Antonenko, D., Flaisch, T., & Flöel, A. (2013). Anodal transcranial direct current stimulation temporarily reverses age-associated cognitive decline and functional brain activity changes. *The Journal of Neuroscience, 33*(30), 12470–12478. https://doi.org/10.1523/JNEUROSCI.5743-12.2013.

Meinzer, M., Lindenberg, R., Phan, M. T., Ulm, L., Volk, C., & Flöel, A. (2015). Transcranial direct current stimulation in mild cognitive impairment: Behavioral effects and neural mechanisms. *Alzheimer's & Dementia, 11*(9), 1032–1040. https://doi.org/10.1016/j.jalz.2014.07.159.

Metais, A., Muller, C. O., Boublay, N., Breuil, C., Guillot, A., Daligault, S., ... Saimpont, A. (2022). Anodal tDCS does not enhance the learning of the sequential finger-tapping task by motor imagery practice in healthy older adults. *Frontiers in Aging Neuroscience, 14*, 1060791. https://doi.org/10.3389/fnagi.2022.1060791.

Meynaghizadeh-Zargar, R., Sadigh-Eteghad, S., Mohaddes, G., Salehpour, F., & Rasta, S. H. (2020). Effects of transcranial photobiomodulation and methylene blue on biochemical and behavioral profiles in mice stress model. *Lasers in Medical Science, 35*(3), 573–584. https://doi.org/10.1007/s10103-019-02851-z.

Morales Fajardo, K., Yan, X., Lungoci, G., Casado Sánchez, M., Mitsis, G. D., & Boudrias, M. H. (2024). The modulatory effects of transcranial alternating current stimulation on brain oscillatory patterns in the beta band in healthy older adults. *Brain Sciences, 14*(12), 1284. https://doi.org/10.3390/brainsci14121284.

Moro, C., Torres, N., El Massri, N., Ratel, D., Johnstone, D. M., Stone, J., ... Benabid, A.-L. (2013). Photobiomodulation preserves behaviour and midbrain dopaminergic cells from MPTP toxicity: Evidence from two mouse strains. *BMC Neuroscience, 14*, 40. https://doi.org/10.1186/1471-2202-14-40.

Mueller, J. K., Ai, L., Bansal, P., & Legon, W. (2017). Numerical evaluation of the skull for human neuromodulation with transcranial focused ultrasound. *Journal of Neural Engineering, 14*(6), 066012. https://doi.org/10.1088/1741-2552/aa843e.

Mukli, P., Pinto, C. B., Owens, C. D., Csipo, T., Lipecz, A., Szarvas, Z., ... Yabluchanskiy, A. (2023). Impaired neurovascular coupling and increased functional connectivity in the frontal cortex predict age-related cognitive dysfunction. *Advanced Science, 11*(10), 2303516. https://doi.org/10.1002/advs.202303516.

Murphy, K. R., Nandi, T., Kop, B., Osada, T., Lueckel, M., N'Djin, W. A., ... Bergmann, T. O. (2025). A practical guide to transcranial ultrasonic stimulation from the IFCN-endorsed ITRUSST consortium. *Clinical Neurophysiology, 171*, 192–226. https://doi.org/10.1016/j.clinph.2025.01.004.

Ngandu, T., Lehtisalo, J., Solomon, A., Levälahti, E., Ahtiluoto, S., Antikainen, R., ... Kivipelto, M. (2015). A 2 year multidomain intervention of diet, exercise, cognitive training, and vascular risk monitoring versus control to prevent cognitive decline in at-risk elderly people (FINGER): A randomised controlled trial. *Lancet, 385*(9984), 2255–2263. https://doi.org/10.1016/S0140-6736(15)60461-5.

Nitsche, M. A., Fricke, K., Henschke, U., Schlitterlau, A., Liebetanz, D., Lang, N., ... Paulus, W. (2003). Pharmacological modulation of cortical excitability shifts induced by transcranial direct current stimulation in humans. *The Journal of Physiology, 553*(Pt 1), 293–301. https://doi.org/10.1113/jphysiol.2003.049916.

Oliviero, A., Profice, P., Tonali, P. A., Pilato, F., Saturno, E., Dileone, M., ... Di Lazzaro, V. (2006). Effects of aging on motor cortex excitability. *Neuroscience Research, 55*(1), 74–77. https://doi.org/10.1016/j.neures.2006.02.002.

Opie, G. M., Sidhu, S. K., Rogasch, N. C., Ridding, M. C., & Semmler, J. G. (2018). Cortical inhibition assessed using paired-pulse TMS-EEG is increased in older adults. *Brain Stimulation, 11*(3), 545–557. https://doi.org/10.1016/j.brs.2017.12.013.

Park, D. C., & Reuter-Lorenz, P. (2009). The adaptive brain: Aging and neurocognitive scaffolding. *Annual Review of Psychology, 60*, 173–196. https://doi.org/10.1146/annurev.psych.59.103006.093656.

Park, T. Y., Koh, H., Lee, W., Park, S. H., Chang, W. S., & Kim, H. (2023). Real-time acoustic simulation framework for tFUS: A feasibility study using navigation system. *Neuroimage, 282*, 120411. https://doi.org/10.1016/j.neuroimage.2023.120411.

Pascual-Leone, A., Valls-Solé, J., Wassermann, E. M., & Hallett, M. (1994). Responses to rapid-rate transcranial magnetic stimulation of the human motor cortex. *Brain, 117*(4), 847–858. https://doi.org/10.1093/brain/117.4.847.

Pchelin, P., Shkarupa, D., Smetanina, N., Grigorieva, T., Lapshin, R., Schelchkova, N., ... Bavrina, A. (2023). Red light photobiomodulation rescues murine brain mitochondrial respiration after acute hypobaric hypoxia. *Journal of Photochemistry and Photobiology. B, Biology, 239*, 112643. https://doi.org/10.1016/j.jphotobiol.2022.112643.

Picca, A., Faitg, J., Auwerx, J., Ferrucci, L., & D'Amico, D. (2023). Mitophagy in human health, ageing and disease. *Nature Metabolism, 5*(12), 2047–2061. https://doi.org/10.1038/s42255-023-00930-8.

Prahl, S. A. (1989). A Monte Carlo model of light propagation in tissue. *Dosimetry of Laser Radiation in Medicine and Biology, 10305*, 105–114. https://doi.org/10.1117/12.2283590.

Prince, J. B., Davis, H. L., Tan, J., Muller-Townsend, K., Markovic, S., Lewis, D. M. G., ... Sohrabi, H. R. (2024). Cognitive and neuroscientific perspectives of healthy ageing. *Neuroscience and Biobehavioral Reviews, 161*, 105649. https://doi.org/10.1016/j.neubiorev.2024.105649.

Qu, X., Li, L., Zhou, X., Dong, Q., Liu, H., Liu, H., ... Niu, H. (2022). Repeated transcranial photobiomodulation improves working memory of healthy older adults: Behavioral outcomes of poststimulation including a three-week follow-up. *Neurophotonics, 9*(3), 035005. https://doi.org/10.1117/1.NPh.9.3.035005.

Reinhart, R. M. G., & Nguyen, J. A. (2019). Working memory revived in older adults by synchronizing rhythmic brain circuits. *Nature Neuroscience, 22*(5), 820–827. https://doi.org/10.1038/s41593-019-0371-x.

Rossi, S., Antal, A., Bestmann, S., Bikson, M., Brewer, C., BrockmÖller, J., … Leocani, L., basis of this article began with a Consensus Statement from the IFCN Workshop on "Present, Future of TMS: Safety, Ethical Guidelines", Siena, October 17–20, 2018, updating through April 2020. (2021). Safety and recommendations for TMS use in healthy subjects and patient populations, with updates on training, ethical and regulatory issues: Expert Guidelines. *Clinical Neurophysiology, 132*(1), 269–306. https://doi.org/10.1016/j.clinph.2020.10.003.

Rossi, S., Hallett, M., Rossini, P. M., & Pascual-Leone, A. (2009). Safety, ethical considerations, and application guidelines for the use of transcranial magnetic stimulation in clinical practice and research. *Clinical Neurophysiology, 120*(12), 2008–2039. https://doi.org/10.1016/j.clinph.2009.08.016.

Salehpour, F., Ahmadian, N., Rasta, S. H., Farhoudi, M., Karimi, P., & Sadigh-Eteghad, S. (2017). Transcranial low-level laser therapy improves brain mitochondrial function and cognitive impairment in D-galactose-induced aging mice. *Neurobiology of Aging, 58*, 140–150. https://doi.org/10.1016/j.neurobiolaging.2017.06.025.

Salthouse, T. A. (1996). The processing-speed theory of adult age differences in cognition. *Psychological Review, 103*(3), 403–428. https://doi.org/10.1037/0033-295x.103.3.403.

Salthouse, T. A. (2019). Trajectories of normal cognitive aging. *Psychology and Aging, 34*(1), 17–24. https://doi.org/10.1037/pag0000288.
Samoudi, M. A., Van Renterghem, T., & Botteldooren, D. (2019). Computational modeling of a single-element transcranial focused ultrasound transducer for subthalamic nucleus stimulation. *Journal of Neural Engineering, 16*(2), 026015. https://doi.org/10.1088/1741-2552/aafa38.
Samuel, N., Zeng, K., Harmsen, I. E., Ding, M. Y. R., Darmani, G., Sarica, C., ... Lozano, A. M. (2022). Multi-modal investigation of transcranial ultrasound-induced neuroplasticity of the human motor cortex. *Brain Stimulation, 15*(6), 1337–1347. https://doi.org/10.1016/j.brs.2022.10.001.
Sandrini, M., Manenti, R., Gobbi, E., Rusich, D., Bartl, G., & Cotelli, M. (2019). Transcranial direct current stimulation applied after encoding facilitates episodic memory consolidation in older adults. *Neurobiology of Learning and Memory, 163*, 107037. https://doi.org/10.1016/j.nlm.2019.107037.
Sarica, C., Nankoo, J.-F., Fomenko, A., Grippe, T. C., Yamamoto, K., Samuel, N., ... Chen, R. (2022). Human studies of transcranial ultrasound neuromodulation: A systematic review of effectiveness and safety. *Brain Stimulation, 15*(3), 737–746. https://doi.org/10.1016/j.brs.2022.05.002.
Saucedo, C. L., Courtois, E. C., Wade, Z. S., Kelley, M. N., Kheradbin, N., Barrett, D. W., & Gonzalez-Lima, F. (2021). Transcranial laser stimulation: Mitochondrial and cerebrovascular effects in younger and older healthy adults. *Brain Stimulation, 14*(2), 440–449. https://doi.org/10.1016/j.brs.2021.02.011.
Seidler, R. D., Bernard, J. A., Burutolu, T. B., Fling, B. W., Gordon, M. T., Gwin, J. T., ... Lipps, D. B. (2010). Motor control and aging: Links to age-related brain structural, functional, and biochemical effects. *Neuroscience and Biobehavioral Reviews, 34*(5), 721–733. https://doi.org/10.1016/j.neubiorev.2009.10.005.
Seidler, R., Erdeniz, B., Koppelmans, V., Hirsiger, S., Mérillat, S., & Jäncke, L. (2015). Associations between age, motor function, and resting state sensorimotor network connectivity in healthy older adults. *Neuroimage, 108*, 47–59. https://doi.org/10.1016/j.neuroimage.2014.12.023.
Shahdadian, S., Wang, X., & Liu, H. (2024). Directed physiological networks in the human prefrontal cortex at rest and post transcranial photobiomodulation. *Scientific Reports, 14*(1), 10242. https://doi.org/10.1038/s41598-024-59879-7.
Shaw, V. E., Peoples, C., Spana, S., Ashkan, K., Benabid, A.-L., Stone, J., ... Mitrofanis, J. (2012). Patterns of cell activity in the subthalamic region associated with the neuroprotective action of near-infrared light treatment in MPTP-treated mice. *Parkinson's Disease, 2012*, 296875. https://doi.org/10.1155/2012/296875.
Stern, Y., Albert, M., Barnes, C. A., Cabeza, R., Pascual-Leone, A., & Rapp, P. R. (2023). A framework for concepts of reserve and resilience in aging. *Neurobiology of Aging, 124*, 100–103. https://doi.org/10.1016/j.neurobiolaging.2022.10.015.
Stern, Y., Arenaza-Urquijo, E. M., Bartrés-Faz, D., Belleville, S., Cantilon, M., Chetelat, G., ... the Reserve, Resilience and Protective Factors PIA Empirical Definitions and Conceptual Frameworks Workgroup (2020). Whitepaper: Defining and investigating cognitive reserve, brain reserve, and brain maintenance. *Alzheimer's & Dementia, 16*(9), 1305–1311. https://doi.org/10.1016/j.jalz.2018.07.219.
Stern, Y., Barnes, C. A., Grady, C., Jones, R. N., & Raz, N. (2019). Brain reserve, cognitive reserve, compensation, and maintenance: Operationalization, validity, and mechanisms of cognitive resilience. *Neurobiology of Aging, 83*, 124–129. https://doi.org/10.1016/j.neurobiolaging.2019.03.022.
Struckmann, W., Bodén, R., Gingnell, M., Fällmar, D., & Persson, J. (2022). Modulation of dorsolateral prefrontal cortex functional connectivity after intermittent theta-burst stimulation in depression: Combining findings from fNIRS and fMRI. *NeuroImage. Clinical, 34*, 103028. https://doi.org/10.1016/j.nicl.2022.103028.

Sun, H., Wang, Y., Yuan, D., Duan, M., Chen, Z., & Fu, Y. (2024). Aftereffect of single transcranial direct and alternating current stimulation on spontaneous home-cage and open-field EEG activities in a mouse model of Alzheimer's disease. *Frontiers in Aging Neuroscience, 16*, 1492838. https://doi.org/10.3389/fnagi.2024.1492838.

Talelli, P., Waddingham, W., Ewas, A., Rothwell, J. C., & Ward, N. S. (2008). The effect of age on task-related modulation of interhemispheric balance. *Experimental Brain Research, 186*(1), 59–66. https://doi.org/10.1007/s00221-007-1205-8.

Taylor, J. L., Bhatt, P., Hernandez, B., Iv, M., Adamson, M. M., Heath, A., ... McNerney, M. W. (2025). Network-targeted transcranial magnetic stimulation (TMS) for mild cognitive impairment (MCI). *NeuroImage. Clinical, 47*, 103819. https://doi.org/10.1016/j.nicl.2025.103819.

Thielscher, A., Antunes, A., & Saturnino, G. B. (2015). Field modeling for transcranial magnetic stimulation: A useful tool to understand the physiological effects of TMS? *37th Annual International Conference of the IEEE Engineering in Medicine and Biology Society (EMBC), 2015*, 222–225. https://doi.org/10.1109/EMBC.2015.7318340.

Tromp, D., Dufour, A., Lithfous, S., Pebayle, T., & Després, O. (2015). Episodic memory in normal aging and Alzheimer disease: Insights from imaging and behavioral studies. *Ageing Research Reviews, 24*(Pt B), 232–262. https://doi.org/10.1016/j.arr.2015.08.006.

Truong, N. C. D., Wang, X., & Liu, H. (2023). Temporal and spectral analyses of EEG microstate reveals neural effects of transcranial photobiomodulation on the resting brain. *Frontiers in Neuroscience, 17*, 1247290. https://doi.org/10.3389/fnins.2023.1247290.

Tufail, Y., Matyushov, A., Baldwin, N., Tauchmann, M. L., Georges, J., Yoshihiro, A., ... Tyler, W. J. (2010). Transcranial pulsed ultrasound stimulates intact brain circuits. *Neuron, 66*(5), 681–694. https://doi.org/10.1016/j.neuron.2010.05.008.

Valero-Cabré, A., Amengual, J. L., Stengel, C., Pascual-Leone, A., & Coubard, O. A. (2017). Transcranial magnetic stimulation in basic and clinical neuroscience: A comprehensive review of fundamental principles and novel insights. *Neuroscience and Biobehavioral Reviews, 83*, 381–404. https://doi.org/10.1016/j.neubiorev.2017.10.006.

Vidal-Piñeiro, D., Martin-Trias, P., Arenaza-Urquijo, E. M., Sala-Llonch, R., Clemente, I. C., Mena-Sánchez, I., ... BartrÉs-Faz, D. (2014). Task-dependent activity and connectivity predict episodic memory network-based responses to brain stimulation in healthy aging. *Brain Stimulation, 7*(2), 287–296. https://doi.org/10.1016/j.brs.2013.12.016.

Wang, J., Zhou, Y., He, Y., Li, Q., Zhang, W., Luo, Z., ... Lou, M. (2022). Impact of different white matter hyperintensities patterns on cognition: A cross-sectional and longitudinal study. *NeuroImage. Clinical, 34*, 102978. https://doi.org/10.1016/j.nicl.2022.102978.

Wang, P., Wang, X., Wang, Q., Jiao, Y., Wang, X., Chen, C., ... Song, T. (2023). Cognitive improvement via a modulated rhythmic pulsed magnetic field in D-galactose-induced accelerated aging mice. *Brain Research, 1810*, 148372. https://doi.org/10.1016/j.brainres.2023.148372.

Wang, X., Wanniarachchi, H., Wu, A., Gonzalez-Lima, F., & Liu, H. (2021). Transcranial photobiomodulation and thermal stimulation induce distinct topographies of EEG alpha and beta power changes in healthy humans. *Scientific Reports, 11*(1), 18917. https://doi.org/10.1038/s41598-021-97987-w.

Wang, X., Wanniarachchi, H., Wu, A., & Liu, H. (2022). Combination of group singular value decomposition and eLORETA identifies human EEG networks and responses to transcranial photobiomodulation. *Frontiers in Human Neuroscience, 16*, 853909. https://doi.org/10.3389/fnhum.2022.853909.

Wessel, M. J., Beanato, E., Popa, T., Windel, F., Vassiliadis, P., Menoud, P., ... Hummel, F. C. (2023). Noninvasive theta-burst stimulation of the human striatum enhances striatal activity and motor skill learning. *Nature Neuroscience, 26*(11), 2005–2016. https://doi.org/10.1038/s41593-023-01457-7.

Wischnewski, M., Alekseichuk, I., & Opitz, A. (2023). Neurocognitive, physiological, and biophysical effects of transcranial alternating current stimulation. *Trends in Cognitive Sciences, 27*(2), 189–205. https://doi.org/10.1016/j.tics.2022.11.013.

Wiseman, S. J., Booth, T., Ritchie, S. J., Cox, S. R., Muñoz Maniega, S., Valdés Hernández, M. del C., ... Bastin, M. E. (2017). Cognitive abilities, brain white matter hyperintensity volume, and structural network connectivity in older age. *Human Brain Mapping, 39*(2), 622–632. https://doi.org/10.1002/hbm.23857.

Xia, X., Wang, Z., Zeng, K., Nankoo, J.-F., Darmani, G., Tran, S., ... Chen, R. (2024). Effects of the motor cortical theta-burst transcranial-focused ultrasound stimulation on the contralateral motor cortex. *The Journal of Physiology, 602*(12), 2931–2943. https://doi.org/10.1113/JP285139.

Xie, K., El Khoury, H., Mitrofanis, J., & Austin, P. J. (2023). A systematic review of the effect of photobiomodulation on the neuroinflammatory response in animal models of neurodegenerative diseases. *Reviews in the Neurosciences, 34*(4), 459–481. https://doi.org/10.1515/revneuro-2022-0109.

Yang, C., Xu, Y., Feng, X., Wang, B., Du, Y., Wang, K., ... Liu, Y. (2025). Transcranial temporal interference stimulation of the right globus pallidus in Parkinson's disease. *Movement Disorders, 40*(6), 1061–1069. https://doi.org/10.1002/mds.29967.

Yang, Q., Qu, X., Sheng, C., Zhao, X., Chen, G., Wang, X., ... Han, Y. (2025). Transcranial photobiomodulation improves functional brain networks and working memory in healthy older adults: An fNIRS study. *Neuroimage, 316*, 121305. https://doi.org/10.1016/j.neuroimage.2025.121305.

Yang, Z., Wen, J., Erus, G., Govindarajan, S. T., Melhem, R., Mamourian, E., ... Davatzikos, C. (2024). Brain aging patterns in a large and diverse cohort of 49,482 individuals. *Nature Medicine, 30*(10), 3015–3026. https://doi.org/10.1038/s41591-024-03144-x.

Yi, F., Yuan, J., Somekh, J., Peleg, M., Zhu, Y.-C., Jia, Z., ... Huang, Z. (2025). Genetically supported targets and drug repurposing for brain aging: A systematic study in the UK Biobank. *Science Advances, 11*(11), eadr3757. https://doi.org/10.1126/sciadv.adr3757.

Yoon, K., Lee, W., Croce, P., Cammalleri, A., & Yoo, S.-S. (2018). Multi-resolution simulation of focused ultrasound propagation through ovine skull from a single-element transducer. *Physics in Medicine and Biology, 63*(10), 105001. https://doi.org/10.1088/1361-6560/aabe37.

Yuan, Y., Cassano, P., Pias, M., & Fang, Q. (2020). Transcranial photobiomodulation with near-infrared light from childhood to elderliness: Simulation of dosimetry. *Neurophotonics*, 7(1), 015009. https://doi.org/10.1117/1.NPh.7.1.015009.

Zhu, J., Xian, Q., Hou, X., Wong, K. F., Zhu, T., Chen, Z., ... Sun, L. (2023). The mechanosensitive ion channel Piezo1 contributes to ultrasound neuromodulation. *Proceedings of the National Academy of Sciences of the United States of America, 120*(18), e2300291120. https://doi.org/10.1073/pnas.2300291120.

Zomorrodi, R., Loheswaran, G., Pushparaj, A., & Lim, L. (2019). Pulsed near infrared transcranial and intranasal photobiomodulation significantly modulates neural oscillations: A pilot exploratory study. *Scientific Reports, 9*(1), 6309. https://doi.org/10.1038/s41598-019-42693-x.

Zonneveld, H. I., Pruim, R. H., Bos, D., Vrooman, H. A., Muetzel, R. L., Hofman, A., ... Vernooij, M. W. (2019). Patterns of functional connectivity in an aging population: The rotterdam study. *Neuroimage, 189*, 432–444. https://doi.org/10.1016/j.neuroimage.2019.01.041.

CHAPTER TWO

# Remote photobiomodulation: A systemic approach to neuroprotection

**Daniel M. Johnstone**[a,*] **and Luke C. Gordon**[b]
[a]School of Biomedical Sciences & Pharmacy, University of Newcastle, Callaghan, New South Wales, Australia
[b]School of Biomedical Engineering, University of Sydney, Sydney, New South Wales, Australia
*Corresponding author. e-mail address: Daniel.johnstone@newcastle.edu.au

## Contents



## Abstract

Photobiomodulation (PBM) therapy is rapidly gaining traction in clinical settings, with an expanding body of clinical trials investigating its efficacy as a non-invasive, non-pharmacological intervention to enhance tissue resilience and mitigate disease pathology. Among the most promising developments is the application of PBM in treating neurological conditions, offering new hope to millions affected by these intractable diseases.

The prevailing approach in this field is **transcranial PBM**, where light is administered directly to the head under the assumption that photons can penetrate the scalp and skull to stimulate neural cells relevant to the targeted condition. Encouragingly, this approach has yielded positive outcomes in many studies, reinforcing the assumption and shaping subsequent research.

However, in the enthusiasm to pioneer these potentially transformative therapies, a critical aspect is often overlooked: PBM can also exert indirect, systemic effects.

*International Review of Neurobiology*, Volume 189
ISSN 0074-7742, https://doi.org/10.1016/bs.irn.2026.01.018

Delivering therapeutic doses of light to one part of the body can produce pronounced benefits in distal tissues – a phenomenon supported by numerous animal experiments and emerging human data. Drawing a parallel to remote ischemic conditioning – where brief, controlled ischemia in a limb can induce resilience in vital organs like the heart and brain – we introduced the term "**remote photobiomodulation**" to describe this modality.

This Chapter explores the concept of remote photobiomodulation: its origins, proposed mechanisms, and its relationship to transcranial PBM. By examining current evidence and theoretical frameworks, we aim to demonstrate why this underappreciated approach warrants deeper investigation and how it could broaden the therapeutic landscape for neurological disorders.

## 1. Beyond transcranial PBM

There is compelling evidence from animal studies that direct irradiation of the brain with therapeutic doses of red and/or near-infrared light can improve both pathological and functional outcomes across a wide spectrum of neurological disorders. These include acute injuries such as stroke and traumatic brain injury, neurodegenerative diseases like Alzheimer's and Parkinson's, and neuropsychiatric conditions including depression. As these findings will be explored in detail in other chapters of this book, we will not elaborate on them here.

Translating these promising results from animal models – primarily rodents – to human applications presents significant challenges due to the physical interactions between light and biological tissues. The scalp and skull of a human are considerably thicker than those of a rodent, creating a greater barrier to transcranial light penetration, with a substantial proportion of emitted light energy being absorbed or scattered before reaching the brain. As a result of this scale disparity, PBM parameters that are effective in small animal models may be insufficient to achieve therapeutic doses in the clinical setting.

Theodore Henderson has extensively reviewed the absorption characteristics of light across the hair, scalp, skull and brain during transcranial delivery (Henderson, 2024), synthesising his own measurements with data from numerous other studies. Light penetration through the skin is highly dependent on its thickness. For example, the relatively thin skin of the forehead (~2 mm) allows approximately 10% of light energy to pass through. In contrast, the thicker skin of the scalp permits minimal penetration unless high-powered lasers (> 6 W) are used.

Once past the skin, light must then penetrate the skull. Unlike the thin skull of a mouse (~0.3 mm), which allows ~40 % of near-infrared light to pass through, the thicker human skull (up to ~7 mm) has penetration rates typically below 5 % (Lapchak et al., 2015) and in some cases is reported to be less than 1 % (Castano-Castano et al., 2024). Light must then traverse the meninges, including the dura mater, subdural space, arachnoid mater, cerebrospinal fluid-filled subarachnoid space and pia mater. Each successive layer introduces additional photon losses through tissue-specific absorption and scattering properties, further diminishing the already severely attenuated light energy reaching the brain surface.

These estimates assume ideal conditions – namely, that light is applied to hairless regions or that hair is removed prior to treatment. However, when light is delivered through hair, more than 97 % of the emitted energy is lost (Henderson, 2024). Additionally, skin pigmentation significantly affects light penetration, with absorption coefficients in individuals with darker skin being up to 74 % higher than those with lighter skin across the 400 – 1000 nm spectrum (Setchfield et al., 2024), further compounding the challenge of effective transcranial light application.

Even under optimal circumstances – application to a hairless area with thin skin such as the forehead, and assuming the highest reported penetration rates – the proportion of light energy reaching the brain's surface is estimated to be less than 0.5 % of that emitted. This generous estimate does not account for additional losses due to scattering and reflectance at tissue interfaces.

The upshot is that transcranial PBM is unlikely to deliver therapeutic doses of light to most brain regions, unless high-power lasers (> 6 W) are used. However, such high-power devices pose significant safety risks and are unsuitable for home-use consumer products, limiting their use to professional medical settings. Most clinical trials to date have not used high-power lasers, suggesting that any direct effects of PBM observed in these studies are likely confined to superficial cortical areas. Indeed, functional changes have been observed in these regions – for example, PBM applied to the forehead to stimulate the prefrontal cortex results in measurable changes in cytochrome c oxidase and cerebral hemodynamics (Saucedo et al., 2021) and resting brain state networks (Ghaderi et al., 2021). These findings suggest that some superficial neural populations may receive sufficient light energy to elicit physiological responses. However, for brain structures that lie beneath the superficial cortex, achieving direct irradiation at therapeutic levels is highly unlikely. Therefore, any neuroprotection of these regions is unlikely to be due to direct stimulation.

## 2. Early observations of indirect effects of PBM

Despite the vast majority of PBM research having focused on treating damaged cells or tissues via direct irradiation, there is growing evidence that a localised irradiation can also have an effect at distal, non-irradiated sites (Table 1). In other words, PBM has indirect as well as direct effects.

The initial observations that PBM can have indirect effects date back to the 1980s. For example, in investigating the effect of PBM on the healing of open skin wounds in rats, Kana and colleagues discovered a significant increase in collagen synthesis in both the irradiated wound and a contralateral, non-irradiated wound (Kana et al., 1981). Similarly, Braverman and colleagues explored the effects of PBM on the healing of skin wounds in rabbits, whereby two incisions were made on the back of each rabbit and only one of the two wounds was irradiated, with either 632.8 nm laser PBM (1.65 J/cm$^2$), 904 nm laser PBM (8.25 J/cm$^2$) or both PBM protocols. Compared to wounds in control animals that had not been subjected to PBM treatment, both irradiated and non-irradiated wounds showed an increase in tensile strength, with the investigators speculating that PBM might induce the release of "circulating factors that affect the contralateral wound" (Braverman et al., 1989).

Arguably the most comprehensive of the earlier studies reporting an indirect effect of PBM was published by Rochkind and colleagues, which systematically demonstrated that PBM can induce healing beyond the site of irradiation across multiple different injuries (Rochkind et al., 1989). Their landmark 1989 paper investigated the indirect effects of helium-neon laser PBM (632.8 nm) using rat models of cutaneous wounds, burns and nerve injuries. In rats with bilateral skin incisions along the back, daily PBM (7.6 J/cm$^2$) applied to the margins of one wound significantly accelerated healing of both treated and untreated wounds, relative to control animals receiving no PBM. Similarly, in rats with burn injuries on both hind feet, daily PBM (10 J/cm$^2$) applied to one foot enhanced healing in both feet relative to non-irradiated controls. In another model involving bilateral sciatic nerve crush injuries, unilateral PBM (10 J/cm$^2$) promoted functional recovery in both nerves. Remarkably, this unilateral treatment also conferred neuroprotection to motor neurons in both anterior horns of the spinal cord, mitigating degeneration induced by sciatic nerve injury. Notably, wound healing and nerve recovery persisted well beyond the cessation of PBM treatment, leading the authors to hypothesise that PBM may stimulate the release of "growth-promoting factors".

**Table 1** Preclinical investigations evaluating remote photobiomodulation effectiveness in animal models of neurodegenerative disorders.

| Study | Model | Treatment groups | PBM parameters (per treatment) | Key results |
|---|---|---|---|---|
| Stone et al. (2013) | MPTP mouse PD model | Whole-body PBM and remote PBM (head covered) | 670 nm LED, 50 mW/cm$^2$, 4 J/cm$^2$ | Whole-body PBM mitigated approximately 30 % of MPTP-induced dopaminergic cell loss, while the remote group showed 20 % mitigation. |
| Johnstone et al. (2014) | MPTP mouse PD model | Transcranial PBM and remote PBM (dorsum) | 670 nm LED, 50 mW/cm$^2$, 4 J/cm$^2$ | Both tPBM and rPBM mitigated MPTP-induced dopaminergic cell loss at the 50 mg/kg dose but not higher doses. |
| Farfara et al. (2015) | 5XFAD mouse AD model | Remote PBM (exposed tibia) | Laser (unspecified wavelength), 1 J/cm$^2$ | Remote PBM improved cognitive capacity and spatial learning. PBM reduced amyloid-β plaque burden in the hippocampus by 68 %. |
| Saliba et al. (2015) | Diabetic retinopathy mouse model (streptozotocin) | Whole-body PBM and remote PBM (head covered) | 670 nm LED, 20 mW, 5 J/cm$^2$, 2–3 cm from animal | Body-only remote PBM showed beneficial effects with respect to superoxide generation, leukostasis, expression of inflammatory proteins, and visual function in a diabetic retinopathy mouse model. |
| Kim et al. (2018) | MPTP mouse PD model | Remote PBM (dorsum) | 670 nm LED, 50 mW/cm$^2$, 9 J/cm$^2$ | Remote PBM mitigated MPTP-induced nigral dopaminergic cell loss (40 %). |
| Ganeshan et al. (2019) | MPTP mouse PD model | Remote preconditioning (body) | 670 nm LED, 50 mW/cm$^2$, 4 J/cm$^2$ daily for 2, 5, or 10 days | 10 days of remote preconditioning significantly reduced dopaminergic cell loss. |

*(continued)*

**Table 1** Preclinical investigations evaluating remote photobiomodulation effectiveness in animal models of neurodegenerative disorders. *(cont'd)*

| Study | Model | Treatment groups | PBM parameters (per treatment) | Key results |
|---|---|---|---|---|
| Wang et al. (2020) | APP/PS1 mouse AD model | Whole-body PBM | Mid-infrared light, 7.7–10 μm, unspecified power density and fluence. Exposed to treatment in a cage. | Whole-body PBM improved learning and memory abilities, and attenuated cortical and hippocampal amyloid-β plaques compared to untreated APP/PS1 mice and altered the gut microbiota to reflect that of the WT control. |
| Chen et al. (2021) | Hippocampal injection of amyloid β peptides (1–42) | Remote PBM (abdomen) | 630, 730, 850 nm LED, 10 $mW/cm^2$, 100 $J/cm^2$ | Remote PBM applied to the abdomen significantly enhanced learning and memory retention, decreased amyloid pathology and tau phosphorylation in the hippocampus, modified the expression of 509 proteins, and altered the composition and abundance of gut microbiota. |
| Tao et al. (2021) | APP/PS1 mouse AD model | Whole-body PBM | 1070 nm LED, 900 $mW/cm^2$, 4.5 $J/cm^2$, light pulse frequencies set to 10 and 40 Hz (duty cycle: 50 %). Exposed to treatment in a cage. | Whole-body PBM ameliorated cognitive and memory impairment by reducing cerebral amyloid β burden. PBM with 10 Hz frequency triggered microglial responses and promoted angiogenesis. |

| | | | | |
|---|---|---|---|---|
| Wu et al. (2022) | APP/PS1 and 3xTg-AD mouse AD models | Remote PBM (abdomen, specifically targeting lymph nodes) | 635 nm Laser, 16.7 $mW/cm^2$, 10 $J/cm^2$ | Remote PBM to lymph nodes improved spatial learning and memory, increased hippocampal neurogenesis, enhanced levels of growth factors in the brain, reduced neuroinflammation, and amyloid-β plaque burden. The mechanism is suggested to be via the promoted expression of IFN-γ and IL-10 in non-parenchymal CD4 + T cells. |
| Gordon et al. (2023) | MPTP mouse PD model | Transcranial, and remote PBM (targeting the abdomen or legs) | 656 nm LED, 50 $mW/cm^2$, 9 $J/cm^2$, shields used to irradiate only targeted regions. | Transcranial and both remote PBM regimens significantly reduced MPTP-induced dopaminergic cell loss in the substantia nigra when applied to the head (57 % protection), abdomen (79 % protection), and legs (79 % protection). Additionally, PBM treatment targeting any of these anatomical regions effectively prevented MPTP-induced degeneration of striatal axonal terminals. |
| Gordon et al. (2023) | MPTP macaque PD model (pilot) | Transcranial, and remote PBM (targeting the abdomen or legs) | 670 nm LED, 50 $mW/cm^2$, 9 $J/cm^2$ | Only one animal was allocated to each treatment regimen. Abdominal remote PBM demonstrated clinical improvement with minimal Parkinsonian signs following MPTP administration. Remote PBM targeting the legs delayed clinical symptom onset by 7 days and provided modest |

*(continued)*

**Table 1** Preclinical investigations evaluating remote photobiomodulation effectiveness in animal models of neurodegenerative disorders. (*cont'd*)

| Study | Model | Treatment groups | PBM parameters (per treatment) | Key results |
|---|---|---|---|---|
| | | | | neuroprotection. Conversely, transcranial PBM offered no therapeutic benefit, with severe symptoms developing that required early euthanasia. Histological analysis revealed that abdominal PBM, and to a lesser extent, leg-directed PBM, provided evidence of neuroprotection in the nigrostriatal pathway. |
| Wu et al. (2025) | Repeated closed-head injury (rCHI) rat | Remote PBM (targeting the lungs) | 808 nm laser, 350 $mW/cm^2$, 42 $J/cm^2$ | Remote PBM reduced depressive like behaviour, improved motor function, enhanced social interaction, and reduced anxiety-like behaviors. No effects were seen on spatial or recognition memory. Irradiation of the lungs also reduced neuronal damage, neuroinflammation, reduced tau hyperphosphorylation. Lung microbiota changes were observed, suggesting this as a potential mechanism of action for neuroprotective observations. |

Abbreviations: *3xTg-AD,* Triple transgenic Alzheimer's disease; *5XFAD,* Five familial Alzheimer's disease mutations; *AD,* Alzheimer's Disease; *APP/PS1,* Amyloid Precursor Protein/Presenilin 1; *CD4+,* Cluster of Differentiation 4 positive; *Hz,* Hertz; *IFN-γ,* Interferon-gamma; *IL-10,* Interleukin-10; $J/cm^2$, joules per square centimeter; *Laser,* Light amplification by simulated emission of radiation; *LED,* Light Emitting Diode; *MPTP,* 1-methyl-4-phenyl-1,2,3,6-tetrahydropyridine; *mW,* milliwatts; $mW/cm^2$, milliwatts per square centimeter; *nm,* nanometers; *PBM,* Photobiomodulation; *PD,* Parkinson's Disease; *rCHI,* repeated closed-head injury; *rPBM,* remote Photobiomodulation; *tPBM,* transcranial Photobiomodulation; *WT,* Wild Type; *μm,* micrometers.

Regardless of the precise mechanisms, these foundational studies offered compelling early evidence that PBM can exert indirect – likely systemic – therapeutic effects, laying the groundwork for the development of a novel treatment approach: remote PBM.

# 3. Remote PBM for CNS disorders

## 3.1 Parkinson's disease

At a neuropathological level, Parkinson's disease (PD) predominantly affects the nigrostriatal pathway – a neural circuit embedded deep within the brain. Approximately 80 mm of tissue lies between this anatomical location and the surface of the scalp, making it a prime candidate for remote PBM as a means to overcome the limitations of transcranial PBM. Pioneering work on PBM for PD was led by John Mitrofanis, who, alongside colleagues from Australia and France, published a series of studies testing the efficacy of transcranial PBM in various rodent models of PD. These included primarily MPTP (1-methyl-4-phenyl-1,2,3,6-tetrahydropyridine) neurotoxin mouse models, as well as the K369I tau transgenic mouse model (Moro et al., 2013; Peoples et al., 2012; Purushothuman et al., 2013; Shaw et al., 2010; Shaw et al., 2012). These studies have been reviewed in detail elsewhere (Johnstone et al., 2021; Petrucco et al., 2020). In brief, transcranial PBM in mice mitigated dopaminergic cell loss in the midbrain, reduced neuronal hyperactivity in the striatum and alleviated motor deficits – clearly demonstrating neuroprotective effects from direct brain irradiation with therapeutic PBM doses.

Recognising the significant clinical challenge of delivering therapeutic doses of light to deep brain structures (e.g. the midbrain) in human patients, Mitrofanis teamed with PD clinician and innovator Alim-Louis Benabid to develop an implantable device for intracranial PBM delivery. This device demonstrated safety and efficacy in preclinical rodent models, including the MPTP mouse model (Moro et al., 2014) and the 6-hydroxydopamine rat model (Reinhart et al., 2016) of PD, before being tested in a non-human primate model.

In this study, an optical fibre connected to a 670 nm laser diode was surgically implanted into the midline of the midbrain of adult macaques. After a 14-day recovery period, the animals received intramuscular injections of MPTP to induce parkinsonism. A subset of these macaques underwent intermittent intracranial PBM treatment (5 s ON / 60 s OFF) throughout the 5- or 7-day injection period (Darlot et al., 2016). Daily clinical assessments revealed significantly reduced parkinsonian symptoms and motor dysfunction in PBM-treated macaques compared to sham-treated controls.

Following a 3-week survival period, immunohistochemical analysis revealed substantial neuroprotection in the midbrain of PBM-treated animals, with dopaminergic cell counts comparable to those of saline-injected controls. Although the protective effect in the striatum was more modest, it remained statistically significant (Darlot et al., 2016). Additionally, intracranial PBM markedly reduced MPTP-induced astrogliosis in both the midbrain and the striatum and mitigated pathological changes in microglial morphology (El Massri et al., 2016). This pioneering work laid the foundation for the clinical translation of intracranial PBM and has since progressed to a human trial, detailed in Chapter 14.

### *3.1.1 Animal studies*

In parallel with the development of intracranial PBM, Johnstone, Stone and Mitrofanis explored the alternative treatment modality of remote PBM (Stone et al., 2013). To test whether remote PBM could induce neuroprotective effects, they used an MPTP mouse model of PD (Balb/c mice; two 25 mg/kg MPTP doses administered over two consecutive days) to directly compare the efficacy of transcranial PBM with remote PBM. For remote PBM treatment, the dorsum of each animal was irradiated with light while the head was shielded using aluminium foil. This initial experiment used 670 nm LED light (50 $mW/cm^2$, 4 $J/cm^2$ emitted light per treatment) as a per-conditioning intervention, delivered twice daily on each MPTP injection day. Neuroprotection was assessed by quantifying functional dopaminergic cells in the substantia nigra pars compacta (SNc), using tyrosine hydroxylase (TH) immunolabelling and stereological analysis, seven days after MPTP administration.

As expected, MPTP induced a 35–40 % loss of dopaminergic cells in the SNc. Transcranial PBM mitigated this loss, with treated mice showing dopaminergic cell counts comparable to saline-injected controls. Remarkably, mice receiving PBM exclusively to the body – without any direct irradiation of the head – also exhibited significantly higher SNc dopaminergic cell counts than sham-treated MPTP animals. These findings demonstrated that remote PBM targeting the body can confer neuroprotection to the brain (Stone et al., 2013).

A follow-up study by the same team expanded on the initial findings by including a larger cohort of animals, additional MPTP dosing regimens and assessments of glial cell responses (Johnstone et al., 2014). This more comprehensive investigation reaffirmed that both transcranial and remote PBM significantly protected dopaminergic neurons in the SNc following a 50 mg/kg MPTP insult, as determined by TH immunohistochemistry. While both treatment modalities were effective, transcranial PBM appeared to offer slightly greater neuroprotection, although the difference in cell counts between the two groups did not reach statistical significance. Remote PBM-induced protection of midbrain dopaminergic neurons against 50 mg/kg MPTP insult was subsequently confirmed in an alternative mouse strain, C57BL/6, that is considered particularly susceptible to MPTP (Kim et al., 2018).

At higher MPTP doses (75 – 100 mg/kg), neither transcranial nor remote PBM was effective in preserving dopaminergic neurons in the SNc, suggesting that the neuroprotective capacity of PBM may be limited beyond a certain threshold of neurotoxic insult. Analysis of glial responses in mice receiving 75 mg/kg MPTP showed no evidence of astrocytic activation but a marked increase in microglial activation, which was not attenuated by either form of PBM (Johnstone et al., 2014).

One limitation of these studies was the concurrent timing of PBM and MPTP administration, raising the possibility that PBM may have influenced MPTP metabolism rather than exerting a direct neuroprotective effect. To address this, a subsequent experiment by the same team tested remote PBM as a pre-conditioning strategy. Mice received 670 nm LED light (50 mW/cm$^2$, 4 J/cm$^2$ emitted light per session) applied to the body once daily for 2, 5, or 10 days. MPTP injections (totalling 50 mg/kg) commenced 24 h after the completion of the PBM regimen. Consistent with the per-conditioning results, 10 days of pre-conditioning with remote PBM significantly reduced dopaminergic cell loss in the SNc and dampened abnormal neuronal activity in the caudate-putamen complex, as assessed by FOS immunohistochemistry (Ganeshan et al., 2019).

The observation that PBM applied solely to the body could protect the brain against parkinsonian MPTP insult prompted investigation into whether specific peripheral target sites might offer superior neuroprotective outcomes. To explore this, Gordon and colleagues compared the effects of remote PBM directed at either the abdomen or lower legs (670 nm, 50 mW/cm$^2$, 9 J/cm$^2$ emitted light per treatment) in an n = 1 pilot study in adult macaques exposed to MPTP. Interestingly, transcranial PBM did not improve clinical scores relative to untreated animals, whereas both animals receiving remote PBM appeared healthier. The animal treated to the lower legs showed a 7-day delay in symptom onset, though its clinical scores eventually matched those of untreated controls. In contrast, the animal receiving PBM to the abdomen displayed only mild clinical signs throughout the experimental period and remained behaviourally unaffected by the end of the study. Immunohistochemical analysis of the midbrain supported these observations: the animal treated with remote PBM to the abdomen retained ~50 % more $TH^+$ dopaminergic cells and over 25 % more Nissl-stained neurons in the SNc than untreated controls. Remote PBM to the lower legs yielded a more modest preservation (20 % and 15 %, respectively), while transcranial PBM did not produce measurable neuroprotection. Immunolabelling of TH in the striatum mirrored these findings, with abdominal PBM resulting in a 3.7-fold increase in dopaminergic terminal density relative to untreated MPTP animals, whereas no substantial rescue was observed in the other treatment groups (Gordon et al., 2023).

To build on the promising results from the pilot study in macaques, the team conducted a follow-up investigation in a larger cohort of MPTP-treated mice (each receiving a 50 mg/kg total dose). Using a custom-designed LED device to deliver targeted PBM (670 nm, 50 $mW/cm^2$, 9 $J/cm^2$ emitted light per treatment) to the shaved abdomen, legs or head of C57BL/6 mice, PBM was delivered as a post-conditioning intervention; that is, PBM treatment commenced the day after the final MPTP injection. The results reinforced the findings of the pilot study: PBM delivered to peripheral sites (i.e. abdomen or legs) significantly preserved dopaminergic neurons in the SNc and improved striatal terminal density. Surprisingly, while transcranial PBM also conferred significant neuroprotection, remote PBM to either the abdomen or legs yielded stronger neuroprotective effects (Gordon et al., 2023), further supporting the potential of targeted remote PBM as a viable strategy for mitigating parkinsonian neurodegeneration.

### *3.1.2 Clinical trials*

Given its favourable safety profile, PBM has recently entered early-stage clinical testing for PD. Initial studies have primarily explored transcranial PBM, with mixed but encouraging results. For instance, Santos and colleagues conducted a randomized controlled trial applying 670 nm light to the temples of PD patients, reporting modest improvements in gait (Santos et al., 2019). Similarly, Hamilton and colleagues observed improvements in fine motor skills and self-reported non-motor symptoms in a small pilot study using a transcranial PBM device delivering 670 nm and 810 nm light (Hamilton et al., 2019).

More recent investigations have also begun to explore remote PBM in clinical settings. Liebert and colleagues conducted a 52-week pilot study in 12 PD patients that incorporated remote PBM treatment (targeting the abdomen and neck) alongside transcranial and intranasal PBM. Notably, all forms of PBM were well tolerated, and participants showed significant improvements in mobility, cognition and fine motor control over the course of treatment (Liebert et al., 2021).

Building on this work, Liebert and colleagues conducted a follow-up proof-of-concept study focusing exclusively on remote PBM applied to the neck and abdomen (Liebert et al., 2022). In this trial, seven PD patients received 12 weeks of in-clinic PBM (904 nm, 30 mW) followed by 33 weeks of at-home treatment. Despite significant disruptions due to bushfires and the COVID-19 pandemic, participants showed notable improvements in mobility, balance and cognitive function. Notably, the authors concluded that the magnitude of clinical benefit observed with remote PBM alone was comparable to that seen in their previous study combining remote and transcranial PBM, suggesting that peripheral application may be sufficient to elicit meaningful therapeutic effects. These findings support the continued exploration of remote PBM as a non-invasive, accessible intervention for PD.

## 3.2 Other CNS diseases

In addition to Parkinson's disease, a small number of studies have reported neuroprotective effects of remote PBM in other disorders of the central nervous system. One notable line of investigation has come from Uri Oron's team in Tel Aviv, who developed the concept of bone marrow-targeted PBM as a means of protecting distant tissues, including the brain. Using the 5xFAD transgenic mouse model of Alzheimer's disease, Farfara and colleagues applied 808 nm PBM (1 J/cm$^2$) to the tibia six times over two months and assessed behavioural performance and hippocampal β-amyloid (Aβ) burden (Farfara et al., 2015). At six months of age, treated mice showed significantly improved performance on object recognition and fear conditioning tasks compared to sham-treated controls, alongside a marked reduction in hippocampal Aβ deposits. Complementary in vitro experiments revealed that bone marrow-derived mesenchymal stem cells (MSCs) exposed to PBM exhibited enhanced phagocytosis of $A\beta_{1-42}$ peptide. These findings suggest that bone marrow-derived MSCs may contribute to Aβ clearance and mediate the neuroprotective effects of remote PBM in Alzheimer's disease (Farfara et al., 2015).

Further evidence for the neuroprotective potential of remote PBM has emerged from a recent study investigating its effects in a rat model of repeated closed-head injury (rCHI). In this work, PBM (808 nm laser light, 42 J/cm$^2$ per treatment) was directed at the lungs daily over a 20-day period. In this model, remote PBM significantly improved behavioural outcomes (depressive-like symptoms, motor deficits, social interaction impairment) and increased grip strength. Histological analysis revealed reduced neuronal damage, apoptosis and degeneration in the brain. The authors also reported attenuation of glial hyperactivation and excessive synaptic pruning by microglia and astrocytes, alongside preservation of microglial SIRPα levels and reduced tau pathology. Interestingly, changes in the lung microbiome were observed following PBM, suggesting a role for the lung–brain axis in mediating these neuroprotective effects. Although the study did not directly compare remote and transcranial PBM, the authors noted that behavioural improvements achieved with lung-targeted PBM appeared greater than those reported in their earlier work using transcranial PBM in the same model (Wu et al., 2025).

Remote PBM has also shown promise in protecting the retina, another CNS structure. Saliba and colleagues investigated its effects in a mouse model of streptozotocin-induced diabetic retinopathy (Saliba et al., 2015). Beginning four weeks after diabetes onset, mice received daily PBM (670 nm, ~5 J/cm$^2$) for ten weeks using a large LED panel. Two treatment groups were included: one with full-body and head exposure, and another with the head shielded using a lead covering. Both PBM-treated diabetic groups exhibited reduced retinal superoxide production and leukostasis, along with improved visual function, relative to sham-treated diabetic mice. These results indicate that body-only PBM is sufficient to confer retinal protection. Furthermore, leukocytes isolated from PBM-treated mice were less cytotoxic to endothelial cells in vitro, suggesting that modulation of immune cell function may play a role in the therapeutic effects of remote PBM in this model (Saliba et al., 2015).

These findings were further supported by a subsequent long-term study from the same group, in which streptozotocin-diabetic mice received daily PBM (670 nm) directed at the body for eight months (Cheng et al., 2018). Although the head was not strictly shielded during treatment, the results reinforced earlier observations: PBM significantly inhibited diabetes-induced retinal capillary leakage and degeneration and improved visual function.

Intriguingly, the beneficial effects of remote PBM on the retina may not be restricted to pathological conditions. Jeffery and colleagues recently demonstrated that 850 nm LED light (~9 mW/cm$^2$ at body surface, 15-minute exposure) directed at the back of volunteers without existing vision disorders resulted in improved colour contrast thresholds 24 h post-treatment. Notably, this enhancement was observed even in participants whose heads were wrapped in aluminium foil to prevent any direct light exposure, suggesting that remote PBM may enhance retinal function via systemic mechanisms, even in individuals with normal vision (Jeffery et al., 2025).

## 4. Potential mechanisms

With multiple independent studies having demonstrated that PBM can exert beneficial effects at sites distant from the point of light application, a central question emerges: ***how*** does this happen? What mechanisms transduce signals from the irradiated peripheral tissue to the brain (or other affected organs) and mediate therapeutic outcomes?

The honest answer is that we don't yet fully know. While substantial research has clarified how cells respond to direct light exposure, our understanding of how these local responses translate into systemic signalling remains in its infancy. We anticipate that the neuroprotective benefits of remote photobiomodulation will not stem from a single, predictable pathway. Land-dwelling mammals have been consistently exposed to longer wavelength sunlight throughout evolution and have likely evolved multiple light-sensitive biological systems that could be fundamental to maintaining neurological health. What follows is a discussion of several plausible mechanisms, offered with the caveat that the true picture is likely to be highly complex and context-dependent.

For the purposes of this Chapter, we have divided the mechanisms of PBM into two complementary categories:

1. Intracellular responses to direct light exposure, and
2. Systemic signalling mediators triggered by these responses that influence distant tissues.

The first category has already been extensively reviewed elsewhere, including in other chapters of this book. Therefore, the focus here will be on the second: the potential mediators that may transduce signals from the site of irradiation to remote targets, such as the brain or retina.

## 4.1 Circulating stem cells

One of the leading candidates for mediating the indirect/systemic effects of PBM is stem cells. There is growing evidence that PBM can stimulate stem cells at their site of residence – particularly within the bone marrow – leading to their proliferation and mobilisation into the bloodstream. Once in circulation, these cells are thought to home to sites of tissue damage, where they release paracrine factors that support repair, rather than directly replacing damaged cells through differentiation.

Pioneering work in this area was led by Uri Oron and colleagues, who first demonstrated that bone marrow-derived cells exposed to 804 nm laser light (50 $mW/cm^2$, 1–3 $J/cm^2$) *in vitro* showed a 2-fold increase in proliferation compared to sham-treated cells (Tuby et al., 2007). Building on this, the team sought to determine if PBM affected the biological function of bone marrow-derived stem cells by treating isolated cells with PBM (804 nm, 1 $J/cm^2$) *ex vivo* and transplanting them into a rat model of myocardial infarction. Transplantation of PBM-treated cells significantly reduced infarct size and enhanced angiogenesis and vascular endothelial growth factor levels relative to non-treated cells (Tuby et al., 2009).

To test whether similar benefits could be achieved by stimulating stem cells *in situ*, the team applied PBM directly to the exposed tibia of rats following myocardial infarction. Remarkably, tibial PBM (804 nm, 1 $J/cm^2$) delivered either 20 min or 4 h post-injury led to a significant reduction in infarct size and ventricular dilatation – effects that were even greater than those observed with direct cardiac PBM treatment (Tuby et al., 2011). Treated animals also showed increased numbers of c-kit$^+$ stem cells in the heart (Tuby et al., 2011), suggesting that remote PBM may enhance the homing of stem cells to injured tissue.

This phenomenon has since been replicated in other models. In a rat model of acute kidney injury, remote PBM of the tibia reduced renal pathology and improved biomarkers of kidney function, in association with increased c-kit$^+$ cell density in the kidneys (Oron et al., 2014). In a pig model of myocardial infarction, PBM applied to the tibia at 30 min, 2 days and 7 days post-injury (808 nm, 1 $J/cm^2$ per treatment) increased circulating c-kit$^+$ cell levels and reduced infarct volume at 90 days, alongside improved cardiac function (Blatt et al., 2016). As discussed earlier, remote PBM of the tibia also improved cognition and reduced neuropathology in a mouse model of Alzheimer's disease (Farfara et al., 2015).

More recently, this approach has been explored in human studies. In healthy volunteers, non-invasive PBM of the tibia increased circulating CD-34$^{+}$ cells, with levels rising around 2 h post-treatment and peaking between 2–4 days. A concurrent rise in circulating macrophages was also observed (Oron et al., 2022). In a small randomised controlled trial assessing safety and feasibility in myocardial infarction patients, tibial PBM was well tolerated and associated with reduced circulating troponin-T levels, although this study was underpowered to detect changes in cardiac function (Elbaz-Greener et al., 2018). These findings suggest that tibial PBM may directly stimulate stem cells in the bone marrow, enabling a technical advance on the low-yield and costly process of stem cell isolation, *in vitro* expansion and reinfusion.

Importantly, the effects of PBM on stem cells may not be limited to those residing within bone marrow. *In vitro* studies have shown that PBM can also enhance viability and proliferation and reduce apoptosis in stem cells isolated from human adipose tissue (Mvula et al., 2010; Zare et al., 2019), suggesting that multiple stem cell populations may contribute to the systemic benefits of PBM.

## 4.2 Mitokines and other circulating molecules

One class of signalling molecules that warrants greater attention in the context of remote PBM – yet has received little – is mitokines. These are factors secreted into the bloodstream in response to localized mitochondrial stress, capable of travelling to distant tissues to induce adaptive mitochondrial responses and enhance cellular resilience (Durieux et al., 2011). While mitokine signalling has not yet been directly investigated in the context of PBM, it remains a highly plausible mechanism, particularly given that mitochondria are the primary intracellular targets of PBM. The molecular effects of PBM – such as transient increases in ROS – are consistent with mild mitochondrial stress. Like PBM, mitokine signalling appears to follow the principles of hormesis, whereby mild or transient mitochondrial stress triggers a mitokine response that induces beneficial adaptations, while chronic or severe mitochondrial stress is associated with a mitokine profile that is deleterious (Burtscher et al., 2023; Ristow & Schmeisser, 2014; Yun & Finkel, 2014). Notably, mitokine release is triggered by lifestyle interventions such as exercise and dietary restriction – interventions that, as discussed elsewhere (Stone et al., 2018), share mechanistic overlap with PBM as "stresses of everyday life" that promote acquired resilience.

The best-characterised mitokines in mammals include fibroblast growth factor 21 (FGF21), growth differentiation factor 15 (GDF15), and mitochondrial-encoded peptides such as humanin and MOTS-c (Burtscher et al., 2023; Zhang et al., 2023). To date, most research on these molecules has focused on metabolic regulation, with limited exploration of their roles in the CNS. One exception is humanin, which has been shown to prevent synapse loss in hippocampal neurons (Zarate et al., 2019). Additionally, plasma levels of FGF21 and humanin are reportedly lower in individuals with Alzheimer's disease than in healthy controls (Conte et al., 2021). Importantly, there is evidence that receptors for mitokines are expressed in neural tissues – for example, the main receptor for GDF15 (GFRAL) is predominantly expressed in the CNS (Burtscher et al., 2023). These findings support the hypothesis that mitokines released from peripheral tissues in response to PBM could mediate protective effects in the brain. However, no studies to date have directly examined whether PBM alters circulating mitokine levels, highlighting a key gap in the literature and an important direction for future research.

In addition to mitokines, there is evidence that PBM can influence the levels of circulating cytokines and chemokines. For example, whole-body PBM treatment of healthy mice significantly reduced plasma levels of eotaxin (Shamloo et al., 2023), a chemokine known to cross the blood-brain barrier and associated with ageing and cognitive impairment (Erickson et al., 2014; Hoefer et al., 2017; Villeda et al., 2011). In mice with systemic inflammation induced by lipopolysaccharide (LPS), pre-conditioning with whole-body PBM attenuated the LPS-driven rise in pro-inflammatory cytokines interleukin (IL)−1β, IL-18 and IL-22, while increasing levels of the anti-inflammatory cytokine IL-10 (Shamloo et al., 2023). Similarly, in an experimental autoimmune encephalomyelitis (EAE) mouse model of multiple sclerosis, whole-body PBM down-regulated pro-inflammatory cytokines such as interferon-γ and tumor necrosis factor-α, and up-regulated anti-inflammatory cytokines including IL-4 and IL-10, in parallel with improvements in disease features (Muili et al., 2012).

## 4.3 Microbiomes

The observation that PBM applied to the abdomen can confer neuroprotection in models of Parkinson's disease (Gordon et al., 2023) has prompted investigation into whether this effect may be mediated, at least in part, by PBM modulation of the gut microbiome. There is already substantial evidence that the gut microbiome influences brain function and contributes to the pathogenesis of neurological disorders, including PD (Houser & Tansey, 2017; Morais & Schreiber, 2021; Sampson et al., 2016; Scheperjans et al., 2015).

Initial work in this area by Bicknell and colleagues demonstrated that abdominal PBM in mice alters microbial diversity in a wavelength-dependent manner, increasing the abundance of species associated with a healthy microbiome (Bicknell et al., 2019; Liebert et al., 2019). This line of research was extended to a mouse model of Alzheimer's disease, where abdominal PBM was shown to reverse disease-associated changes in gut microbial composition, in parallel with improvements in cognitive function and reductions in neuropathology (Chen et al., 2021). In a rat model of type 2 diabetes, PBM applied to the duodenum using dual-wavelength LED light also modified the gut microbiome, in association with improvements in glucose tolerance and hepatic parameters (Min et al., 2022).

Evidence from human studies is beginning to emerge. In a retrospective analysis of a Parkinson's disease trial cohort described earlier in this Chapter (Liebert et al., 2021), Bicknell and colleagues found that abdominal PBM altered the gut microbiome, including a shift in the Firmicutes/Bacteroidetes ratio – a commonly used proxy for gut health (Bicknell, Liebert, et al., 2022). In a separate case study of a breast cancer patient who had undergone radiotherapy and immunotherapy, abdominal PBM (904 nm) led to marked changes in microbiome diversity. These included increases in several genera commonly associated with gut health (e.g. *Akkermansia, Faecalibacterium* and *Roseburia*) and reductions in genera considered potentially pathogenic (Bicknell, Laakso, et al., 2022).

Beyond the gut, PBM may also influence microbiomes in other tissues. As discussed earlier in this Chapter, a study in a rat model of repeated closed-head injury found that PBM targeted at the lungs not only conferred neuroprotection but also altered the lung microbiome, including enrichment of genetic pathways related to lipopolysaccharide biosynthesis – changes the authors speculated may prime microglia and dampen neuroinflammation (Wu et al., 2025). While not yet explored in the context brain disorders, PBM has also been shown to modulate microbial communities in the skin and oral mucosa (Munoz Declara et al., 2024; Zanotta et al., 2020).

## 4.4 Other potential mechanisms

In addition to effects on circulating cells, soluble factors and microbiomes, several other plausible – though less extensively studied – mechanisms may contribute to the systemic effects of remote PBM.

One intriguing possibility involves circulating mitochondria. Al Amir Dache and colleagues discovered functional cell-free mitochondria that circulate in the bloodstream, raising the possibility that these organelles could act as systemic messengers (Al Amir Dache et al., 2020). Given that mitochondria are the primary targets of PBM, it is plausible that circulating mitochondria play an important role in mediating the systemic effects of remote PBM.

Another emerging area of interest is the role of extracellular vesicles (EVs). PBM has been shown to increase the release of EVs from various cell types. For example, Weihrauch and colleagues demonstrated that PBM enhances EV release from endothelial cells, facilitating the export of vasodilatory molecules and promoting smooth muscle relaxation (Weihrauch et al., 2021). Similarly, PBM has been reported to stimulate EV release from mesenchymal stem cells (Chang et al., 2024; Gholami et al., 2022), in association with increased cell viability and migratory capacity (Chang et al., 2024). Given the proposed role of mesenchymal stem cells in mediating the systemic effects of remote PBM, these findings suggest that EVs may serve as a critical link between peripheral stem cell activation and neuroprotection in the brain.

Neural pathways may also contribute to the systemic signalling initiated by PBM. Although this possibility has not been directly explored in PBM research, precedent exists in a related intervention: remote ischemic conditioning (RIC). In RIC, brief cycles of limb ischemia and reperfusion confer protection to distant organs, including the heart and brain. As reviewed elsewhere (Kim et al., 2017), both humoral and neural pathways appear to be involved in mediating these protective effects. One hypothesis is that RIC activates peripheral nerves, which in turn triggers the release of circulating protective factors. Supporting this, disruption of vagus nerve signalling has been shown to impair RIC-induced cardioprotection (Mastitskaya et al., 2016).

## 5. The paradox of transcranial PBM: Is it remote PBM by another name?

As discussed earlier in this Chapter, transcranial PBM has demonstrated strong neuroprotective effects in small animal models of brain disorders, presumably due to the direct irradiation of vulnerable or damaged cells by photons penetrating the scalp, skull and superficial neural tissue. Early clinical trials suggest that transcranial PBM may also benefit human patients with neurological conditions (Table 2). However, a paradox arises: in many of these trials, the irradiation parameters used appear insufficient to deliver therapeutic doses of light to the target brain regions (Henderson, 2024). This raises a critical question – how is transcranial PBM exerting its effects?

One possibility, yet to be formally tested, is that the observed benefits are indirect. Rather than acting directly on neurons or glia, transcranial PBM may stimulate cells within the overlying tissues – such as the scalp or skull – that then initiate signalling cascades that transduce protective effects to the brain. In this sense, transcranial PBM may operate as a form of remote PBM.

The skull is more than a passive barrier; it plays an active role in neuroimmune communication. For example, Herisson and colleagues identified vascular channels connecting the skull brain marrow to the brain's surface, allowing immune cells to migrate directly from the marrow to sites of brain injury or inflammation (Herisson et al., 2018). While their study focused on myeloid cells, it is plausible that stem cells within the skull marrow exhibit similar migratory behaviour, homing to damaged brain regions and releasing neurotrophic factors. If skull-resident stem cells respond to PBM in a manner analogous to those in the tibia, then irradiation of the skull during transcranial PBM may enhance local stem cell activity and contribute to neuroprotection.

Further evidence that transcranial PBM may exert its effects through non-neuronal mechanisms comes from recent work on meningeal lymphatic vessels (mLVs), a network within the dura mater responsible for clearing macromolecular waste and inflammatory mediators from the brain. In aged mice, transcranial PBM using 808 nm light enhanced mLV drainage and distribution, leading to improvements in spatial learning, memory and object recognition. Similar benefits were observed in Alzheimer's disease models, where PBM reduced amyloid burden and neuroinflammation. Crucially, these cognitive improvements were abolished in mice with ablated mLVs, indicating that functional mLV drainage is necessary for PBM-induced neuroprotection (Wang et al., 2024). These findings suggest that transcranial PBM may act, at least in part, by modulating extracerebral structures such as the mLVs, further supporting the hypothesis that, in humans, its therapeutic effects might be mediated through indirect pathways rather than direct stimulation of neural cells.

**Table 2** Clinical research assessing photobiomodulation therapy in individuals with Alzheimer's and Parkinson's disease.

| Study | Study design | Patient population | Type of PBM | PBM parameters (per treatment) | Treatment regimen | Key results |
|---|---|---|---|---|---|---|
| Saltmarche et al. (2017) | Case series study, no control group | Cases of mild to moderately severe dementia (n = 5) | Transcranial & Remote (Intranasal) PBM | 810 nm, 10 Hz pulsed, LED. Transcranial: 41 $mW/cm^2$, 25 $J/cm^2$, Intranasal: 14–23 $mW/cm^2$, 11–14 $J/cm^2$ | Daily home use; 20 min for 12 weeks with an intranasal device. Weekly clinic visit with transcranial and intranasal PBM. | Significant cognitive improvements with mean MMSE scores increasing 2.6 points ($p < 0.003$) and ADAS-cog scores improving 7.73 points ($p < 0.023$). Benefits were reported to decline after treatment was discontinued. |
| Berman et al. (2017) | Double-blind, randomized, placebo-controlled trial | Patients with AD, n = 9 (6 tPBM and 3 sham) | Transcranial PBM | tPBM: 1072 nm LED, unspecified irradiance and fluence. | 6-minute treatment daily for 28 days. | After 28 days, no statistically significant difference as compared to the control group. |
| Hamilton et al. (2019) | Case series study, no control group | PD patients (n = 6) | Transcranial & Remote (Intranasal) PBM | tPBM: 670, 810 and 850 nm LED, rPBM: 660 nm LED, unspecified irradiance and fluence. | 1–2 daily treatment sessions of 10–30 min | Observed improvements, 55 % of initial signs and symptoms showed overall improvement, 43 % stayed the same, and 2 % got worse. Improvements dependent on individual, both motor and non-motor symptoms affected. |

| | | | | | | |
|---|---|---|---|---|---|---|
| Salehpour et al. (2019) | Single case report, no control group | Single case of a person with MCI | Transcranial & Remote (Body, targeting lower back and intranasal) | tPBM: 635 and 810 nm LED, Remote (body): same as tPBM, Remote (intranasal): 810 nm LED | All three devices were applied simultaneously, twice daily, for 25 min each | Significant improvement in Montreal Cognitive Assessment score from 18 to 24 and Working Memory Questionnaire score from 53 to 10. Reversal of olfactory dysfunction (Alberta Smell Test). QoL measures improved. |
| Chao (2019) | Open, randomized, placebo-controlled trial | Patients with AD, n = 8 (4 tPBM and 4 standard care) | Transcranial and remote (intranasal) PBM | tPBM: 810 nm, 75–100 $mW/cm^2$, 40–60 $J/cm^2$. rPBM: 25 $mW/cm^2$, 15 $J/cm^2$ | 20-minute treatment, 3 days per week for 12 weeks | After 12 weeks of LED-based PBM treatment, the intervention group showed significantly better outcomes compared to usual care in cognitive function (ADAS-cog), behavioral symptoms (NPI scores), and cerebral blood flow, with no adverse effects reported. |
| Santos et al. (2019) | Randomized, placebo-controlled trial | Patients with PD, n = 35 (17 tPBM and 18 sham) | Transcranial PBM | tPBM: 670 nm LED, 60 $mW/cm^2$, 8 $J/cm^2$ | Applied to left and right temples, alternating in 6 blocks of 1 min each with a 30-second rest between for each session. Sessions were provided twice a week for 9 weeks. | No significant effect on the primary outcome measure, MDS-UPDRS motor scores. Significant improvement in fast rhythm walking speed ($p = 0.001$). No changes detected in static posturography, preferred walking speed, or other motor function measures. |

*(continued)*

**Table 2** Clinical research assessing photobiomodulation therapy in individuals with Alzheimer's and Parkinson's disease. (*cont'd*)

| Study | Study design | Patient population | Type of PBM | PBM parameters (per treatment) | Treatment regimen | Key results |
|---|---|---|---|---|---|---|
| Liebert et al. (2021) | Pilot study, six were assigned to active treatment, and another six were waitlisted for 14 weeks before commencing the same treatment. | PD patients (n = 12) | Remote (abdomen/ neck/ intranasal) + Transcranial + Intranasal | tPBM: 810 nm LED, 75–100 $mW/cm^2$, ~45–180 $J/cm^2$. rPBM: 904 nm Laser, 47 $mW/cm^2$, ~6–15 $J/cm^2$ | Variable; between 1–3 times per week. rPBM for 330–660 s, tPBM for 2100 s per treatment | Participants demonstrated significant improvements in mobility, cognition, and fine motor control over 12 weeks, with sustained benefits up to one year. |
| Nagy et al. (2021) | Open, randomized, placebo-controlled trial | Patients with AD, n = 60 (30 rPBM, 30 sham) | rPBM (intranasal + wrist watch) | rPBM: Laser 650 nm, unspecified irradiance and fluence | A 30-minute treatment, twice daily, 3 days a week, for 12 weeks. Associated with physical exercise. | At 12 weeks, compared to the sham group, those treated with rPBM showed improved cognition and quality of life. |
| Nizamutdinov et al. (2021) | Double-blind, randomized, placebo-controlled trial | Patients with AD, n = 57 (40 PBM and 17 sham) | Transcranial and ocular PBM | tPBM: 1060-1080 nm LED, 23 $mW/cm^2$, 8 $J/cm^2$ | 6-minute treatment, twice daily, for 8 weeks. | After 8 weeks of PBM treatment, the intervention group showed significant improvements in multiple |

| | | | | | | |
|---|---|---|---|---|---|---|
| | | | | | | cognitive domains including general cognition (MMSE), memory (logical and auditory verbal learning tests), executive function (TMT A&B), and language (Boston naming test), along with patient-reported improvements in sleep, energy, and daily activities, and caregiver-reported improvements in anxiety and mood, while the sham group showed no significant changes. |
| Bullock-Saxton et al. (2021) | Double-blind randomized, placebo-controlled trial | Patients with PD, n = 20 (10 group 1 and 10 group 2). 2 lost to follow-up. | Transcranial and intraoral | PBM: 4 cranial and 1 intraoral points of stimulation, 904 nm laser, 60 mW, 2 J per point. | 33 s per point (21 points/session). Crossover trial: 4 weeks active vs. sham (3/week), 4-week washout, then 4 weeks with varying active/sham frequencies (1–2/week active + sham). | No statistically significant differences between groups on any outcome measures. Some clinically important differences in the 9-hole peg test (dominant hand) and spiral writing test. It was suggested that applying 2–3 times per week for more than 4 weeks was required to show effects. |

*(continued)*

**Table 2** Clinical research assessing photobiomodulation therapy in individuals with Alzheimer's and Parkinson's disease. (*cont'd*)

| Study | Study design | Patient population | Type of PBM | PBM parameters (per treatment) | Treatment regimen | Key results |
|---|---|---|---|---|---|---|
| Liebert et al. (2022) | Pilot study, no control group | PD patients ($n = 7$) | Remote PBM targeting abdomen and neck (no transcranial) | 904 nm Laser, 47 $mW/cm^2$, 11.3 $J/cm^2$ | 3 treatment sessions per week; 12 weeks with clinic treatments, 33 weeks of home treatment. | Improvements in PD clinical signs included enhanced functional mobility, cognitive function, sense of smell, and fine motor skills. Some improvements were maintained at 45 weeks despite COVID-19 lockdowns. |
| Kheradmand et al. (2022) | Double-blind, randomized, placebo-controlled trial | Patients with dementia; $n = 16$ PBM group, $n = 16$ sham group | Transcranial & Remote (Intranasal) PBM | tPBM: 630 and 810 nm, 12 diodes of each in a helmet configuration. 90 $mW/cm^2$ per diode, total fluence of 57 $J/cm^2$. Pulsed at 75 Hz. | 3 sessions/week for 2 weeks. 10 min per session. Total of 6 treatment sessions. | Significant improvements in the PBM group vs the sham control: MMSE scores increased at weeks 2 and 8, while the sham group showed a decline in MMSE scores. Improvements sustained 6 weeks post-treatment, no adverse effects reported. |
| Blivet et al. (2022) | Double-blind, randomized, placebo-controlled trial | Patients with AD, $n = 53$ (27 PBM vs. 26 sham groups) | Transcranial & Remote (abdomen) PBM | tPBM and rPBM: 660 and 850 nm, laser + LED in helmet and abdominal belt configuration. | 25-minute sessions, 5 times weekly for 8 weeks (40 total sessions) | Improved comprehension and executive function. Sustained benefits 4 weeks post-treatment of enhanced working memory. Well-tolerated with minimal adverse events reported. |

| | | | | | | |
|---|---|---|---|---|---|---|
| | | | | Laser: 21 mW/cm$^2$, LED: 29 mW/cm$^2$ (infrared) and 25 mW/cm$^2$ (red), total fluence of 16–22 J/cm$^2$. Pulsed at 10 Hz. | | |
| Herkes et al. (2023) | Double-blind, randomized, placebo-controlled feasibility trial | Patients with PD, n = 38 (20 tPBM and 18 sham) | Transcranial PBM | tPBM: 635 and 810 nm LED, 20 × 52 mW and 27 mW diodes, respectively. This produced ~ 10 J of infrared and ~ 5 J of red light per session. | A 12-minute treatment, 6 days a week, for 12 weeks. | Treatments were well tolerated, with only minor, transient adverse events. On average, participants receiving tPBM showed improved PD motor symptoms; however, no significant differences were detected between the active and sham groups after 12 weeks. |
| McGee et al. (2023) | Double-blind randomized, placebo-controlled trial | Patients with PD, n = 40 (20 PBM vs 20 sham) | Transcranial PBM | tPBM: 635 + 810 nm, unspecified irradiance and fluence. | 24-minute treatment, 6 days per week for 12 weeks | After 12 weeks of treatment, no statistically significant differences were observed in MDS-UPDRS-III scores or subscores between groups, with no safety concerns or adverse events reported, except for occasional mild, transient dizziness. |

*(continued)*

**Table 2** Clinical research assessing photobiomodulation therapy in individuals with Alzheimer's and Parkinson's disease. (*cont'd*)

| Study | Study design | Patient population | Type of PBM | PBM parameters (per treatment) | Treatment regimen | Key results |
|---|---|---|---|---|---|---|
| Liebert et al. (2024) | Case series study, no control group | Patients with PD, n = 7 | Remote PBM and exercise specific to PD | rPBM: 904 nm Laser 30 mW across two diodes, with fluence of 5.64 $J/cm^2$ per point. Each treatment consisted of 9 points on the abdomen and 1 on the neck. | 20-minute treatment, 3 days per week. Two participants received treatment for 6 weeks, while the remainder received treatment for 2–4 years. | All participants reported anecdotal improvements in some of their Parkinsonian symptoms, including both motor and non-motor symptoms. |

Abbreviations: *AD*, Alzheimer's Disease; *ADAS-cog*, Alzheimer's Disease Assessment Scale-cognitive subscale; *COVID-19*, Coronavirus Disease 2019; *Hz*, Hertz; *J*, Joules; $J/cm^2$, joules per square centimeter; *Laser*, Light amplification by simulated emission of radiation; *LED*, Light Emitting Diode; *MCI*, Mild Cognitive Impairment; *MDS-UPDRS-III*, Movement Disorder Society-Unified Parkinson's Disease Rating Scale Part III; *MMSE*, Mini-Mental State Examination; *mW*, milliwatts; $mW/cm^2$, milliwatts per square centimeter; *nm*, nanometers; *NPI*, Neuropsychiatric Inventory; *PBM*, Photobiomodulation; *PD*, Parkinson's Disease; *QoL*, Quality of Life; *rPBM*, remote Photobiomodulation; *TMT A&B*, Trail Making Test A and B; *tPBM*, transcranial Photobiomodulation.

## 6. Conclusion

Remote PBM offers a compelling solution to the limitations of transcranial light delivery in humans, particularly for deep brain targets. Its favourable safety profile, compatibility with home-use devices and emerging clinical evidence position it as a practical and scalable intervention for neurological disorders.

To translate this promise into clinical impact, the next phase of research must focus on standardising protocols, validating efficacy across diverse populations and integrating mechanistic insights into trial design. Remote PBM is not just an alternative to transcranial approaches – it may be the key to unlocking broader therapeutic access and effectiveness.

### Declaration of generative AI and AI-assisted technologies in the writing process

During the preparation of this work the authors used M365 Copilot for feedback on the phrasing of some sections of text. After using this tool, the authors reviewed and edited the content as needed and take full responsibility for the content of the publication.

## References

Al Amir Dache, Z., Otandault, A., Tanos, R., Pastor, B., Meddeb, R., Sanchez, C., ... Thierry, A. R. (2020). Blood contains circulating cell-free respiratory competent mitochondria. *The FASEB Journal, 34*(3), 3616–3630. https://doi.org/10.1096/fj.201901917RR.

Berman, M. H., Halper, J. P., Nichols, T. W., Jarrett, H., Lundy, A., & Huang, J. H. (2017). Photobiomodulation with near infrared light helmet in a pilot, placebo controlled clinical trial in dementia patients testing memory and cognition. *Journal of Neurology and Neuroscience, 8*(1), https://doi.org/10.21767/2171-6625.1000176.

Bicknell, B., Laakso, E. L., Liebert, A., & Kiat, H. (2022). Modifying the microbiome as a potential mechanism of photobiomodulation: A case report. *Photobiomodulation, Photomedicine, and Laser Surgery, 40*(2), 88–97. https://doi.org/10.1089/photob.2021.0057.

Bicknell, B., Liebert, A., Johnstone, D., & Kiat, H. (2019). Photobiomodulation of the microbiome: Implications for metabolic and inflammatory diseases. *Lasers in Medical Science, 34*(2), 317–327. https://doi.org/10.1007/s10103-018-2594-6.

Bicknell, B., Liebert, A., McLachlan, C. S., & Kiat, H. (2022). Microbiome changes in humans with Parkinson's disease after photobiomodulation therapy: A retrospective study. *Journal of Personalized Medicine, 12*(1), https://doi.org/10.3390/jpm12010049.

Blatt, A., Elbaz-Greener, G. A., Tuby, H., Maltz, L., Siman-Tov, Y., Ben-Aharon, G., ... Oron, U. (2016). Low-level laser therapy to the bone marrow reduces scarring and improves heart function post-acute myocardial infarction in the pig. *Photomedicine and Laser Surgery, 34*(11), 516–524. https://doi.org/10.1089/pho.2015.3988.

Blivet, G., Relano-Gines, A., Wachtel, M., & Touchon, J. (2022). A randomized, double-blind, and sham-controlled trial of an innovative brain-gut photobiomodulation therapy: Safety and patient compliance. *Journal of Alzheimer's Disease: JAD, 90*(2), 811–822. https://doi.org/10.3233/JAD-220467.

Braverman, B., McCarthy, R. J., Ivankovich, A. D., Forde, D. E., Overfield, M., & Bapna, M. S. (1989). Effect of helium-neon and infrared laser irradiation on wound healing in rabbits. *Lasers in Surgery and Medicine, 9*(1), 50–58. https://www.ncbi.nlm.nih.gov/pubmed/2927230.

Bullock-Saxton, J., Lehn, A., & Laakso, E. L. (2021). Exploring the effect of combined transcranial and intra-oral photobiomodulation therapy over a four-week period on physical and cognitive outcome measures for people with Parkinson's disease: A randomized double-blind placebo-controlled pilot study. *Journal of Alzheimer's Disease: JAD, 83*(4), 1499–1512. https://doi.org/10.3233/JAD-210170.

Burtscher, J., Soltany, A., Visavadiya, N. P., Burtscher, M., Millet, G. P., Khoramipour, K., & Khamoui, A. V. (2023). Mitochondrial stress and mitokines in aging. *Aging Cell, 22*(2), e13770. https://doi.org/10.1111/acel.13770.

Castano-Castano, S., Zorzo, C., Martinez-Esteban, J. A., & Arias, J. L. (2024). Dosimetry in cranial photobiomodulation therapy: effect of cranial thickness and bone density. *Lasers in Medical Science, 39*(1), 76. https://doi.org/10.1007/s10103-024-04024-z.

Chang, C. Y., Avina, A. E., Chang, C. J., Lu, L. S., Chong, Y. Y., Ho, T. Y., & Yang, T. S. (2024). Exploring the biphasic dose-response effects of photobiomodulation on the viability, migration, and extracellular vesicle secretion of human adipose mesenchymal stem cells. *Journal of Photochemistry and Photobiology. B, Biology, 256*, 112940. https://doi.org/10.1016/j.jphotobiol.2024.112940.

Chao, L. L. (2019). Effects of home photobiomodulation treatments on cognitive and behavioral function, cerebral perfusion, and resting-state functional connectivity in patients with dementia: A pilot trial. *Photobiomodulation, Photomedicine, and Laser Surgery, 37*(3), 133–141. https://doi.org/10.1089/photob.2018.4555.

Chen, Q., Wu, J., Dong, X., Yin, H., Shi, X., Su, S., ... Yang, J. (2021). Gut flora-targeted photobiomodulation therapy improves senile dementia in an Ass-induced Alzheimer's disease animal model. *Journal of Photochemistry and Photobiology. B, Biology, 216*, 112152. https://doi.org/10.1016/j.jphotobiol.2021.112152.

Cheng, Y., Du, Y., Liu, H., Tang, J., Veenstra, A., & Kern, T. S. (2018). Photobiomodulation inhibits long-term structural and functional lesions of diabetic retinopathy. *Diabetes, 67*(2), 291–298. https://doi.org/10.2337/db17-0803.

Conte, M., Sabbatinelli, J., Chiariello, A., Martucci, M., Santoro, A., Monti, D., ... Salvioli, S. (2021). Disease-specific plasma levels of mitokines FGF21, GDF15, and Humanin in type II diabetes and Alzheimer's disease in comparison with healthy aging. *Geroscience, 43*(2), 985–1001. https://doi.org/10.1007/s11357-020-00287-w.

Darlot, F., Moro, C., El Massri, N., Chabrol, C., Johnstone, D. M., Reinhart, F., ... Benabid, A. L. (2016). Near-infrared light is neuroprotective in a monkey model of Parkinson disease. *Annals of Neurology, 79*(1), 59–75. https://doi.org/10.1002/ana.24542.

Durieux, J., Wolff, S., & Dillin, A. (2011). The cell-non-autonomous nature of electron transport chain-mediated longevity. *Cell, 144*(1), 79–91. https://doi.org/10.1016/j.cell.2010.12.016.

El Massri, N., Moro, C., Torres, N., Darlot, F., Agay, D., Chabrol, C., ... Mitrofanis, J. (2016). Near-infrared light treatment reduces astrogliosis in MPTP-treated monkeys. *Experimental Brain Research, 234*(11), 3225–3232. https://doi.org/10.1007/s00221-016-4720-7.

Elbaz-Greener, G., Sud, M., Tzuman, O., Leitman, M., Vered, Z., Ben-Dov, N., ... Blatt, A. (2018). Adjunctive laser-stimulated stem-cells therapy to primary reperfusion in acute myocardial infarction in humans: Safety and feasibility study. *Journal of Interventional Cardiology, 31*(6), 711–716. https://doi.org/10.1111/joic.12539.

Erickson, M. A., Morofuji, Y., Owen, J. B., & Banks, W. A. (2014). Rapid transport of CCL11 across the blood-brain barrier: Regional variation and importance of blood cells. *The Journal of Pharmacology and Experimental Therapeutics, 349*(3), 497–507. https://doi.org/10.1124/jpet.114.213074.

Farfara, D., Tuby, H., Trudler, D., Doron-Mandel, E., Maltz, L., Vassar, R. J., ... Oron, U. (2015). Low-level laser therapy ameliorates disease progression in a mouse model of Alzheimer's disease. *Journal of Molecular Neuroscience: MN, 55*(2), 430–436. https://doi.org/10.1007/s12031-014-0354-z.

Ganeshan, V., Skladnev, N. V., Kim, J. Y., Mitrofanis, J., Stone, J., & Johnstone, D. M. (2019). Pre-conditioning with remote photobiomodulation modulates the brain transcriptome and protects against MPTP insult in mice. *Neuroscience, 400*, 85–97. https://doi.org/10.1016/j.neuroscience.2018.12.050.

Ghaderi, A. H., Jahan, A., Akrami, F., & Moghadam Salimi, M. (2021). Transcranial photobiomodulation changes topology, synchronizability, and complexity of resting state brain networks. *Journal of Neural Engineering, 18*(4), https://doi.org/10.1088/1741-2552/abf97c.

Gholami, L., Khorsandi, K., Nooshabadi, V. T., Shahabi, S., Jazaeri, M., Esfahani, H., ... Mostafa, N. (2022). Effect of photobiomodulation on structure and function of extracellular vesicle secreted from mesenchymal stem cells. *Photochemistry and Photobiology, 98*(6), 1447–1458. https://doi.org/10.1111/php.13633.

Gordon, L. C., Martin, K. L., Torres, N., Benabid, A. L., Mitrofanis, J., Stone, J., ... Johnstone, D. M. (2023). Remote photobiomodulation targeted at the abdomen or legs provides effective neuroprotection against parkinsonian MPTP insult. *The European Journal of Neuroscience, 57*(9), 1611–1624. https://doi.org/10.1111/ejn.15973.

Hamilton, C. L., El Khoury, H., Hamilton, D., Nicklason, F., & Mitrofanis, J. (2019). "Buckets": Early observations on the use of red and infrared light helmets in Parkinson's disease patients. *Photobiomodulation, Photomedicine, and Laser Surgery, 37*(10), 615–622. https://doi.org/10.1089/photob.2019.4663.

Henderson, T. A. (2024). Can infrared light really be doing what we claim it is doing? Infrared light penetration principles, practices, and limitations. *Frontiers in Neurology, 15*, 1398894. https://doi.org/10.3389/fneur.2024.1398894.

Herisson, F., Frodermann, V., Courties, G., Rohde, D., Sun, Y., Vandoorne, K., ... Nahrendorf, M. (2018). Direct vascular channels connect skull bone marrow and the brain surface enabling myeloid cell migration. *Nature Neuroscience, 21*(9), 1209–1217. https://doi.org/10.1038/s41593-018-0213-2.

Herkes, G., McGee, C., Liebert, A., Bicknell, B., Isaac, V., Kiat, H., & McLachlan, C. S. (2023). A novel transcranial photobiomodulation device to address motor signs of Parkinson's disease: a parallel randomised feasibility study. *EClinicalMedicine, 66*, 102338. https://doi.org/10.1016/j.eclinm.2023.102338.

Hoefer, J., Luger, M., Dal-Pont, C., Culig, Z., Schennach, H., & Jochberger, S. (2017). The "Aging Factor" Eotaxin-1 (CCL11) is detectable in transfusion blood products and increases with the donor's age. *Frontiers in Aging Neuroscience, 9*, 402. https://doi.org/10.3389/fnagi.2017.00402.

Houser, M. C., & Tansey, M. G. (2017). The gut-brain axis: Is intestinal inflammation a silent driver of Parkinson's disease pathogenesis? *NPJ Parkinsons Disease, 3*, 3. https://doi.org/10.1038/s41531-016-0002-0.

Jeffery, G., Fosbury, R., Barrett, E., Hogg, C., Carmona, M. R., & Powner, M. B. (2025). Longer wavelengths in sunlight pass through the human body and have a systemic impact which improves vision. *Science Reports, 15*(1), 24435. https://doi.org/10.1038/s41598-025-09785-3.

Johnstone, D. M., el Massri, N., Moro, C., Spana, S., Wang, X. S., Torres, N., ... Mitrofanis, J. (2014). Indirect application of near infrared light induces neuroprotection in a mouse model of parkinsonism - An abscopal neuroprotective effect. *Neuroscience, 274*, 93–101. https://doi.org/10.1016/j.neuroscience.2014.05.023.

Johnstone, D. M., Hamilton, C., Gordon, L. C., Moro, C., Torres, N., Nicklason, F., ... Mitrofanis, J. (2021). Exploring the use of intracranial and extracranial (remote) photobiomodulation devices in Parkinson's disease: A comparison of direct and indirect systemic stimulations. *Journal of Alzheimer's Disease: JAD, 83*(4), 1399–1413. https://doi.org/10.3233/JAD-210052.

Kana, J. S., Hutschenreiter, G., Haina, D., & Waidelich, W. (1981). Effect of low-power density laser radiation on healing of open skin wounds in rats. *Archives of Surgery, 116*(3), 293–296. https://www.ncbi.nlm.nih.gov/pubmed/7469766.

Kheradmand, A., Donboli, S., Tanjani, P. T., Farhadinasab, A., Tabeie, F., Qutbi, M., & Kordmir, T. (2022). Therapeutic effects of low-level laser therapy on cognitive symptoms of patients with dementia: A double-blinded randomized clinical trial. *Photobiomodulation, Photomedicine, and Laser Surgery, 40*(9), 632–638. https://doi.org/10.1089/photob.2021.0135.

Kim, B., Brandli, A., Mitrofanis, J., Stone, J., Purushothuman, S., & Johnstone, D. M. (2017). Remote tissue conditioning - An emerging approach for inducing body-wide protection against diseases of ageing. *Ageing Research Reviews, 37*, 69–78. https://doi.org/10.1016/j.arr.2017.05.005.

Kim, B., Mitrofanis, J., Stone, J., & Johnstone, D. M. (2018). Remote tissue conditioning is neuroprotective against MPTP insult in mice. *IBRO Reports, 4*, 14–17.

Lapchak, P. A., Boitano, P. D., Butte, P. V., Fisher, D. J., Holscher, T., Ley, E. J., ... Rajput, P. S. (2015). Transcranial near-infrared laser transmission (NILT) profiles (800 nm): systematic comparison in four common research species. *PLoS One, 10*(6), e0127580. https://doi.org/10.1371/journal.pone.0127580.

Liebert, A., Bicknell, B., Johnstone, D. M., Gordon, L. C., Kiat, H., & Hamblin, M. R. (2019). "Photobiomics": Can light, including photobiomodulation, alter the microbiome? *Photobiomodulation, Photomedicine, and Laser Surgery, 37*(11), 681–693. https://doi.org/10.1089/photob.2019.4628.

Liebert, A., Bicknell, B., Laakso, E. L., Heller, G., Jalilitabaei, P., Tilley, S., ... Kiat, H. (2021). Improvements in clinical signs of Parkinson's disease using photobiomodulation: A prospective proof-of-concept study. *BMC Neurology, 21*(1), 256. https://doi.org/10.1186/s12883-021-02248-y.

Liebert, A., Bicknell, B., Laakso, E. L., Jalilitabaei, P., Tilley, S., Kiat, H., & Mitrofanis, J. (2022). Remote photobiomodulation treatment for the clinical signs of Parkinson's disease: A case series conducted during COVID-19. *Photobiomodulation, Photomedicine, and Laser Surgery, 40*(2), 112–122. https://doi.org/10.1089/photob.2021.0056.

Liebert, A., Bicknell, B., Laakso, E. L., Tilley, S., Heller, G., Kiat, H., & Herkes, G. (2024). Improvements in clinical signs and symptoms of Parkinson's disease using photobiomodulation: a five-year follow-up. *BMC Neurology, 24*(1), 381. https://doi.org/10.1186/s12883-024-03857-z.

Mastitskaya, S., Basalay, M., Hosford, P. S., Ramage, A. G., Gourine, A., & Gourine, A. V. (2016). Identifying the source of a humoral factor of remote (pre)conditioning cardioprotection. *PLoS One, 11*(2), e0150108. https://doi.org/10.1371/journal.pone.0150108.

McGee, C., Liebert, A., Bicknell, B., Pang, V., Isaac, V., McLachlan, C. S., ... Herkes, G. (2023). A randomized placebo-controlled study of a transcranial photobiomodulation helmet in Parkinson's disease: Post-hoc analysis of motor outcomes. *Journal of Clinical Medicine, 12*(8), https://doi.org/10.3390/jcm12082846.

Min, S. H., Kwon, J., Do, E. J., Kim, S. H., Kim, E. S., Jeong, J. Y., ... Park, D. H. (2022). Duodenal dual-wavelength photobiomodulation improves hyperglycemia and hepatic parameters with alteration of gut microbiome in type 2 diabetes animal model. *Cells, 11*(21), https://doi.org/10.3390/cells11213490.

Morais, L. H., Schreiber, H. L. T., & Mazmanian, S. K. (2021). The gut microbiota-brain axis in behaviour and brain disorders. *Nature Reviews. Microbiology, 19*(4), 241–255. https://doi.org/10.1038/s41579-020-00460-0.

Moro, C., Massri, N. E., Torres, N., Ratel, D., De Jaeger, X., Chabrol, C., ... Benabid, A. L. (2014). Photobiomodulation inside the brain: A novel method of applying near-infrared light intracranially and its impact on dopaminergic cell survival in MPTP-treated mice. *Journal of Neurosurgery, 120*(3), 670–683. https://doi.org/10.3171/2013.9.JNS13423.

Moro, C., Torres, N., El Massri, N., Ratel, D., Johnstone, D. M., Stone, J., ... Benabid, A. L. (2013). Photobiomodulation preserves behaviour and midbrain dopaminergic cells from MPTP toxicity: Evidence from two mouse strains. *BMC Neuroscience, 14*, 40. https://doi.org/10.1186/1471-2202-14-40.

Muili, K. A., Gopalakrishnan, S., Meyer, S. L., Eells, J. T., & Lyons, J. A. (2012). Amelioration of experimental autoimmune encephalomyelitis in C57BL/6 mice by photobiomodulation induced by 670 nm light. *PLoS One, 7*(1), e30655. https://doi.org/10.1371/journal.pone.0030655.

Munoz Declara, S., D'Alessandro, A., Gori, A., Cerasuolo, B., Renzi, S., Berlanda, M., ... Zanna, G. (2024). Evaluation of the impact of near-infrared multiwavelength locked system laser therapy on skin microbiome in atopic dogs. *Animals (Basel), 14*(6), https://doi.org/10.3390/ani14060906.

Mvula, B., Moore, T. J., & Abrahamse, H. (2010). Effect of low-level laser irradiation and epidermal growth factor on adult human adipose-derived stem cells. *Lasers in Medical Science, 25*(1), 33–39. https://doi.org/10.1007/s10103-008-0636-1.

Nagy, E. N., Ali, A. Y., Behiry, M. E., Naguib, M. M., & Elsayed, M. M. (2021). Impact of combined photo-biomodulation and aerobic exercise on cognitive function and quality-of-life in elderly Alzheimer patients with anemia: A randomized clinical trial. *International Journal of General Medicine, 14*, 141–152. https://doi.org/10.2147/IJGM.S280559.

Nizamutdinov, D., Qi, X., Berman, M. H., Dougal, G., Dayawansa, S., Wu, E., ... Huang, J. H. (2021). Transcranial near infrared light stimulations improve cognition in patients with dementia. *Aging and Disease, 12*(4), 954–963. https://doi.org/10.14336/AD.2021.0229.

Oron, A., Efrati, S., Doenyas-Barak, K., Tuby, H., Maltz, L., & Oron, U. (2022). Photobiomodulation therapy to autologous bone marrow in humans significantly increases the concentration of circulating stem cells and macrophages: A pilot study. *Photobiomodulation, Photomedicine, and Laser Surgery, 40*(3), 178–182. https://doi.org/10.1089/photob.2021.0123.

Oron, U., Tuby, H., Maltz, L., Sagi-Assif, O., Abu-Hamed, R., Yaakobi, T., ... Efrati, S. (2014). Autologous bone-marrow stem cells stimulation reverses post-ischemic-reperfusion kidney injury in rats. *American Journal of Nephrology, 40*(5), 425–433. https://doi.org/10.1159/000368721.

Peoples, C., Spana, S., Ashkan, K., Benabid, A. L., Stone, J., Baker, G. E., & Mitrofanis, J. (2012). Photobiomodulation enhances nigral dopaminergic cell survival in a chronic MPTP mouse model of Parkinson's disease. *Parkinsonism & Related Disorders, 18*(5), 469–476. https://doi.org/10.1016/j.parkreldis.2012.01.005.

Petrucco, C., Benson, P., Gordon, L., Stone, J., & Johnstone, D. M. (2020). Photobiomodulation as a neuroprotective strategy for Parkinson's disease. In C. R. Martin, & V. R. Preedy (Eds.). *The neuroscience of Parkinson's disease: Diagnosis and management in Parkinson's disease* (pp. 697–712)Elsevier.

Purushothuman, S., Nandasena, C., Johnstone, D. M., Stone, J., & Mitrofanis, J. (2013). The impact of near-infrared light on dopaminergic cell survival in a transgenic mouse model of parkinsonism. *Brain Research, 1535*, 61–70. https://doi.org/10.1016/j.brainres.2013.08.047.

Reinhart, F., Massri, N. E., Chabrol, C., Cretallaz, C., Johnstone, D. M., Torres, N., ... Moro, C. (2016). Intracranial application of near-infrared light in a hemi-parkinsonian rat model: the impact on behavior and cell survival. *Journal of Neurosurgery, 124*(6), 1829–1841. https://doi.org/10.3171/2015.5.JNS15735.

Ristow, M., & Schmeisser, K. (2014). Mitohormesis: Promoting health and lifespan by increased levels of reactive oxygen species (ROS). *Dose Response, 12*(2), 288–341. https://doi.org/10.2203/dose-response.13-035.Ristow.

Rochkind, S., Rousso, M., Nissan, M., Villarreal, M., Barr-Nea, L., & Rees, D. G. (1989). Systemic effects of low-power laser irradiation on the peripheral and central nervous system, cutaneous wounds, and burns. *Lasers in Surgery and Medicine, 9*(2), 174–182. https://www.ncbi.nlm.nih.gov/pubmed/2716462.

Salehpour, F., Hamblin, M. R., & DiDuro, J. O. (2019). Rapid reversal of cognitive decline, olfactory dysfunction, and quality of life using multi-modality photobiomodulation therapy: Case report. *Photobiomodulation, Photomedicine, and Laser Surgery, 37*(3), 159–167. https://doi.org/10.1089/photob.2018.4569.

Saliba, A., Du, Y., Liu, H., Patel, S., Roberts, R., Berkowitz, B. A., & Kern, T. S. (2015). Photobiomodulation mitigates diabetes-induced retinopathy by direct and indirect mechanisms: Evidence from intervention studies in pigmented mice. *PLoS One, 10*(10), e0139003. https://doi.org/10.1371/journal.pone.0139003.

Saltmarche, A. E., Naeser, M. A., Ho, K. F., Hamblin, M. R., & Lim, L. (2017). Significant improvement in cognition in mild to moderately severe dementia cases treated with transcranial plus intranasal photobiomodulation: Case series report. *Photomedicine and Laser Surgery, 35*(8), 432–441. https://doi.org/10.1089/pho.2016.4227.

Sampson, T. R., Debelius, J. W., Thron, T., Janssen, S., Shastri, G. G., Ilhan, Z. E., ... Mazmanian, S. K. (2016). Gut microbiota regulate motor deficits and neuroinflammation in a model of Parkinson's disease. *Cell, 167*(6), 1469–1480.e1412. https://doi.org/10.1016/j.cell.2016.11.018.

Santos, L., Olmo-Aguado, S. D., Valenzuela, P. L., Winge, K., Iglesias-Soler, E., Arguelles-Luis, J., ... Lucia, A. (2019). Photobiomodulation in Parkinson's disease: A randomized controlled trial. *Brain Stimul, 12*(3), 810–812. https://doi.org/10.1016/j.brs.2019.02.009.

Saucedo, C. L., Courtois, E. C., Wade, Z. S., Kelley, M. N., Kheradbin, N., Barrett, D. W., & Gonzalez-Lima, F. (2021). Transcranial laser stimulation: Mitochondrial and cerebrovascular effects in younger and older healthy adults. *Brain Stimulation, 14*(2), 440–449. https://doi.org/10.1016/j.brs.2021.02.011.

Scheperjans, F., Aho, V., Pereira, P. A., Koskinen, K., Paulin, L., Pekkonen, E., ... Auvinen, P. (2015). Gut microbiota are related to Parkinson's disease and clinical phenotype. *Movement Disorders: Official Journal of the Movement Disorder Society, 30*(3), 350–358. https://doi.org/10.1002/mds.26069.

Setchfield, K., Gorman, A., Simpson, A., Somekh, M. G., & Wright, A. J. (2024). Effect of skin color on optical properties and the implications for medical optical technologies: A review. *Journal of Biomedical Optics, 29*(1), 010901. https://doi.org/10.1117/1.JBO.29.1.010901.

Shamloo, S., Defensor, E., Ciari, P., Ogawa, G., Vidano, L., Lin, J. S., ... Barron, A. E. (2023). The anti-inflammatory effects of photobiomodulation are mediated by cytokines: Evidence from a mouse model of inflammation. *Frontiers in Neuroscience, 17*, 1150156. https://doi.org/10.3389/fnins.2023.1150156.

Shaw, V. E., Peoples, C., Spana, S., Ashkan, K., Benabid, A. L., Stone, J., ... Mitrofanis, J. (2012). Patterns of cell activity in the subthalamic region associated with the neuroprotective action of near-infrared light treatment in MPTP-treated mice. *Parkinsons Disease, 2012*, 296875. https://doi.org/10.1155/2012/296875.

Shaw, V. E., Spana, S., Ashkan, K., Benabid, A. L., Stone, J., Baker, G. E., & Mitrofanis, J. (2010). Neuroprotection of midbrain dopaminergic cells in MPTP-treated mice after near-infrared light treatment. *The Journal of Comparative Neurology, 518*(1), 25–40. https://doi.org/10.1002/cne.22207.

Stone, J., Johnstone, D., & Mitrofanis, J. (2013). The helmet experiment in Parkinson's disease: an observation of the mechanism of neuroprotection by near infra-red light Proceedings of the 9th World Association for Laser Therapy Congress,

Stone, J., Mitrofanis, J., Johnstone, D. M., Falsini, B., Bisti, S., Adam, P., ... Eells, J. (2018). Acquired resilience: An evolved system of tissue protection in mammals. 1559325818803428 *Dose Response, 16*(4), https://doi.org/10.1177/1559325818803428.

Tao, L., Liu, Q., Zhang, F., Fu, Y., Zhu, X., Weng, X., ... Wei, X. (2021). Microglia modulation with 1070-nm light attenuates Abeta burden and cognitive impairment in Alzheimer's disease mouse model. *Light: Science & Applications, 10*(1), 179. https://doi.org/10.1038/s41377-021-00617-3.

Tuby, H., Maltz, L., & Oron, U. (2007). Low-level laser irradiation (LLLI) promotes proliferation of mesenchymal and cardiac stem cells in culture. *Lasers in Surgery and Medicine, 39*(4), 373–378. https://doi.org/10.1002/lsm.20492.

Tuby, H., Maltz, L., & Oron, U. (2009). Implantation of low-level laser irradiated mesenchymal stem cells into the infarcted rat heart is associated with reduction in infarct size and enhanced angiogenesis. *Photomedicine and Laser Surgery, 27*(2), 227–233. https://doi.org/10.1089/pho.2008.2272.

Tuby, H., Maltz, L., & Oron, U. (2011). Induction of autologous mesenchymal stem cells in the bone marrow by low-level laser therapy has profound beneficial effects on the infarcted rat heart. *Lasers in Surgery and Medicine, 43*(5), 401–409. https://doi.org/10.1002/lsm.21063.

Villeda, S. A., Luo, J., Mosher, K. I., Zou, B., Britschgi, M., Bieri, G., ... Wyss-Coray, T. (2011). The ageing systemic milieu negatively regulates neurogenesis and cognitive function. *Nature, 477*(7362), 90–94. https://doi.org/10.1038/nature10357.

Wang, M., Cao, J., Amakye, W. K., Gong, C., Li, Q., & Ren, J. (2020). Mid infrared light treatment attenuates cognitive decline and alters the gut microbiota community in APP/PS1 mouse model. *Biochemical and Biophysical Research Communications, 523*(1), 60–65. https://doi.org/10.1016/j.bbrc.2019.12.015.

Wang, M., Yan, C., Li, X., Yang, T., Wu, S., Liu, Q., ... Zhou, F. (2024). Non-invasive modulation of meningeal lymphatics ameliorates ageing and Alzheimer's disease-associated pathology and cognition in mice. *Nature Communications, 15*(1), 1453. https://doi.org/10.1038/s41467-024-45656-7.

Weihrauch, D., Keszler, A., Lindemer, B., Krolikowski, J., & Lohr, N. L. (2021). Red light stimulates vasodilation through extracellular vesicle trafficking. *Journal of Photochemistry and Photobiology. B, Biology, 220*, 112212. https://doi.org/10.1016/j.jphotobiol.2021.112212.

Wu, C., Li, M., Chen, Z., Feng, S., Deng, Q., Duan, R., ... Yang, L. (2025). Remote photobiomodulation ameliorates behavioral and neuropathological outcomes in a rat model of repeated closed head injury. *Translational Psychiatry, 15*(1), 8. https://doi.org/10.1038/s41398-025-03228-3.

Wu, X., Shen, Q., Chang, H., Li, J., & Xing, D. (2022). Promoted CD4(+) T cell-derived IFN-gamma/IL-10 by photobiomodulation therapy modulates neurogenesis to ameliorate cognitive deficits in APP/PS1 and 3xTg-AD mice. *Journal of Neuroinflammation, 19*(1), 253. https://doi.org/10.1186/s12974-022-02617-5.

Yun, J., & Finkel, T. (2014). Mitohormesis. *Cell Metabolism, 19*(5), 757–766. https://doi.org/10.1016/j.cmet.2014.01.011.

Zanotta, N., Ottaviani, G., Campisciano, G., Poropat, A., Bovenzi, M., Rupel, K., ... Zacchigna, S. (2020). Photobiomodulation modulates inflammation and oral microbiome: A pilot study. *Biomarkers: Biochemical Indicators of Exposure, Response, and Susceptibility to Chemicals, 25*(8), 677–684. https://doi.org/10.1080/1354750X.2020.1825812.

Zarate, S. C., Traetta, M. E., Codagnone, M. G., Seilicovich, A., & Reines, A. G. (2019). Humanin, a mitochondrial-derived peptide released by astrocytes, prevents synapse loss in hippocampal neurons. *Frontiers in Aging Neuroscience, 11*, 123. https://doi.org/10.3389/fnagi.2019.00123.

Zare, F., Moradi, A., Fallahnezhad, S., Ghoreishi, S. K., Amini, A., Chien, S., & Bayat, M. (2019). Photobiomodulation with 630 plus 810 nm wavelengths induce more in vitro cell viability of human adipose stem cells than human bone marrow-derived stem cells. *Journal of Photochemistry and Photobiology. B, Biology, 201*, 111658. https://doi.org/10.1016/j.jphotobiol.2019.111658.

Zhang, H., Li, X., Fan, W., Pandovski, S., Tian, Y., & Dillin, A. (2023). Inter-tissue communication of mitochondrial stress and metabolic health. *Life Metabolism, 2*(1), https://doi.org/10.1093/lifemeta/load001.

CHAPTER THREE

# Light talks: The role of biophoton in cellular communication

**Jaimie Hoh Kam[a,*], Ifigeneia Kalampouka[b], and Tigrane Cantat-Moltrecht[a]**
[a]Université Grenoble Alpes, Fonds de Dotation Clinatec, Grenoble, France
[b]University of Westminster, London, United Kingdom
*Corresponding author. e-mail address: mariejaimie.hohkam@cea.fr

## Contents



## Abstract

Biophotons represent ultra-weak light emission (200–800 nm) spontaneously generated by living cells through oxidative metabolic processes, particularly within mitochondria. In 1923, Alexander Gurwitsch demonstrated non-chemical cellular communication in pioneering experiments on what he called 'mitogenetic radiation', which transmitted through quartz but not through glass: it was later confirmed to involve light and especially UV. Since then, this field has evolved to encompass sophisticated detection technologies and theoretical frameworks. Evidence suggests that biophotons, produced mainly by reactive oxygen species (ROS) and excited molecular intermediates during cellular respiration, may serve functional roles in

*International Review of Neurobiology*, Volume 189
ISSN 0074-7742, https://doi.org/10.1016/bs.irn.2026.01.014

intercellular signalling beyond being mere metabolic byproducts. In neural systems, biophotonic emissions correlate with brain activity, metabolic states, and oxidative stress levels, with emerging applications in neurodegenerative disease monitoring and cognitive assessment. Advanced detection methodologies using photomultiplier tubes (PMT) and electron-multiplying charge-coupled device (CCD) cameras have revealed species-specific spectral signatures and stress-responsive emission patterns. While quantum mechanical explanations remain debated, accumulating evidence supports biophotons as potential biomarkers for cellular health and novel communication pathways. Future therapeutic applications may integrate biophotonic monitoring with photobiomodulation (PBM) interventions for personalized medical treatments.

## 1. Introduction and historical perspective

Cells are known to respond to extracellular light from external sources, such as in photobiomodulation (PBM) therapy, which uses red and near-infrared light on body tissues (Hamblin, 2016). However, emerging evidence indicates that cells can also generate and respond to endogenous light produced by intracellular metabolic processes (Cifra & Pospíšil, 2014). This spontaneous biophotonic emission suggests a previously unrecognized optical communication network operating within and between cells, potentially complementing established electrical and chemical signalling pathways (Tang & Dai, 2014a).

The discovery of biophotons, also known as ultraweak photon emission (UPE), represents one of the most intriguing chapters at the intersection of physics and biology. In the early 1920s, as quantum mechanics was revolutionising our understanding of light and matter, Einstein's proposal that light radiation could exhibit both wave-like and particle-like behaviour introduced the concept of "photon" (Marais et al., 2018). This fundamental shift in physics would soon find unexpected applications in biology.

Just a few years after Einstein's groundbreaking work, Russian embryologist Alexander Gurwitsch made a remarkable discovery that would challenge conventional understanding of cellular communication. In 1923, Gurwitsch demonstrated the possibility of non-contact, non-chemical cellular communication through what he then termed as "mitogenetic radiation". His simple yet elegant experiment used two onion rootlets: an inductor and a detector. When the inductor rootlet was in a high cell division state, Gurwitsch observed that a nearby detector rootlet, separated by quartz glass (which allows ultraviolet [UV] light to pass through), showed an increased rate of cell division, too. However, when the rootlets were separated by regular glass, which blocks most light in the UV spectrum, no change in division rate was observed (Fig. 1; Gurwitsch, 1923).

This observation could not be explained by traditional biochemical and physical laws of the scientific theories of that time. The implications of Gurwitsch's experiment were revolutionary: onion root cells appeared capable of influencing each other through non-chemical means, most possibly through light emission, suggesting that photonic emission from living tissue could stimulate biological processes. However, right after this discovery, the Second World War erupted and among all other detrimental consequences, research in this area was largely abandoned for the next few years.

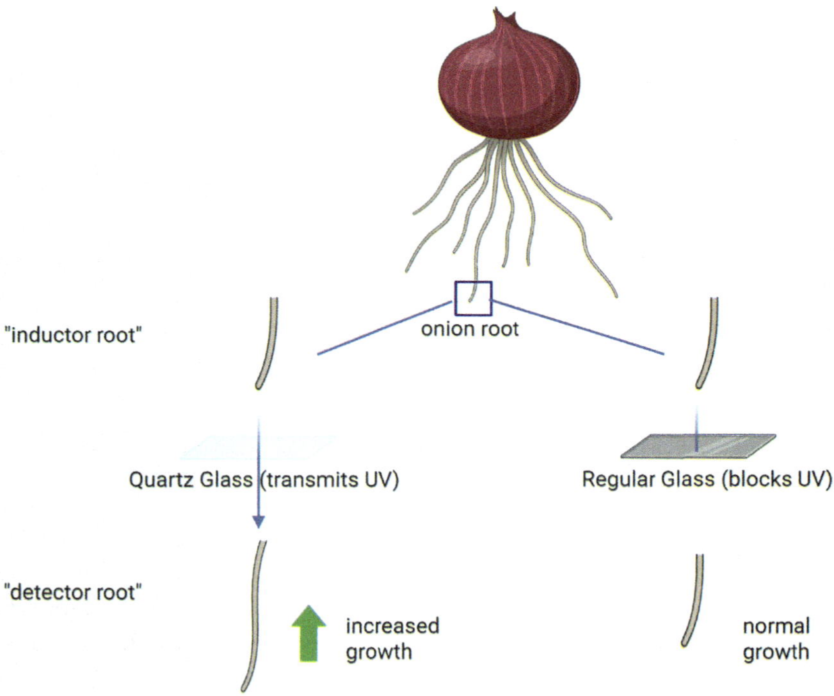


**Fig. 1** Gurwitsch's Onion root experiment (1923). Experimental setup demonstrating non-chemical cellular communication between onion roots. The inductor root stimulated increased growth in the detector root when separated by UV-transparent quartz glass, but not through UV-opaque regular glass, providing first evidence of light-mediated intercellular signaling. *Created in BioRender. Kalampouka, I. (2025) https://BioRender.com/0jag4f4.*

Fortunately, in the early 1980s, researchers started re-exploring the potential of living tissue communicating non-chemically. Kaznacheev and colleagues demonstrated that dying cells could send distress signals to healthy neighbouring cells, observing that pathological changes in treated cultures were mirrored in adjacent untreated cultures when separated by quartz (UV-transparent) but not glass (UV-opaque) barriers (Kaznacheev et al., 1980). Similarly, four years later, Bat'yanov showed that isolated mitochondria from rat liver cells could influence the oxygen consumption of physically separated mitochondria through quartz barriers, concluding that this non-chemical interaction occurred specifically within the ultra-violet range (around 200–400 nm) (Bat'yanov, 1984). Both researchers inferred light-based communication from their barrier experiments, establishing that cellular communication could occur through electro-magnetic radiation rather than chemical contact alone.

The field remained relatively dormant until German biophysicist Fritz-Albert Popp revived interest in the phenomenon over 60 years after Gurwitsch's initial experiments. Popp introduced the term "biophotons," describing them as low-intensity photons emitted by biological systems in the UV range (Popp et al., 1984). His contribution went beyond nomenclature, as Popp proposed that biophotons are produced in a coherent electromagnetic field and are influenced by cellular activities (Popp, 1992, 2003). This hypothesis suggested for the first time that bio-photons might carry information and serve functional roles in biological systems rather than being mere byproducts of metabolism.

Early evidence for non-chemical cellular communication through light continued to accumulate throughout the later part of the 20th century. Albrecht-Buehler's work with baby hamster kidney cells demonstrated that the distribution and orientation of newly grown cells could be modulated by cultivated cells placed on the opposite surface of glass slides. When a thin piece of metal that blocked light passage was added between cell populations, the effect disappeared (Albrecht-Buehler, 1992). Similarly, Shen and colleagues demonstrated cellular responses in neutrophils from pig blood after non-chemical, optical contact with a second neutrophil population exposed to an oxidative burst (Shen et al., 1994).

More sophisticated experimental setups emerged in the 21st century. (Farhadi et al., 2007) used intestinal epithelial cells separated by a custom-made container, while (Rossi et al., 2011) developed a "dish-on-dish" design containing endothelial cell populations (Farhadi et al., 2007; Rossi et al., 2011). In both cases, physically disconnected cell populations influenced each other by changing protein content and altering cellular growth and morphology, supporting the role of biophotons as a method of cellular communication.

This chapter provides a comprehensive examination of biophotons from fundamental principles to cutting-edge neural applications. We begin by exploring the physical and spectral characteristics of these ultra-weak emissions, including their production mechanisms through mitochondrial oxidative processes and the technological challenges in their detection. We later focus on the fascinating world of neural biophotonics, where we examine how brain cells generate and potentially utilize these photonic signals for communication. The chapter culminates with emerging clinical applications, from novel diagnostic tools for neurodegenerative diseases to potential therapeutic interventions combining biophotonic monitoring with targeted light therapies. By the end of the chapter, we hope readers will understand why these barely detectable biological light emissions — once dismissed as metabolic noise — may represent an ancient and fundamental communication system that could revolutionize our understanding of cellular signalling and open new frontiers in neuroscience and medicine.

## 2. Fundamental properties and mechanisms

Understanding biophotons requires examining both their physical characteristics and the biological processes that generate them. This section explores the spectral properties of these ultra-weak emissions, their cellular sources — particularly mitochondria — and the molecular mechanisms that transform metabolic energy into optical signals.

### 2.1 Physical and spectral characteristics

Today, compelling evidence confirms that biophotons are spontaneous, ultra-weak emissions of light produced by living cells of microbial, plant, or animal origin (Cifra & Pospišil, 2014; Prasad & Pospíšil, 2013). Understanding their physical properties is crucial for both detection and potential applications. The field has employed various terms over the years to describe this phenomenon, including biological auto-luminescence, metabolic photon emission, and UPE (Mould et al., 2024). The recent publication of "Ultra-Weak Photon Emission from Biological Systems" by field experts further emphasises the use of "UPE" to characterise this effect

(Volodyaev et al., 2023). This terminology arises from the extremely low photon count rather than the energy of individual photons themselves—while the number of emitted photons is indeed ultra-weak (ranging from several to tens of photons per second per square centimetre), each photon carries the same quantum of electromagnetic energy as any other photon of its wavelength, making the phenomenon "weak" only in terms of intensity, not in the fundamental nature of the light itself (Cifra & Pospíšil, 2014; Niggli, 1992).

### 2.1.1 Wavelength and intensity

The spectral range of biophotons spans from 200 to 800 nm, encompassing the ultraviolet, visible, and near-infrared portions of the electromagnetic spectrum (Cifra & Pospíšil, 2014). This broad range suggests multiple chromophores and emission mechanisms within biological systems. The intensity of biophotonic emission under normal conditions is remarkably low, typically ranging from several to tens of photons per second per square centimetre (photons $sec^{-1}$ $cm^{-2}$) (Cifra & Pospíšil, 2014).

However, biological systems under stress show dramatically increased biophotonic intensity. Evidence indicates that abiotic or biotic stress can enhance emission to hundreds of photons.$sec^{-1}$.$cm^{-2}$ (Burgos et al., 2017; Kamal & Komatsu, 2016; Kobayashi et al., 2009). This stress-induced enhancement occurs because biophotons are produced during oxidative metabolic reactions (Bókkon et al., 2010; Prasad & Pospíšil, 2013; Van Wijk et al., 2008). Both biotic stressors (such as viral or bacterial infection) and abiotic stressors (including temperature changes, ionizing radiation, mechanical, light damage or chemical stress) increase oxidative stress, leading to greater reactive oxygen species (ROS) production and explaining the observed changes in biophotonic intensity under stress conditions (Cifra & Pospíšil, 2014).

The relationship between mechanical injury/stress and biophotons has been particularly well-characterized. Prasad and Pospíšil conducted comprehensive analyses of the spectral distribution of biophotons following wounding in plants, providing detailed characterization of the origins of wound-induced emission and establishing that mechanical damage triggers a cascade of oxidative reactions that significantly elevate photon emission (Prasad & Pospíšil, 2013). Later, the importance of using intact living plants for valid biophotonic measurements was further demonstrated by showing that the cutting process itself induces stress responses that dramatically elevate photon emission levels through lipid peroxidation-associated chemiluminescence (Havaux & Ksas, 2022). Recent imaging studies have expanded on these findings, with (Salari et al., 2025) showing that injured plant tissues exhibited significantly higher biophotonic intensity, particularly when treated with chemical agents like benzocaine, and that temperature increases up to 36 °C progressively enhanced emission before thermal stress caused a decline in photon output.

## 2.2 Cellular sources and metabolic origins

### *2.2.1 Mitochondrial and metabolic mechanisms*

The production of biophotons is fundamentally linked to cellular metabolism, with mitochondria playing a central role. Understanding these mechanisms provides insight into how metabolic states translate into optical signals.

Early studies such as those by (Cadenas et al., 1980), demonstrated that biophotonic emissions are associated with electronically excited molecular species generated during mitochondrial oxidative metabolism. They used sensitizers and quenchers to identify mitochondria as a principal source of cellular biophotons.

ROS and reactive nitrogen species (RNS), metabolic by-products critical to neural function, are major contributors to biophotonic generation (Tuszyński & Dixon, 2001). These reactive intermediates are integral to cerebral signalling pathways, including synaptic plasticity, neurotransmitter release, hippocampal long-term potentiation (LTP), and memory consolidation. While physiological ROS levels maintain homeostasis, excessive ROS precipitate oxidative stress and neuronal apoptosis. Consequently, biophotonic emission may serve as a sensitive, non-invasive biomarker of oxidative metabolic dynamics.

### *2.2.2 Electron transport chain origins and ROS production*

Biophotonic production is characterized by electronically excited species, including singlet oxygen ($^1O_2$) and carbonyl species ($^3R{=}O$), transitioning from high to lower energy states and releasing photons in the process (Cifra & Pospíšil, 2014). Research suggests that biophotonic emission is primarily associated with regulated intracellular redox processes, particularly those occurring in mitochondria (Bókkon et al., 2010).

The main sites of ROS generation and other redox reactions—mitochondria, cell membranes, and chloroplasts—are thought to be the principal cellular sources of biophotons (Pospíšil et al., 2019). Within mitochondria, the electron transport chain represents a major source of

ROS. Under normal conditions, approximately 1–2 % of electrons passing through the chain leak prematurely, combining with oxygen to form superoxide radicals. This leakage occurs primarily at complexes I and III of the electron transport chain (Turrens, 2003).

The mechanistic pathways of biophotonic emission during oxidative metabolism involve two parallel processes. The first pathway involves the transition of carbonyl radicals (R=O•) to ground state carbonyls (R=O) with photon emission, initiated by the reaction of two peroxyl radicals (2ROO•) forming alcohol (R-OH), carbonyl radical (R=O•), and triplet oxygen ($^3O_2$). The second pathway involves the conversion of singlet oxygen ($^1O_2$) to ground state triplet oxygen ($^3O_2$) with photon emission, generated from the reaction of superoxide radicals ($O_2^-$) with protons ($H^+$) via hydrogen peroxide ($H_2O_2$) formation (Mould et al., 2024).

## 2.3 Molecular mechanisms

### *2.3.1 Cytochrome c oxidase and other chromophores*

At the molecular level, mitochondria contain numerous chromophores capable of absorbing and emitting light across a broad spectrum. The electron transport chain components, including cytochromes, ubiquinone, and NADPH, have absorption spectra spanning from UV to visible light (Nunn et al., 2020). Additionally, flavins and opsins can absorb light in the blue and green spectral regions (Hamblin, 2017).

Cytochrome c oxidase (CCO), located in the inner mitochondrial membrane, is widely recognized as a principal photoreceptor for red to near-infrared (NIR) light in the range of 600–850 nm. Absorption of these wavelengths by CCO can potentially increase electron transport chain activity, leading to enhanced adenosine triphosphate (ATP) production and altered metabolic homeostasis (Hamblin, 2016; Karu, 2008). This bidirectional relationship between light and mitochondrial function suggests that mitochondria can both emit and respond to photonic signals.

Recent research has highlighted the role of nanoscopic interfacial water layers within and around mitochondria in NIR light absorption (Sommer et al., 2015). These water layers, particularly near hydrophilic surfaces, may have their structure (volume and viscosity) that can be changed by NIR light, potentially affecting the efficiency of the ATP synthase motor. Furthermore, the inner and outer mitochondrial membranes, containing various proteins and lipids, can absorb NIR light, potentially influencing membrane potential, permeability, and the function of membrane-bound proteins (Passarella & Karu, 2014).

### 2.3.2 Metabolic correlations and stress responses

The concentration of biophotons within cells could considerably exceed that of the extracellular environment, reflecting the compartmentalised nature of metabolic processes (Bókkon et al., 2010). Mitochondrial respiration, lipid peroxidation, peroxisome activity, and catecholamine biochemistry all contribute to biophotonic emission through ROS generation and excited state formation.

Experimental evidence supports the metabolic origin of biophotons. Treatment with antimycin, a complex III inhibitor that disrupts mitochondrial oxidative phosphorylation (OXPHOS), leads to decreased oxygen consumption while increasing ROS production through electron leakage (Kim et al., 1999). This creates an ideal model for studying stress-induced biophotonic emission, as increased ROS levels are associated with enhanced biophotonic production (Cifra & Pospíšil, 2014). The fundamental link between metabolism and biophotonic emission is further demonstrated by (Salari et al., 2025), who showed that live mice consistently exhibited significantly higher biophotonic intensity compared to recently euthanised mice, with emission nearly extinguished upon death despite maintaining body temperature at 37 °C. These findings, together with the ones mentioned before regarding chemical treatments, dramatically modulate emission patterns, confirming that biophotonic emission directly reflects active metabolic processes and can be chemically and physically modulated through alterations in cellular oxidative state.

The dynamic nature of biophotonic emission reflects cellular metabolic states. Under normal conditions, the low level of emission corresponds to basal metabolic activity. However, various stressors can dramatically increase emission intensity, providing a potential non-invasive window into cellular stress responses and metabolic dysfunction.

### 2.3.3 Other cellular sources and pathways

While mitochondria serve as primary generators of biophotons through their oxidative phosphorylation processes, the phenomenon extends throughout multiple cellular compartments. Beyond mitochondrial sources, cellular membranes represent significant sites of photon generation. Membrane phospholipids rich in polyunsaturated fatty acids undergo oxidative degradation, forming peroxyl radicals that subsequently yield light-emitting excited carbonyl species (Cifra & Pospíšil, 2014). Peroxisomes contribute through their concentrated oxidative enzyme systems involved in fatty acid β-oxidation and hydrogen peroxide metabolism, while the endoplasmic reticulum participates particularly in cells with high protein synthesis rates, where oxidative protein folding generates reactive intermediates.

According to (Pospíšil et al., 2019), these organelles, along with mitochondria and cell membranes, constitute the principal cellular sources of ROS-mediated photon emission. Imaging studies have revealed organism-specific emission patterns that reflect cellular architecture and composition. Prasad and Pospíšil (2013) documented striking differences in detection requirements across biological systems: yeast cultures required 30-minute exposures, human skin needed 20 min, while plant tissues required only 10 min for clear visualization (Prasad & Pospíšil, 2013). These variations correlate with cellular density, metabolic rate, and the presence of photosensitive pigments that can increase emission through energy transfer processes. Plant wound response studies have provided detailed characterisation of lipid peroxidation-associated emission, with Prasad and Pospí šil mapping the temporal and spectral evolution of photon emission following mechanical injury, revealing distinct phases corresponding to primary damage responses and secondary oxidative cascades (Pospíšil et al., 2019; Prasad & Pospíšil, 2013). Havaux and Ksas (2022) demonstrated that even minimal tissue disruption triggers substantial emission increases, emphasizing the sensitivity of membrane lipids to oxidative perturbation (Havaux & Ksas, 2022).

### *2.3.4 Integration of cellular sources*

Rather than functioning as isolated emitters, cellular biophotonic sources form an interconnected network mediated by ROS diffusion and redox signalling. (Pospíšil et al., 2019) describe how reactive species generated in one compartment can initiate oxidative reactions throughout the cell. The differential mobility of ROS species creates spatial gradients: hydrogen peroxide diffuses across membranes to reach distant targets, while hydroxyl radicals remain confined to their generation sites. This spatial heterogeneity produces complex emission patterns that encode information about cellular redox state. Mitochondrial superoxide production can trigger lipid peroxidation in adjacent organellar membranes, creating cascading photonic signals. The spectral signature varies with the source and cellular environment: naked carbonyl species emit at shorter wavelengths (350–550 nm), while energy transfer to endogenous chromophores shifts emission toward longer wavelengths—chlorophylls at 680–740 nm in plants, melanin at 360–560 nm in animal tissues (Cifra & Pospíšil, 2014). These integrated cellular sources collectively generate the complex biophotonic signatures observed in living systems, providing a real-time optical readout of cellular metabolic and oxidative status.

## 3. Detection technologies and methodologies

The correlation between biophotons and oxidative stress has led to investigations into their potential as a non-invasive tool for detecting and monitoring oxidative stress dynamics (Du et al., 2023). However, detecting and measuring biophotons remains a significant challenge due to their extremely low intensity, as well as the difficulty to control background light and noise (Mould et al., 2024; Hoh Kam et al., 2025). This section examines the evolution of detection technologies from early photomultiplier tubes to modern ultra-sensitive imaging systems. We discuss critical methodological issues including the differentiation of true biophotonic emission from delayed luminescence, the trade-offs between different detector technologies, and recent breakthroughs that have pushed the boundaries of sensitivity.

### 3.1 General detection principles

Historically, researchers have employed various sophisticated detection methods. Two primary technologies have dominated the field: photomultiplier tubes (PMT) and charge-coupled device (CCD) or electron-multiplying CCD (EM-CCD) cameras. A PMT essentially consists in 1. a photocathode which releases an electron upon arrival of a photon (a "photoelectron", in a process similar to photovoltaic cells), 2. a cascaded ensemble of dynodes supplied with high voltages by which the photoelectrons are accelerated and multiplied (each time an electron hits a dynode, it triggers the release of 10–100 more), and 3. an anode through which the flux of incoming electrons can be extracted and measured as an electrical current. PMTs have a very low noise, which makes them very sensitive and useful for low-light applications. However, their quantum efficiency is low (5–50 %, the chance for an incoming photon to trigger the release of an electron) and, perhaps more importantly, they cannot provide spatial information. A CCD camera consists of an array of pixels, each of which stores an electrical charge when it receives photons. The charge is then transferred across the array and read out pixel by pixel. An EM-CCD adds an electron-multiplying stage in the readout circuitry of the pixels, allowing for increased sensitivity. CCD and EM-CCD cameras can be very sensitive as well, and with quantum efficiencies up to > 90 %; but the limited size of each pixel (which is necessary to image with good resolution) means that each of them receives very few photons in low-light applications, sometimes not enough to overcome the intrinsic noise of the device. Recent progress in digital sensors technology has allowed for the development of CMOS sensors which can record down to single photons, provided the right post-processing. In CMOS sensor, the charge generated by incoming photons is read out pixel by pixel and does not have to be transferred along the array. This can allow for a reduction in readout noise and thus for the detection of single photons in the best case. However, current CMOS sensors tend to have smaller pixels than CCD cameras, which means they receive even less photons, and they tend to be more susceptible to dark noise (charges generated even though no photons reach the sensor, and which accumulate over time). Investigations into using ultra-sensitive CMOS sensors for the detection of biophotons are underway, but they have not given tangible results yet (Mould et al., 2024; Hoh Kam et al., 2024).

Regardless of the choice of detector, a critical methodological consideration is differentiating biophotonic emission from delayed luminescence. Delayed luminescence is weak light re-emission induced by exposure to external light sources, occurring from seconds to several hours after light exposure (Cifra & Pospíšil, 2014; Scordino et al., 2014). Additionally, delayed luminescence can occur from biological and non-biological systems and it is not a byproduct of metabolic reaction. Although delayed luminescence has been proposed to serve communication purposes (Fleiss & Sarkisyan, 2019), it differs fundamentally from spontaneous biophotonic emission produced by ROS.

To control for delayed luminescence, studies employ strict dark conditions. (Kobayashi et al., 2009), who imaged diurnal changes in human biophotonic production using a cryogenic CCD camera, dark-adapted subjects for 15 min before 20-minute imaging sessions in a darkroom. (Ortega-Ojeda et al., 2018) measured the spectral distribution of biophotons from the human hand using an electron multiplying CCD sensor and liquid crystal tuneable filter, requiring participants to wear a light-tight sleeve for at least 20 min prior to measurements in a controlled darkroom environment. There is a growing concern that such durations for dark adaptation might not be enough to eliminate delayed luminescence efficiently (Mackenzie et al., 2024).

## 3.2 Evolution of detection methods

A recent breakthrough by (Mackenzie et al., 2024) marks a significant advancement in the detection of biophotons. By employing highly sensitive detection systems and optimising experimental conditions, they successfully detected biophotonic emission in growing mung beans. Their innovations included improvements in light-tight experimental chambers, reduction of background noise, and careful management of delayed luminescence effects. Their setup demonstrated the ability to detect changes as small as 0.1 average counts per second using a photomultiplier tube with an average dark count of 15–17 counts per second, allowing measurement of signal changes of fractions of a count per second by averaging 1000 s of data (Mackenzie et al., 2024). And with a similar setup, researchers from the University of Westminster in London are currently working on detecting biophotons from mammalian whole cells and isolated mitochondria.

Detection techniques still face challenges, including the need for complete darkness during experiments, the challenge of discriminating biophotons from other light sources, and the difficulty in obtaining fully resolved spectra due to low photon count. And despite these technical difficulties that continue to shape experimental design and interpretation in the field, ongoing technological advances coupled with deeper understanding of phenomena such as delayed luminescence are progressively refining our ability to study these elusive biological light emissions.

# 4. Biophotons in neural systems

Although the precise physiological role of biophotons remains to be fully elucidated, accumulating evidence suggests that these emissions reflect oxidative metabolic states taking place in the mitochondria and may participate in cellular signalling and tissue repair processes. Research efforts have been historically constrained by technological limitations in the detection and quantification of such low-intensity emissions.

This section explores the unique challenges and methodologies for detecting biophotons specifically in neural tissues. We begin by examining detection approaches across different experimental models—from cultured neurons to whole brain imaging—highlighting the technical innovations required at each level of biological complexity. We then discuss how these detection methods have revealed the distinctive properties of neuronal biophoton emission, including their relationship to neural activity and potential role in brain communication networks. In vivo imaging studies have already demonstrated associations between spontaneous biophotonic emission and markers of cerebral energy metabolism and oxidative stress (Kataoka et al., 2001; Kobayashi et al., 1999a), suggesting functional significance. The section concludes by examining the correlation between biophotonic signatures and brain states, setting the stage for understanding their potential clinical applications.

## 4.1 Neural-specific detection methods

The detection and measurement of biophotons is a challenge in general, due to the extremely low number of photons emitted outside the cells or organisms. Targeting specifically the biophotons from the brain adds a layer of complexity, especially in vivo as a number of barriers have to be overcome, the first of which are the skull and scalp which themselves emit their own biophotons. Below we examine the progressive strategies researchers have developed to address these challenges, from simple cell culture systems to sophisticated in vivo imaging techniques.

### *4.1.1 In cell cultures*

A first strategy to explore "brain biophotons" is to look at biophotons from cultures of brain or neuron cells, where confounding factors can be minimized. However, even this simplified system presents challenges. While recording biophotonic emission from cultured brain cells typically involves placing either a PMT or a CCD/EM-CCD camera close to the culture vessel in a dark enclosure, maintaining physiological conditions during extended dark adaptation periods and for long recording times remains problematic (Fig. 2).

To date, no published studies have successfully recorded biophotons from brain cells maintained in optimal culture conditions with controlled temperature, humidity and pH throughout the measurement period. This represents a significant limitation, as cellular stress from suboptimal conditions may confound the interpretation of biophotonic signatures. Another limitation to such studies is the possibility to obtain recordings from a sample representative of a full and functional brain, starting with a full set of brain cells: neurons, astrocytes and microglia all together. Moreover, culture of brain cells has been historically challenging and the detection of brain biophotons started with ex vivo samples.

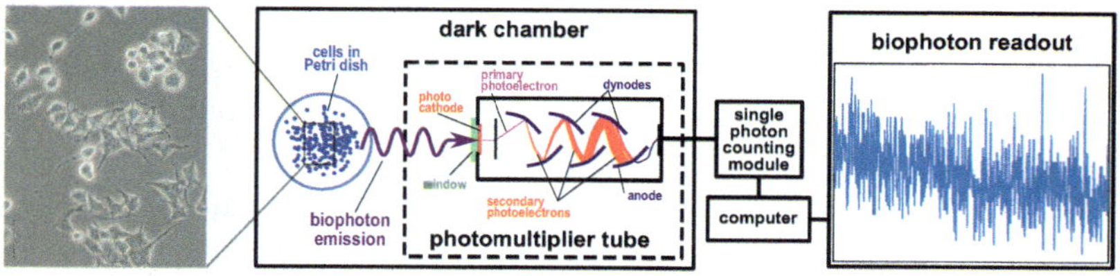


**Fig. 2** Schematic of an experimental setup to measure biophoton emission from cells in culture. Cells (Neuro-2a; mouse neuroblasts from brain tissue, shown as a photomicrograph) are in a petri dish, inside of a dark chamber. They emit biophotons which are detected by a photomultiplier tube: when a photon impinges on the photocathode, it is absorbed and releases a 'primary photoelectron', which is accelerated and multiplied through a cascade of dynodes. At the end of the cascade, the avalanche of 'secondary photoelectrons' on the anode generates an electrical current which is discriminated by a single photon counting module. The output signal is finally interpreted by a computer, as a number of detected photons per time window (typ. 1 s), as represented in the graph.

### *4.1.2 In brain slices*

(Isojima et al., 1995) measured the biophotons from hippocampal slices from rats: a tapered optical fibre was used to collect biophotons in the sample and bring it to with a single-photon detector using a silicon avalanche photodiode kept in liquid nitrogen. (Kobayashi et al., 1999a) reports a brief study of biophotons of mouse brain slices under different conditions modulating energy metabolism and cellular respiration. The brain slices were placed in a quartz chamber with circulating artificial cerebro-spinal fluid bubbled with $O_2$ and $CO_2$. The intensity of biophotonic emission was recorded through a PMT and a photon counting system.

Fifteen years later, the team of Jiapei Dai at the Wuhan Institute for Neuroscience had developed a more modern in vitro imaging system UBIS, for Ultraweak Biophoton Imaging System, see (Tang & Dai, 2014b) that allowed for 2D imaging of rat and mouse brain slices. It consists of an EMCCD camera, a wide aperture lens and a stand, all placed inside a dark box, with fluidic connections allowing for perfusion to keep the brain slices alive. With this imaging setup, they were able to explore many aspects of biophotons in the brain.

A notable evolution of this system is the capability to explore spectral properties of biophotons from brain slices, through what they called a Biophoton Spectral Analysis Device (BSAD). It adds to the "UBIS" a slit and a diffraction grating, which diffracts biophotons to a first order fringe whose position is wavelength-dependent. Careful calibration against sources at known wavelengths allows for a coarse estimation of the spectral properties of the biophotons from brain slices of different animals bullfrog, mouse, chicken, pig, monkey and post-mortem human brain materials, see (Wang et al., 2016).

The same team also tried a technique based on the photoreduction of dissolved Ag+ ions into Ag deposits, in a process similar to silver-based photography, termed "in-situ biophoton autography" or IBA (Sun et al., 2010). They used this technique on green bean seeds and on rat brain slices, sciatic nerves and spinal nerve root. The samples are biophoton-imaged with an EMCCD for 30 min, after which the Ag deposits are in turn imaged under a microscope. The same technique was used by (Zangari et al., 2021), on the sciatic nerve of rats. The isolated nerves were electrically stimulated, after which deposits of Ag granules, indicative of photonic activity, were detected in the nodes of Ranvier. Even though this technique seems to be very sensitive and to allow for some localization of biophotonic trajectories or regions of more intense emission, there is concern on whether it could affect the living tissue through the presence of silver ions, and on whether this could in turn affect the biophoton count itself. Further exploration is also needed to validate its accuracy (Hoh Kam et al., 2024). These studies were all, however, still done in vitro/ex vivo and a natural extension to this is to go in vivo.

### 4.1.3 In whole rodents

The Dai team modified their setup to accommodate for whole animals to be placed in the dark box, with mice being maintained in a stereotactic holder. A "window" to their brains was opened: the animals were shaved, their scalp cut and their parietal bones removed to expose the cerebral cortex (Li et al., 2021). The rest of the imaging system is the same as in their previous studies.

It is worth mentioning that similar studies on exposed animal brains were conducted twenty years earlier by the team of Humio Inaba in Yamagata, Japan. At the time, EMCCD technology was still in its infancy, but photomultiplying tubes were already an established tool for low-level light detection. In their studies (Kobayashi et al., 1999a; Kobayashi et al., 1999b), whole anesthetized rats are placed in a light-tight enclosure, artificially ventilated and immobilized. Their brains are exposed by cutting the skin and removing the parietal bones. They are imaged, through a high aperture lens system, (0.45–0.5 NA, magnification 1.0) by a 2D photon-counting tube. This tube consists of a photocathode; a photoelectron multiplier made of micro-channel plates and a position-sensitive anode. This imaging system provides a spatial resolution around 200 μm together with the capability to time-tag the detected photons, providing a spatio-temporal monitoring of biophotons emitted by the brain. In a study by (Kobayashi et al., 1999a), spatial resolution is reduced to 2.3 mm due to image post-processing (binning and smoothing), but the time monitoring allows them to explore possible correlations between biophotonic emission intensity and electroencephalographic (EEG) activity in the -wave component.

### 4.1.4 In human subjects

Much more recently, studies have been carried out on human subjects, building on the history of exploration exposed above. The team of Nirosha Murugan at Wilfried Laurier University recruited twenty adult humans, who were seated in a dark room with multiple PMTs pointed at their heads and EEG devices placed on them (Casey et al., 2025). Biophotonic emission and EEG records were acquired during different phases of cognitive significance, with the goal to correlate biophotonic intensity to different cognitive tasks and to EEG activity. Noticeably, task-dependent variations of biophotonic intensity, therefore attributed to biophotons emitted from inside the brain since they somehow correlate to brain activity, can be detected even through the skull, scalp and hair of the subjects.

### 4.1.5 What's next for measuring brain biophotons?

Even though there is evidence to support the presence of biophotons from the brain, progress has been slow in measuring them. As with biophotons in general, the main obstacle is the very small amount of emitted biophotons and, therefore, the difficulty to detect them. The history of devices presented here does not claim to be exhaustive but aims at presenting the different avenues to date that have explored biophotons in the brain: most of the studies so far have focused either on photomultiplier tubes or CCD cameras.

It should be noted as well that most of these studies mention dark adaptation periods of their samples of a few tens of minutes. (Mackenzie et al., 2024) has demonstrated, however, that delayed luminescence from the sample's environment only can be significant over much longer, typically 24 h, after 60 s exposure to diffuse daylight. Extra care should be taken henceforth to ensure that delayed luminescence, which emits at least 2 orders of magnitude more photons that UPE, has decayed sufficiently to allow for the detection of proper biophotons. This requirement is especially critical when trying to detect biophotons that must first go through tissue layers such as skull and scalp. Cultured cells, brain slices, anesthetized animals as well as human subjects should then, ideally, be kept in the dark for 24 h before measurement, which for each case poses different but always potentially challenging issues.

Several avenues can be explored however, to progress on the long-term and delayed luminescence-free measurement of brain biophotons. For cultured cells or brain slices, one could design a light-tight experiment chamber in which a measurement device can be operated safely while the cells can be kept in physiological culture conditions for several days.

In the case of animals, one could imagine a new type of measurement device (based on, e.g., microscopic plasmon antennae or near-field scanning optical microscope fibre tips) which could be inserted inside the brain. Hopefully, such a device, if it reaches a sufficient sensitivity, could record brain biophotons from animals in different states of wakefulness, and would be shielded from external light just as much as the animal's brain cells usually are. Such a device could also be used on 3D brain organoids or organotypic samples, provided they are satisfactorily kept in the dark. Finally, a small enough version of such a detector would be instrumental in probing the density of photons inside of brain cells versus the number of emitted photons, as there is research hypothesizing that the emitted biophotons are but a fraction of the ones being constantly emitted and absorbed inside cells, especially brain cells. Furthermore, with small enough detectors, one could be able to spatially probe where inside the cells are the biophotons primarily emitted.

## 4.2 Neuronal biophotonic emission

Numerous cells in the human body, including neurones that are deeply embedded within the cranial cavity—regions typically inaccessible to external light—exhibit the capacity to respond to photonic stimuli. The presence of photoreceptive molecules in these light-shielded neural cells raises important questions about their functional significance. One compelling hypothesis posits that these cells generate endogenous biophotons, which may serve as a modality of intra- and intercellular communication.

Neuronal biophotonic interactions are mediated by a variety of endogenous chromophores, including porphyrin rings, flavinic and pyridinic rings, lipid chromophores, and aromatic amino acids (Karu, 1999; Kato et al., 1981; Mazhul' & Shcherbin, 1999; Salari et al., 2015; Thar & Kühl, 2004).

These photosensitive biomolecules, once excited, can transfer the absorbed biophotonic energy to adjacent biomolecules by resonant energy transfer and may lead to conformational changes. Such structural alterations can subsequently initiate or modulate downstream biochemical signalling cascades, influencing processes such as gene expression, enzymatic activation, and synaptic plasticity.

Sun et al. (Sun et al., 2010) showed that neurones can conduct photonic signals. Due to their high metabolic demands, neurones emit biophotons constitutively without any external excitation, with emission intensity modulated by neurotransmitters and electrical activity. For example, high-potassium environments enhance neuronal biophotonic emission via induced depolarization, while extracellular calcium removal attenuates this effect (Kataoka et al., 2001). It has been proposed that depolarization-associated energy is harnessed to generate biophotons, which may facilitate microtubule-mediated information processing and transmission to adjacent neurones.

If biophotons mediate nerve signal transmission and processing, it follows that they must be transmittable along nerve fibres. Biophotonic transmission is hypothesized to involve mitochondrial and microtubular networks within axons. Mitochondrial morphology and spatial distribution may form a functionally coupled system enabling intracellular signal coordination (Dedov & Roufogalis, 1999; Müller et al., 2005; Skulachev, 2001). Microtubules contribute to signal transduction and information encoding (Craddock et al., 2009), with theoretical models proposing that filamentous mitochondria and microtubules function as neuronal "optical fibres" for biophotonic propagation (Jibu et al., 1994). Empirical studies further demonstrate that nerve fibre orientation affects biophotonic propagation in brain tissue (Hebeda et al., 1994).

The dual-component model proposed by Tang and Dai conceptualizes biophotonic emission as comprising of: (1) background biophotons produced constitutively during metabolism, maintaining a "pre-action" molecular state, and (2) signal biophotons actively propagated along dendrites and axons to encode and transmit neural information (Tang & Dai, 2014a). Supporting this, (Kumar et al., 2016) provided theorectical evidence for myelinated axons serving as photonic waveguides facilitating optical communication.

Biophotonic transmission in the brain appears to occur via two principal mechanisms: long-distance transmission through white matter tracts connecting distant cortical and subcortical regions (including the brainstem and spinal cord), and short-range signalling within the extracellular matrix over micrometre scales (Grass et al., 2004). Monoaminergic pathways involving dopamine, serotonin, and noradrenaline have been implicated as prominent conduits for biophoton-related activity. Tang and Dai (2014a) employed spatiotemporal imaging to demonstrate robust glutamate-induced biophotonic activity at axonal cut terminals in mouse cortical slices. They further showed that glutamate administration amplifies biophotonic emissions, whereas metabolic inhibitors such as sodium azide and oxygen-glucose deprivation suppress these signals. Notably, blockade of action potentials or calcium chelation only partially inhibits emission, suggesting that biophotonic production is partially independent of conventional electrophysiological activity. This finding supports the hypothesis that neuronal networks utilize intrinsic biophotonic signalling for information processing. (Li et al., 2021) employed a highly sensitive photon-counting imaging system to capture two-dimensional in vivo images of spontaneous biophotonic emission from the entire mouse brain. The resulting data revealed heterogeneous spatial distribution of biophotonic emission with distinct brain regions exhibiting elevated photon intensities, indicative of region-specific metabolic or oxidative activity. It is important to note that the current biophotonic detection methodologies primarily quantify extracellular emissions, suggesting that intracellular biophoton dynamics may be substantially underestimated.

Emerging interdisciplinary research has revealed a potential functional relationship between biophotons and EEG activity, suggesting that endogenously emitted light may contribute to neural information processing. Theoretical models propose that mitochondrial biophotons interact with neuronal microtubules, potentially modulating the transition of their quantum states between coherent and incoherent, thereby influencing electrophysiological rhythms, particularly within the alpha frequency range (Rahnama et al., 2011). Studies in both human and animal models have demonstrated significant correlations between EEG activity and biophotonic emission intensity, linking biophoton levels to brain activity, cerebral energy metabolism, blood flow, and oxidative processes (Dotta et al., 2012; Kobayashi et al., 1999a; Van Wijk et al., 2020). Notably, Dotta and colleagues (Dotta et al., 2012) reported that mental imagery of light in a dark environment elicited a significant increase in right hemispheric biophotonic emission, which was strongly correlated with changes in EEG power. These findings lend support to the hypothesis that biophotons may serve as an additional modality of neural signalling, complementing established electrical and chemical pathways. Further evidence of this relationship is observed in altered states of consciousness, such as near-death experiences, during which transient increases in biophotonic production have been hypothesized to co-occur with surges in high-frequency gamma oscillations recorded via EEG (Crone et al., 2006). Additionally, photobiological neuromodulation studies have demonstrated that exposure to 40 Hz pulsed violet light can entrain EEG activity, enhancing cross-frequency coupling between alpha-phase and gamma-amplitude oscillations (Noda et al., 2021). Collectively, these findings suggest that biophotonic emissions are not merely metabolic by-products but may constitute a functional component of brain dynamics, particularly in contexts involving visual cognition and non-ordinary states of awareness.

## 5. Spectral signatures and information processing

Beyond detecting biophotons, analysing their spectral characteristics has revealed a lot of biological information. The wavelength distribution and spectral shifts of biophotonic emission provide unique insights into cellular processes, metabolic states, and even cognitive function. This section examines how different biological systems produce distinctive spectral "fingerprints" that can distinguish between species, cell types, and physiological conditions. We explore the controversial findings linking spectral properties to intelligence and ageing, and critically evaluate the theoretical frameworks proposed to explain these phenomena, including quantum coherence and holographic brain theories. Understanding these spectral signatures and their underlying mechanisms is crucial for developing diagnostic applications and determining whether biophotons truly represent a functional communication system or remain an intriguing epiphenomenon of cellular metabolism.

### 5.1 Species-specific spectral characteristics

Spectral analysis of biophotonic emissions has provided valuable insights into neural activity, aging, and cognitive function. Biophotons are increasingly recognized as potential non-local information carriers that respond to environmental stressors, exhibiting species-specific spectral signatures. For instance, (Tessaro et al., 2019) reported distinctive spectral characteristics in bacterial cultures, suggesting functional specificity in biophotonic emission profiles.

Extending these findings to cancer research, (Murugan et al., 2018, 2020) demonstrated that biophotonic emission patterns differ markedly between cancerous and non-cancerous cells in both in vivo and in vitro models. Specifically, the ratio of emissions within the infrared and ultraviolet spectral ranges effectively discriminated malignant from benign cell populations. Furthermore, spectral power density analyses revealed particular intensity modulation frequencies near 20 Hz and a narrow optical band centered around 500 nm as promising biomarkers for tumour identification (Murugan et al., 2018, 2020). These results underscore the potential of biophotonic spectral profiling as a non-invasive tool for early cancer detection.

In the domain of cognitive neuroscience, Wang et al. (Wang et al., 2016) reported a spectral redshift in glutamate-induced biophotonic activity from lower animals to humans. This redshift was proposed as a biophysical correlate of increased cognitive capacity and intelligence, potentially reflecting more efficient biophotonic signal transduction mechanisms in the human brain. In contrast, (Chen et al., 2020) observed a spectral blueshift in biophotonic emissions associated with aging in mouse brains, a phenomenon potentially linked to cognitive decline. However, (Salari et al., 2016) have cautioned against inferring causal relationships from such correlations, particularly regarding intelligence. They emphasized the absence of a mechanistic framework to explain the observed spectral shifts and urged prudence in interpreting these associations without rigorous theoretical and experimental validation. Another study by (Wang et al., 2023) reported a marked reduction in biophotonic emissions accompanied by a spectral blueshift in synaptosomes and brain slices derived from mouse models of Alzheimer's disease and vascular dementia, further implicating altered biophotonic dynamics in cognitive impairment.

Methodologically, most investigations into biophoton spectral properties have utilized band-pass optical filters coupled with PMTs. While these filters enable selective wavelength analysis, they inherently attenuate signal strength, resulting in reduced photon counts. This attenuation poses a significant limitation by diminishing detection sensitivity and potentially obscuring subtle but biologically relevant features.

## 5.2 Theoretical frameworks

The intersection of quantum mechanics and biology in biophotonic phenomena has generated both excitement and controversy. Several theoretical frameworks have been proposed to explain how quantum effects might play a role in biophotonic communication.

### 5.2.1 Microtubules as quantum optical waveguides

One of the most intriguing proposals involves microtubules serving as quantum optical waveguides. (Kurian et al., 2017) suggested that these cylindrical proteins, which play crucial roles in cellular structure and mitochondrial dynamics, could transmit light signals over long distances within cells. According to their model, biophotons generated by mitochondrial metabolic processes could excite the uniquely arranged aromatic amino acids, especially tryptophan, within tubulin proteins. This excitation could then propagate along the microtubule structure through quantum coherent energy transfer between neighbouring tryptophan residues (Kurian et al., 2017).

This quantum coherent energy transfer mechanism aligns with earlier work by (Craddock et al., 2014), who suggested that the arrangement of aromatic amino acids in tubulin could support quantum effects. The close association of microtubules with mitochondria could potentially facilitate the exchange of biophotonic signals between these structures within cells (Craddock et al., 2014).

### 5.2.2 Mitochondria as optical waveguides

Complementing the microtubule hypothesis, Thar and Kühl (2004) proposed that filamentous mitochondria themselves could act as "optical waveguides." Due to their higher refractive index compared to the surrounding cytoplasm, mitochondria could potentially direct the transmission of biophotons along the mitochondrial network. They described mitochondria as a "multi-layer optical system" with alternating refractive indices, particularly in their metabolically active state. This dynamic structure could potentially support coherent propagation and even amplification of light, allowing for the enhancement of low biophotonic signals and potentially increasing the range and efficacy of biophotonic communication (Thar & Kühl, 2004).

### 5.2.3 Coherence and information processing

Popp's original hypothesis of coherent biophotonic emission suggests that biophotons are not random thermal emissions but carry information through their coherence properties (Popp, 1992, 2003). This coherence could theoretically allow for more efficient information transfer than would be possible with incoherent light. In his 1992 analysis, Popp rejected the hypothesis that biophotonic emission originates from spontaneous chemiluminescence, arguing that all experimental data—including correlations to biological processes, temperature dependence, spectral distribution, and photocount statistics—align with the alternative hypothesis of a coherent field as the source of biophotonic emission (Popp, 1992).

However, a comprehensive critical review by (Cifra et al., 2015) evaluated decades of research on biophoton coherence and found that while the phenomenon of biophotonic emission from biological systems is experimentally well-established, no reliable evidence for coherence or non-classical states of biophotons has been achieved. They noted that the observation of Poisson statistics, often cited by Popp and colleagues as evidence for coherence, can arise from many sources including thermal light with many modes, and that extraordinary coherence times claimed by some authors (ranging from hours to days) lack rigorous experimental support (Cifra et al., 2015).

Despite these controversies, theoretical frameworks continue to develop. (Grass et al., 2004) proposed that the brain might function as a "holographic computer" via biophotons and microtubules. They noted that neurons possess ideal conditions for photonic communication: they are large colourless cells with wide arborizations, contain little pigment, have prominent cytoskeletal microtubules, and are protected from environmental light by bone and connective tissue. They suggested that neurotransmitters with strong fluorescence properties (serotonin, dopamine, and norepinephrine) might modulate biophotonic communication through their optical characteristics.

Tang and Dai (2014a) expanded on these concepts by reviewing evidence for biophotonic involvement in neural signal transmission. They proposed that biophotons may consist of two components: background biophotons generated by metabolic activity, and signal biophotons that transmit along neural circuits carrying neural information. Their review suggested that biophotonic transmission might be executed through quantum computation mechanisms, potentially explaining higher brain functions that traditional bioelectrical theories struggle to address.

Most recently, (Kumar et al., 2016) demonstrated through theoretical modelling that myelinated axons could serve as photonic waveguides. They proposed that the myelin sheath, with its higher refractive index than surrounding tissues, could guide photons efficiently over significant distances in the brain. This work provides a potential physical substrate for targeted optical communication between neurons, addressing a crucial gap in earlier biophotonic theories.

While these theoretical frameworks offer intriguing possibilities for understanding neural communication and information processing, (Cifra et al., 2015) emphasize that standard quantum optics methods could deliver more reliable information on coherence properties of biophotons, such as measuring light interference or correlation functions. However, the extremely low intensity of biophotonic emission and inherent non-stationarity of biological signals make such experiments highly challenging. Thus, while coherence remains a compelling theoretical possibility, it awaits definitive experimental verification.

## 6. Clinical applications and future directions

The potential clinical applications of biophotons extend beyond basic research, offering promising avenues for both diagnostic and therapeutic interventions in neurological and medical conditions. This section explores how biophotonic detection technologies are being translated into practical clinical tools for disease monitoring and how understanding endogenous light emission may inform novel treatment strategies.

### 6.1 Diagnostic applications

Recently, Casey and colleagues (2025) introduced "photoencephalography," a novel non-invasive technique utilizing biophotonic emission detection for real-time brain activity monitoring. Their data demonstrated task-dependent modulation of biophoton spectral and entropy features correlating with neural oscillations in alpha and beta frequency bands, supporting a functional link between biophotonic emissions and brain rhythms.

Biophotonic emission also holds promise as a biomarker in neurodegenerative diseases. (Sefati et al., 2024) observed elevated hippocampal biophotonic emissions in a rat model of sporadic Alzheimer's disease (AD), correlating with oxidative stress and cognitive decline. Pharmacological intervention with donepezil reduced both biophotonic emission and oxidative markers. Similarly, (Wang et al., 2023) reported diminished biophotonic activity and spectral blueshift in AD and vascular dementia models, associating these changes with synaptic dysfunction and cognitive deficits.

These findings advocate for the continued exploration of biophotonic emission as a tool for understanding neural function, encoding, and network dynamics. Future research may focus on wavelength-specific biophotonic signatures related to distinct neurological conditions, advancing non-invasive diagnostics and therapeutic monitoring.

## 6.2 Therapeutic implications

### 6.2.1 Biophotons and photobiomodulation: Interactions and therapeutic implications

If biophotons contribute to cellular communication and repair mechanisms, it becomes increasingly important to understand how external light stimuli may interact with or influence this endogenous photonic network. Most neural cells are shielded from external light by cranial bone and connective tissue; nevertheless, research has demonstrated that neurones and glial cells can respond to externally applied red and NIR light. This response forms the basis of PBM, a non-invasive therapeutic approach known to enhance cellular metabolism, promote tissue repair, and attenuate inflammation (Moro et al., 2021, 2023; Hoh Kam et al., 2024).

Recent studies suggest a complex relationship between PBM and endogenous biophotonic emissions. PBM delivers photons in the red to near-infrared spectral range which are absorbed by mitochondrial chromophores, particularly CCO. This photonic absorption enhances mitochondrial respiration by increasing ATP synthesis and restoring electron transport chain function, notably they have been shown to influence complex I activity (Le et al., 2018). Biophotons, albeit significantly weaker in intensity, have been implicated in modulating these critical metabolic processes (Moro et al., 2021, 2023; Hoh Kam et al., 2024).

Therefore, a consequence of improved mitochondrial efficiency is the regulation of biophotonic emission patterns. Increased and more coherent biophotonic activity may reflect healthier mitochondrial dynamics and improved cellular communication within neuronal networks. Although PBM utilizes light intensities several orders of magnitude higher than those emitted by cells intrinsically (Hamblin, 2022), both modalities involve overlapping spectral ranges—particularly in the red and NIR regions—and appear to produce similar functional outcomes.

While CCO has traditionally been considered the primary photoacceptor in PBM, emerging evidence points to additional photophysical mechanisms that may underly its therapeutic effects. These include potential indirect interactions with endogenous biophotons and engagement of mechanotransductive signalling pathways (Liebert et al., 2023). Such mechanisms may modulate neural micro-oscillatory behaviour and influence broader network dynamics, offering a plausible explanation for PBM-associated alterations in brainwave activity.

Moreover, exposure to LED light has been shown to enhance biophotonic emission at specific frequencies, which may serve as biomarkers of cellular metabolic state and viability (Ravindran et al., 2023). These findings support the hypothesis that externally applied light not only interacts with cellular structures but may also modulate the biophotonic communication landscape, thereby influencing both cell function and survival.

Importantly, cells retain the ability to communicate via biophotonic signalling even in the absence of external light exposure. It is hypothesized that cells distinguish between endogenous and exogenous light sources based on specific parameters, including wavelength, intensity, temporal coherence, and potentially other yet unidentified characteristics. Despite growing interest, the mechanisms underlying biophotonic signalling remain insufficiently characterized. Future research employing advanced optical imaging, photonic sensing, and computational modelling will be critical for elucidating the biological significance and therapeutic potential of this underexplored communication modality.

### 6.2.2 Biophotons in clinical applications: Therapeutic implications in neurodegenerative diseases

While the precise physiological significance of biophotonic emissions remains incompletely understood, emerging evidence suggests they may hold substantial therapeutic potential. A promising avenue for future development involves the creation of a non-invasive transcranial device capable of detecting biophotonic signals originating from brain tissue (Hoh Kam et al., 2024). To achieve reliable signal acquisition, such a device would require advanced optical filtering and calibration mechanisms to account for attenuation and scattering caused by intervening anatomical layers—specifically, the scalp, skull, and connective tissue. Additionally, the device could be engineered with variable depth sensitivity to enable selective monitoring of both superficial cortical regions and deeper subcortical structures, such as the thalamus and brainstem. At its core, the system would detect the spectral properties—namely, the wavelengths and intensities—of biophotonic emissions, which are hypothesized to reflect patterns of oxidative metabolism and mitochondrial function at the cellular level. Such spectral profiling could serve as a real-time biomarker for neuronal health and dysfunction, offering novel insights into disease progression and treatment efficacy.

The development of an ultra-sensitive biophoton detection device holds considerable promise for guiding therapeutic interventions tailored to specific neurological conditions. Such interventions may include pharmacological treatment, targeted red and near-infrared light therapy, or neurosurgical procedures, depending on the nature and severity of the disorder—be it stroke, traumatic brain injury, psychiatric illness, or neurodegenerative diseases such as Alzheimer's and Parkinson's. In this context, a novel and emerging strategy involves the integration of photonic brain-computer interface (BCI) chips. These minimally invasive devices would be designed to monitor biophotonic emissions in real time and, based on the detected optical signals, facilitate personalized therapeutic guidance. Functioning analogously to the aforementioned transcranial detection system, BCI photonic chips could offer a more localized and precise means of assessing mitochondrial function and oxidative metabolism at the cellular level within the brain (Moro et al., 2021, 2023; Hoh Kam et al., 2024).

## 7. Conclusion and perspectives

The study of biophotons has evolved from Gurwitsch's pioneering observations to a multidisciplinary field incorporating quantum physics, biochemistry, and cellular biology. The evidence for ultra-weak photon emission from biological systems is now well-established, with clear connections to oxidative metabolism and cellular stress responses (Cifra & Pospíšil, 2014; Pospíšil et al., 2019). Recent experimental evidence, including Mould et al. (2023) demonstrating biophotonic signalling between isolated mitochondria, supports the fundamental concept of light-mediated biological communication. However, theoretical frameworks proposing quantum mechanical effects remain largely speculative, facing challenges from environmental decoherence in biological systems.

Progress in biophoton research has been constrained by technological limitations, primarily the extremely low intensity of emissions—often comprising only a few photons per cell per second. This presents considerable barriers to experimental reproducibility and comprehensive analysis. Nonetheless, ongoing innovations in photodetection, including enhanced PMTs, CCDs, and single-photon detectors, are progressively improving sensitivity and spatial resolution, holding promise for elucidating the functional role of biophotonic emissions in cellular physiology (Hoh Kam et al., 2024).

The refinement of ultra-sensitive detection technologies could have profound clinical implications. Real-time, non-invasive monitoring of mitochondrial health and oxidative stress could revolutionize diagnostic and therapeutic strategies for neurological disorders ranging from acute conditions such as strokes to chronic neurodegenerative diseases. Future therapeutic strategies may leverage PBM to deliberately modulate cellular metabolism, thereby influencing endogenous biophotonic production patterns. By integrating external light therapy with real-time biophoton monitoring, clinicians could optimize treatment protocols based on the individual's cellular responses, creating a feedback loop between applied phototherapy and endogenous optical biomarkers.

As our understanding deepens, biophotons may reveal themselves to be more than metabolic byproducts—perhaps serving as an ancient communication system that complements electrical and chemical signalling. The journey from Gurwitsch's onion roots to modern quantum biological theories illustrates how phenomena at the edge of detectability can open new windows into the fundamental processes of life.

## References

Albrecht-Buehler, G. (1992). Rudimentary form of cellular "vision". *Proceedings of the National Academy of Sciences, 89*(17), 8288–8292. https://doi.org/10.1073/pnas.89.17.8288.

Bat'yanov, A. P. (1984). Distant-optical interaction of mitochondria through quartz. *Bulletin of Experimental Biology and Medicine, 97*(6), 740–742. https://doi.org/10.1007/BF00804160.

Bókkon, I., Salari, V., Tuszynski, J. A., & Antal, I. (2010). Estimation of the number of biophotons involved in the visual perception of a single-object image: Biophoton intensity can be considerably higher inside cells than outside. *Journal of Photochemistry and Photobiology B: Biology, 100*(3), 160–166. https://doi.org/10.1016/j.jphotobiol.2010.06.001.

Burgos, R. C. R., Schoeman, J. C., Winden, L. J. van, Červinková, K., Ramautar, R., Van Wijk, E. P. A., ... Greef, J. van der (2017). Ultra-weak photon emission as a dynamic tool for monitoring oxidative stress metabolism. *Scientific Reports,* 7(1), 1229. https://doi.org/10.1038/s41598-017-01229-x.

Cadenas, E., Arad, I. D., Boveris, A., Fisher, A. B., & Chance, B. (1980). Partial spectral analysis of the hydroperoxide-induced chemiluminescence of the perfused lung. *FEBS Letters, 111*(2), 413–418. https://doi.org/10.1016/0014-5793(80)80839-8.

Casey, H., DiBerardino, I., Bonzanni, M., Rouleau, N., & Murugan, N. J. (2025). Exploring ultraweak photon emissions as optical markers of brain activity. *IScience, 28*(3), 112019. https://doi.org/10.1016/j.isci.2025.112019.

Chen, L., Wang, Z., & Dai, J. (2020). Spectral blueshift of biophotonic activity and transmission in the ageing mouse brain. *Brain Research, 1749*, 147133. https://doi.org/10.1016/j.brainres.2020.147133.

Cifra, M., & Pospíšil, P. (2014). Ultra-weak photon emission from biological samples: Definition, mechanisms, properties, detection and applications. *Journal of Photochemistry and Photobiology B: Biology, 139*, 2–10. https://doi.org/10.1016/j.jphotobiol.2014.02.009.

Cifra, M., Brouder, C., Nerudová, M., & Kučera, O. (2015). Biophotons, coherence and photocount statistics: A critical review. *Journal of Luminescence, 164*, 38–51. https://doi.org/10.1016/j.jlumin.2015.03.020.

Craddock, T. J. A., Beauchemin, C., & Tuszynski, J. A. (2009). Information processing mechanisms in microtubules at physiological temperature: Model predictions for experimental tests. *Bio Systems, 97*(1), 28–34. https://doi.org/10.1016/j.biosystems.2009.04.001.

Craddock, T. J. A., Friesen, D., Mane, J., Hameroff, S., & Tuszynski, J. A. (2014). The feasibility of coherent energy transfer in microtubules. *Journal of The Royal Society Interface, 11*(100), 20140677. https://doi.org/10.1098/rsif.2014.0677.

Crone, N. E., Sinai, A., & Korzeniewska, A. (2006). High-frequency gamma oscillations and human brain mapping with electrocorticography. *Progress in Brain Research, 159*, 275–295. https://doi.org/10.1016/S0079-6123(06)59019-3.

Dedov, V. N., & Roufogalis, B. D. (1999). Organisation of mitochondria in living sensory neurons. *FEBS Letters, 456*(1), 171–174. https://doi.org/10.1016/s0014-5793(99)00951-5.

Dotta, B. T., Saroka, K. S., & Persinger, M. A. (2012). Increased photon emission from the head while imagining light in the dark is correlated with changes in electroencephalographic power: Support for Bókkon's biophoton hypothesis. *Neuroscience Letters, 513*(2), 151–154. https://doi.org/10.1016/j.neulet.2012.02.021.

Du, J., Deng, T., Cao, B., Wang, Z., Yang, M., & Han, J. (2023). The application and trend of ultra-weak photon emission in biology and medicine. *Frontiers in Chemistry, 11*. https://doi.org/10.3389/fchem.2023.1140128.

Farhadi, A., Forsyth, C., Banan, A., Shaikh, M., Engen, P., Fields, J. Z., & Keshavarzian, A. (2007). Evidence for non-chemical, non-electrical intercellular signaling in intestinal epithelial cells. *Bioelectrochemistry (Amsterdam, Netherlands), 71*(2), 142–148. https://doi.org/10.1016/j.bioelechem.2007.03.001.

Fleiss, A., & Sarkisyan, K. S. (2019). A brief review of bioluminescent systems (2019). *Current Genetics, 65*(4), 877–882. https://doi.org/10.1007/s00294-019-00951-5.

Grass, F., Klima, H., & Kasper, S. (2004). Biophotons, microtubules and CNS, is our brain a "Holographic computer"? *Medical Hypotheses, 62*(2), 169–172. https://doi.org/10.1016/S0306-9877(03)00308-6.

Gurwitsch, A. (1923). Die Natur des spezifischen Erregers der Zellteilung. *Archiv Für Mikroskopische Anatomie Und Entwicklungsmechanik, 100*(1–2), 11–40. https://doi.org/10.1007/BF02111053.

Hamblin, M. R. (2016). Photobiomodulation or low-level laser therapy. *Journal of Biophotonics, 9*(11–12), 1122–1124. https://doi.org/10.1002/jbio.201670113.

Hamblin, M. R. (2017). Mechanisms and applications of the anti-inflammatory effects of photobiomodulation. *AIMS Biophysics, 4*(3), 337–361. https://doi.org/10.3934/biophy.2017.3.337.

Hamblin, M. R. (2022). Do biophotons play any role in transcranial photobiomodulation of the brain? *Photobiomodulation, Photomedicine, and Laser Surgery, 40*(11), 731–733. https://doi.org/10.1089/photob.2022.0095.

Havaux, M., & Ksas, B. (2022). Imaging of lipid peroxidation-associated chemiluminescence in plants: Spectral features, regulation and origin of the signal in leaves and roots. *Antioxidants, 11*(7), 1333. https://doi.org/10.3390/antiox11071333.

Hebeda, K. M., Menovsky, T., Beek, J. F., Wolbers, J. G., & van Gemert, M. J. (1994). Light propagation in the brain depends on nerve fiber orientation. *Neurosurgery, 35*(4), 720–724. https://doi.org/10.1227/00006123-199410000-00019.

Hoh Kam, J., Billeres, M., Herault, L., Cali, C., Sarmiento, B., Cassano, P., ... Mitrofanis, J. (2024). Exploring current and future technologies to make sense of the biophoton phenomenon: A narrative review. *Advanced Technology in Neuroscience, 1*(2) https://journals.lww.com/atn/fulltext/2024/12000/exploring_current_and_future_technologies_to_make.5.aspx.

Hoh Kam, J., Clément, R., Cantat-Moltrecht, T., Billères, M., & Mitrofanis, J. (2025). Red and near-infrared light treatment can change the intensity of biophoton emissions in cell culture. *Scientific Reports, 15*(1), 38541. https://doi.org/10.1038/s41598-025-22344-0.

Isojima, Y., Isoshima, T., Nagai, K., Kikuchi, K., & Nakagawa, H. (1995). Ultraweak biochemiluminescence detected from rat hippocampal slices. *Neuroreport, 6*(4), 658–660. https://doi.org/10.1097/00001756-199503000-00018.

Jibu, M., Hagan, S., Hameroff, S. R., Pribram, K. H., & Yasue, K. (1994). Quantum optical coherence in cytoskeletal microtubules: Implications for brain function. *Bio Systems, 32*(3), 195–209. https://doi.org/10.1016/0303-2647(94)90043-4.

Kamal, A. H. M., & Komatsu, S. (2016). Proteins involved in biophoton emission and flooding-stress responses in soybean under light and dark conditions. *Molecular Biology Reports, 43*(2), 73–89. https://doi.org/10.1007/s11033-015-3940-4.

Karu, T. (1999). Primary and secondary mechanisms of action of visible to near-IR radiation on cells. *Journal of Photochemistry and Photobiology. B, Biology, 49*(1), 1–17. https://doi.org/10.1016/S1011-1344(98)00219-X.

Karu, T. I. (2008). Mitochondrial signaling in mammalian cells activated by red and Near-IR radiation. *Photochemistry and Photobiology, 84*(5), 1091–1099. https://doi.org/10.1111/j.1751-1097.2008.00394.x.

Kataoka, Y., Cui, Y., Yamagata, A., Niigaki, M., Hirohata, T., Oishi, N., & Watanabe, Y. (2001). Activity-dependent neural tissue oxidation emits intrinsic ultraweak photons. *Biochemical and Biophysical Research Communications, 285*(4), 1007–1011. https://doi.org/10.1006/bbrc.2001.5285.

Kato, M., Shinzawa, K., & Yoshikawa, S. (1981). Cytochrome oxidase is a possible photoreceptor in mitochondria. *Photobiochemistry and Photobiophysics, 2*(4), 263–269. https://doi.org/10.1016/S0165-8646(24)00252-6.

Kaznacheev, A. V. P., Mikhailova, L. P., & Kartashov, N. B. (1980). Distant intercellular electromagnetic interaction between two tissue cultures. *Bulletin of Experimental Biology and Medicine, 89*(3), 345–348. https://doi.org/10.1007/BF00834249.

Kim, H., Esser, L., Hossain, M. B., Xia, D., Yu, C.-A., Rizo, J., ... Deisenhofer, J. (1999). Structure of antimycin A1, a specific electron transfer inhibitor of ubiquinol−cytochrome c oxidoreductase. *Journal of the American Chemical Society, 121*(20), 4902–4903. https://doi.org/10.1021/ja990190h.

Kobayashi, M., Kikuchi, D., & Okamura, H. (2009). Imaging of ultraweak spontaneous photon emission from human body displaying diurnal rhythm. *PLoS One, 4*(7), e6256. https://doi.org/10.1371/journal.pone.0006256.

Kobayashi, M., Takeda, M., Sato, T., Yamazaki, Y., Kaneko, K., Ito, K., ... Inaba, H. (1999a). In vivo imaging of spontaneous ultraweak photon emission from a rat's brain correlated with cerebral energy metabolism and oxidative stress. *Neuroscience Research, 34*(2), 103–113. https://doi.org/10.1016/s0168-0102(99)00040-1.

Kobayashi, M., Takeda, M., Ito, K., Kato, H., & Inaba, H. (1999b). Two-dimensional photon counting imaging and spatiotemporal characterization of ultraweak photon emission from a rat's brain in vivo. *Journal of Neuroscience Methods, 93*(2), 163–168. https://doi.org/10.1016/s0165-0270(99)00140-5.

Kumar, S., Boone, K., Tuszyński, J., Barclay, P., & Simon, C. (2016). Possible existence of optical communication channels in the brain. *Scientific Reports, 6*(1), 36508. https://doi.org/10.1038/srep36508.

Kurian, P., Obisesan, T. O., & Craddock, T. J. A. (2017). Oxidative species-induced excitonic transport in tubulin aromatic networks: Potential implications for neurodegenerative disease. *Journal of Photochemistry and Photobiology B: Biology, 175*, 109–124. https://doi.org/10.1016/j.jphotobiol.2017.08.033.

Le, M., McNeill, F. E., Seymour, C. B., Rusin, A., Diamond, K., Rainbow, A. J., ... Mothersill, C. E. (2018). Modulation of oxidative phosphorylation (OXPHOS) by radiation-induced biophotons. *Environmental Research, 163*, 80–87. https://doi.org/10.1016/j.envres.2018.01.027.

Li, J., Xia, C., Wang, Y., Chen, L., & Dai, J. (2021). In vivo imaging of biophoton emission in the whole brain of mice. *Natural Science, 13*(09), 407–411. https://doi.org/10.4236/ns.2021.139033.

Liebert, A., Capon, W., Pang, V., Vila, D., Bicknell, B., McLachlan, C., & Kiat, H. (2023). Photophysical mechanisms of photobiomodulation therapy as precision medicine. *Biomedicines, 11*(2), https://doi.org/10.3390/biomedicines11020237.

Mackenzie, A. M., Smith, H. E., Mould, R. R., Bell, J. D., Nunn, A. V. W., & Botchway, S. W. (2024). Rooting out ultraweak photon emission a-mung bean sprouts. *Journal of Photochemistry and Photobiology, 19*, 100224. https://doi.org/10.1016/j.jpap.2023.100224.

Marais, A., Adams, B., Ringsmuth, A. K., Ferretti, M., Gruber, J. M., Hendrikx, R., ... van Grondelle, R. (2018). The future of quantum biology. *Journal of The Royal Society Interface, 15*(148), 20180640. https://doi.org/10.1098/rsif.2018.0640.

Mazhul, V. M., & Shcherbin, D. G. (1999). Phosphorescent analysis of lipid peroxidation products in liposomes. *Biofizika, 44*(4), 676–681.

Moro, C., Liebert, A., Hamilton, C., Pasqual, N., Jeffery, G., Stone, J., & Mitrofanis, J. (2021). The code of light: Do neurones generate light to communicate and repair? *Neural Regeneration Research, 17*(6), 1251–1252. https://doi.org/10.4103/1673-5374.327332.

Moro, C., Valverde, A., Dole, M., Hoh Kam, J., Hamilton, C., Liebert, A., ... Mitrofanis, J. (2023). The effect of photobiomodulation on the brain during wakefulness and sleep. *Frontiers in Neuroscience, 16*. https://www.frontiersin.org/articles/10.3389/fnins.2022.942536.

Mould, R. R., Kalampouka, I., Thomas, E. L., Guy, G. W., Nunn, A. V. W., & Bell, J. D. (2023). Non-chemical signalling between mitochondria. *Frontiers in Physiology, 14*, 1268075. https://doi.org/10.3389/fphys.2023.1268075.

Mould, R. R., Mackenzie, A. M., Kalampouka, I., Nunn, A. V. W., Thomas, E. L., Bell, J. D., & Botchway, S. W. (2024). Ultra weak photon emission—a brief review. *Frontiers in Physiology, 15*. https://doi.org/10.3389/fphys.2024.1348915.

Müller, M., Mironov, S. L., Ivannikov, M. V., Schmidt, J., & Richter, D. W. (2005). Mitochondrial organization and motility probed by two-photon microscopy in cultured mouse brainstem neurons. *Experimental Cell Research, 303*(1), 114–127. https://doi.org/10.1016/j.yexcr.2004.09.025.

Murugan, N. J., Persinger, M. A., Karbowski, L. M., & Dotta, B. T. (2020). Ultraweak photon emissions as a non-invasive, early-malignancy detection tool: An in vitro and in vivo study. *Cancers, 12*(4), https://doi.org/10.3390/cancers12041001.

Murugan, N. J., Rouleau, N., Karbowski, L. M., & Persinger, M. A. (2018). Biophotonic markers of malignancy: Discriminating cancers using wavelength-specific biophotons. *Biochemistry and Biophysics Reports, 13*, 7–11. https://doi.org/10.1016/j.bbrep.2017.11.001.

Niggli, H. J. (1992). Ultraweak photons emitted by cells: Biophotons. *Journal of Photochemistry and Photobiology B: Biology, 14*(1–2), 144–146. https://doi.org/10.1016/1011-1344(92)85090-H.

Noda, Y., Takano, M., Hayano, M., Li, X., Wada, M., Nakajima, S., ... Tsubota, K. (2021). Photobiological neuromodulation of resting-state EEG and steady-state visual-evoked potentials by 40 Hz violet light optical stimulation in healthy individuals. *Journal of Personalized Medicine, 11*(6), https://doi.org/10.3390/jpm11060557.

Nunn, A. V. W., Guy, G. W., Botchway, S. W., & Bell, J. D. (2020). From sunscreens to medicines: Can a dissipation hypothesis explain the beneficial aspects of many plant compounds? *Phytotherapy Research, 34*(8), 1868–1888. https://doi.org/10.1002/ptr.6654.

Ortega-Ojeda, F., Calcerrada, M., Ferrero, A., Campos, J., & Garcia-Ruiz, C. (2018). Measuring the human ultra-weak photon emission distribution using an electron-multiplying, charge-coupled device as a sensor. *Sensors, 18*(4), 1152. https://doi.org/10.3390/s18041152.

Passarella, S., & Karu, T. (2014). Absorption of monochromatic and narrow band radiation in the visible and near IR by both mitochondrial and non-mitochondrial photoacceptors results in photobiomodulation. *Journal of Photochemistry and Photobiology B: Biology, 140*, 344–358. https://doi.org/10.1016/j.jphotobiol.2014.07.021.

Popp, F.-A. (1992). *Some essential questions of biophoton research and probable answers. Recent advances in biophoton research and its applications*. World Scientific 1–46. https://doi.org/10.1142/9789814439671_0001.

Popp, F.-A. (2003). *Biophotons — background, experimental results, theoretical approach and applications. Integrative biophysics*. Springer Netherlands 387–438. https://doi.org/10.1007/978-94-017-0373-4_12.

Popp, F. A., Nagl, W., Li, K. H., Scholz, W., Weingärtner, O., & Wolf, R. (1984). Biophoton emission. *Cell Biophysics, 6*(1), 33–52. https://doi.org/10.1007/BF02788579.

Pospíšil, P., Prasad, A., & Rác, M. (2019). Mechanism of the formation of electronically excited species by oxidative metabolic processes: Role of reactive oxygen species. *Biomolecules, 9*(7), 258. https://doi.org/10.3390/biom9070258.

Prasad, A., & Pospíšil, P. (2013). Towards the two-dimensional imaging of spontaneous ultra-weak photon emission from microbial, plant and animal cells. *Scientific Reports, 3*(1), 1211. https://doi.org/10.1038/srep01211.

Rahnama, M., Tuszynski, J. A., Bókkon, I., Cifra, M., Sardar, P., & Salari, V. (2011). Emission of mitochondrial biophotons and their effect on electrical activity of membrane via microtubules. *Journal of Integrative Neuroscience, 10*(1), 65–88. https://doi.org/10.1142/S0219635211002622.

Ravindran, R., Branigan, K. S., Lefebvre, L. M., & Dotta, B. T. (2023). Effects of patterned electromagnetic fields and light-emitting diodes on cancer cells: Impact on cell density and biophoton emission when applied individually vs. simultaneously. *Applied Biosciences, 2*(4), 542–549. https://doi.org/10.3390/applbiosci2040034.

Rossi, C., Foletti, A., Magnani, A., & Lamponi, S. (2011). New perspectives in cell communication: Bioelectromagnetic interactions. *Seminars in Cancer Biology, 21*(3), 207–214. https://doi.org/10.1016/j.semcancer.2011.04.003.

Salari, V., Bókkon, I., Ghobadi, R., Scholkmann, F., & Tuszynski, J. A. (2016). Relationship between intelligence and spectral characteristics of brain biophoton emission: Correlation does not automatically imply causation. *Proceedings of the National Academy of Sciences of the United States of America, 113*(38), https://doi.org/10.1073/pnas.1612646113 E5540-1.

Salari, V., Seshan, V., Frankle, L., England, D., Simon, C., & Oblak, D. (2025). Imaging ultraweak photon emission from living and dead mice and from plants under stress. *The Journal of Physical Chemistry Letters, 16*(17), 4354–4362. https://doi.org/10.1021/acs.jpclett.4c03546.

Salari, V., Valian, H., Bassereh, H., Bókkon, I., & Barkhordari, A. (2015). Ultraweak photon emission in the brain. *Journal of Integrative Neuroscience, 14*(3), 419–429. https://doi.org/10.1142/S0219635215300012.

Scordino, A., Baran, I., Gulino, M., Ganea, C., Grasso, R., Niggli, J. H., & Musumeci, F. (2014). Ultra-weak delayed luminescence in cancer research: A review of the results by the ARETUSA equipment. *Journal of Photochemistry and Photobiology B: Biology, 139*, 76–84. https://doi.org/10.1016/j.jphotobiol.2014.03.027.

Sefati, N., Esmaeilpour, T., Salari, V., Zarifkar, A., Dehghani, F., Ghaffari, M. K., ... Oblak, D. (2024). Monitoring Alzheimer's disease via ultraweak photon emission. *IScience, 27*(1), 108744. https://doi.org/10.1016/j.isci.2023.108744.

Shen, X., Mei, W., & Xu, X. (1994). Activation of neutrophils by a chemically separated but optically coupled neutrophil population undergoing respiratory burst. *Experientia, 50*(10), 963–968. https://doi.org/10.1007/BF01923488.

Skulachev, V. P. (2001). Mitochondrial filaments and clusters as intracellular power-transmitting cables. *Trends in Biochemical Sciences, 26*(1), 23–29. https://doi.org/10.1016/s0968-0004(00)01735-7.

Sommer, A. P., Haddad, M., Kh, & Fecht, H.-J. (2015). Light effect on water viscosity: Implication for ATP biosynthesis. *Scientific Reports, 5*(1), 12029. https://doi.org/10.1038/srep12029.

Sun, Y., Wang, C., & Dai, J. (2010). Biophotons as neural communication signals demonstrated by in situ biophoton autography. *Photochemical & Photobiological Sciences, 9*(3), 315–322. https://doi.org/10.1039/b9pp00125e.

Tang, R., & Dai, J. (2014a). Biophoton signal transmission and processing in the brain. *Journal of Photochemistry and Photobiology B: Biology, 139*, 71–75. https://doi.org/10.1016/j.jphotobiol.2013.12.008.

Tang, R., & Dai, J. (2014b). Spatiotemporal imaging of glutamate-induced biophotonic activities and transmission in neural circuits. *PLoS One, 9*(1), e85643. https://doi.org/10.1371/journal.pone.0085643.

Tessaro, L. W. E., Dotta, B. T., & Persinger, M. A. (2019). Bacterial biophotons as non-local information carriers: Species-specific spectral characteristics of a stress response. *MicrobiologyOpen, 8*(6), e00761. https://doi.org/10.1002/mbo3.761.

Thar, R., & Kühl, M. (2004). Propagation of electromagnetic radiation in mitochondria? *Journal of Theoretical Biology, 230*(2), 261–270. https://doi.org/10.1016/j.jtbi.2004.05.021.

Turrens, J. F. (2003). Mitochondrial formation of reactive oxygen species. *The Journal of Physiology, 552*(2), 335–344. https://doi.org/10.1113/jphysiol.2003.049478.

Tuszyński, J. A., & Dixon, J. M. (2001). Quantitative analysis of the frequency spectrum of the radiation emitted by cytochrome oxidase enzymes. *Physical Review. E, Statistical, Nonlinear, and Soft Matter Physics, 64*(5Pt 1), 51915. https://doi.org/10.1103/PhysRevE.64.051915.

Van Wijk, R., Van Wijk, E. P. A., Wiegant, F. A. C., & Ives, J. (2008). Free radicals and low-level photon emission in human pathogenesis: State of the art. *Indian Journal of Experimental Biology, 46*(5), 273–309.

Van Wijk, R., Van Wijk, E. P. A., Pang, J., Yang, M., Yan, Y., & Han, J. (2020). Integrating ultra-weak photon emission analysis in mitochondrial research. *Frontiers in Physiology, 11*, 717. https://doi.org/10.3389/fphys.2020.00717.

Volodyaev, I., van Wijk, E., Cifra, M., & Vladimirov, Y. A. (2023). In (I. Volodyaev, E. van Wijk, M. Cifra, & Y. A. Vladimirov (Eds.). *Ultra-weak photon emission from biological systems*. Springer International Publishing. https://doi.org/10.1007/978-3-031-39078-4.

Wang, Z., Wang, N., Li, Z., Xiao, F., & Dai, J. (2016). Human high intelligence is involved in spectral redshift of biophotonic activities in the brain. *Proceedings of the National Academy of Sciences of the United States of America, 113*(31), 8753–8758. https://doi.org/10.1073/pnas.1604855113.

Wang, Z., Xu, Z., Luo, Y., Peng, S., Song, H., Li, T., ... Zhang, J. (2023). Reduced biophotonic activities and spectral blueshift in Alzheimer's disease and vascular dementia models with cognitive impairment. *Frontiers in Aging Neuroscience, 15*, 1208274. https://doi.org/10.3389/fnagi.2023.1208274.

Zangari, A., Micheli, D., Galeazzi, R., Tozzi, A., Balzano, V., Bellavia, G., & Caristo, M. E. (2021). Photons detected in the active nerve by photographic technique. *Scientific Reports, 11*(1), 3022. https://doi.org/10.1038/s41598-021-82622-5.

CHAPTER FOUR

# Photobiomodulation for traumatic brain injury

**Michael R. Hamblin***
Laser Research Centre, Faculty of Health Science, University of Johannesburg, Doornfontein, South Africa
*Corresponding author. e-mail address: hamblin.lab@gmail.com

## Contents



*International Review of Neurobiology,* Volume 189
ISSN 0074-7742, https://doi.org/10.1016/bs.irn.2026.01.016

## Abstract

There is a notable lack of therapeutic alternatives for what is fast becoming a global epidemic of traumatic brain injury (TBI). Photobiomodulation (PBM) employs red or near-infrared (NIR) light (600–1300 nm) from lasers or LEDs to stimulate healing, protect tissue from dying, increase mitochondrial function, improve blood flow and tissue oxygenation. PBM can also act to reduce swelling, increase antioxidants, decrease inflammation, protect against apoptosis, and modulate microglial activation state. All these mechanisms of action strongly suggest that PBM delivered to the head should be beneficial in cases of both acute and chronic TBI. Many studies in small animal models of acute TBI have found positive effects on neurological function, learning and memory, and reduced inflammation and cell death in the brain. There is evidence that PBM can help the brain to repair itself by stimulating neurogenesis, upregulating BDNF synthesis, and encouraging synaptogenesis. Clinical studies have been conducted in patients suffering from acute TBI and the chronic effects of TBI. There have been reports of improvements in executive function, working memory, and improved sleep. Functional magnetic resonance imaging has shown modulation of the activation in intrinsic brain networks likely to be damaged in TBI (default mode network and salience network).

This chapter is an updated version of a previous paper "Hamblin, M. R. (2018). Photobiomodulation for traumatic brain injury and stroke. *Journal of Neuroscience Research*, 96(4), 731–743. https://doi.org/10.1002/jnr.24190".

## 1. Introduction

Photobiomodulation (PBM) formerly known as low-level laser (light) therapy (LLLT) is approaching its 50th anniversary, after being discovered by Endre Mester working in Hungary in 1967 (Hamblin, de Sousa, & Agrawal, 2016). Originally thought to be a property of red lasers (600–700 nm), PBM has broadened to include near-infrared (NIR) wavelengths 760–1300 nm, and even blue and green wavelengths. Moreover the advent of inexpensive and safe light emitting diodes (LEDs) has supplanted the use of expensive lasers in many indications. The better tissue penetration properties of NIR light, together with its good efficacy, has made it the most popular wavelength range overall. The best-known medical applications of PBM have been for indications such as stimulation of wound healing (Hopkins, McLoda, Seegmiller, & David Baxter, 2004;

Kovacs, Mester, & Gorog, 1974), reduction of pain and inflammation in orthopedic and musculoskeletal conditions (Aimbire et al., 2006; Gam, Thorsen, & Lonnberg, 1993), and mitigation of cancer therapy side-effects (Zecha, Raber-Durlacher, Nair, Epstein, Elad, et al., 2016; Zecha, Raber-Durlacher, Nair, Epstein, Sonis, et al., 2016). However in recent years there has been growing interest in the use of PBM for various brain disorders (Hamblin, 2016; Hennessy & Hamblin, 2016; Naeser & Hamblin, 2011, 2015). The almost complete lack of any adverse side-effects of PBM, coupled with growing disillusion with pharmaceutical drugs that affect brain function, have combined together to suggest that PBM could be an alternative physical therapy approach to improving brain function.

Traumatic brain injury (TBI) is caused by many types of traumatic impacts to the head, often resulting from road traffic accidents, assaults, falls, sports injuries, or blast injuries suffered in military conflict. TBI is classified as (i) mild (loss of consciousness 0–30 min; altered mental state <24 h; post-trauma amnesia <1 day); (ii) moderate (loss of consciousness 30 min to 24 h; altered mental state >24 h; post-trauma amnesia >1–7 days), or (iii) severe (loss of consciousness >24 h; altered mental state >24 h; post-trauma amnesia >7 days) (Blennow et al., 2016). Three individuals suffer a TBI each minute in the US (Faul, Xu, Wald, & Coronado, 2010). Repeated mild episodes of TBI (also known as concussions) even without loss of consciousness, may have devastating cumulative effects (Kamins & Giza, 2016). Chronic traumatic encephalopathy is a recently recognized condition resulting from repeated head trauma, often found in boxers, football players, and military personnel (McKee, Alosco, & Huber, 2016; Safinia et al., 2016). There is presently no accepted treatment for TBI, although some investigational approaches have been tested in both the acute (neuroprotection) and chronic (neurorehabilitation) phases (Loane & Faden, 2010). One of these novel approaches is PBM, which has shown considerable promise (Hamblin, 2018; Lim, 2024; Naeser & Hamblin, 2015; Stevens et al., 2023; Thunshelle & Hamblin, 2016).

## 2. Mechanisms of action

Uncertainties about the mechanism of action of PBM at the molecular or cellular level, have undoubtedly held back its acceptance by the wider biomedical community. However in recent years, substantial progress has been made in understanding these mechanisms of action (de Freitas & Hamblin, 2016). In the following section the state-of-the-art knowledge about the mechanisms of PBM is summarized.

## 2.1 Chromophores

The first law of photobiology states that a photon must be absorbed by some molecule within the tissue to have any biological effect. The identity of these chromophores has been the subject of much scientific investigation and speculation. Largely due to the efforts of Tiina Karu in Russia, the enzyme cytochrome c oxidase (CCO) has been identified as a major chromophore of red/NIR light (Karu & Kolyakov, 2005; Karu, 1999; Karu, Pyatibrat, & Afanasyeva, 2004; Karu, Pyatibrat, & Kalendo, 2004). CCO is unit IV in the mitochondrial respiratory chain and has absorption peaks reaching well into the NIR spectral region (up to 900 nm) as well as in the red and blue regions. The most discussed hypothesis to explain exactly how photon absorption can stimulate the activity of CCO involves the photodissociation of inhibitory nitric oxide (NO) that can bind to the copper and heme centers in the enzyme and prevent oxygen from gaining access to the active sites (Lane, 2006). When red/NIR light is applied to laboratory models such as isolated mitochondria, oxygen consumption and ATP production are increased, and the mitochondrial membrane potential is raised (Passarella et al., 1984). Fig. 1 shows the absorption of light by mitochondria and the subsequent activation of signaling pathways and transcription factors.

A less well-appreciated mechanism involves light and heat-gated ion channels. These cation ion channels are thought to be members of the transient receptor potential (TRP) superfamily consisting of over 28 distinct members organized into six subfamilies, based on their primary amino acid structures (Caterina & Pang, 2016). TRPV (vanilloid sub-family) members including TRPV1 (capsaicin receptor) have been shown to be activated by various wavelengths of light including green, red and NIR (Sharma, Sardana, & Hamblin, 2023). When blue or green light is used to activate TRPV channels it is thought the primary chromophore is an opsin protein (either melanopsin OPN4 or neuropsin OPN5, which rely on the photoisomeriztion of 11-cis retinal to all-trans retinal. Opsins are G-protein coupled receptors, which can activate TRPV channels via diacylglycerol or kinase phosphorylation. On the other hand, when NIR or FIR is used to activate TRPV channels it is thought that a nanostructured water cluster

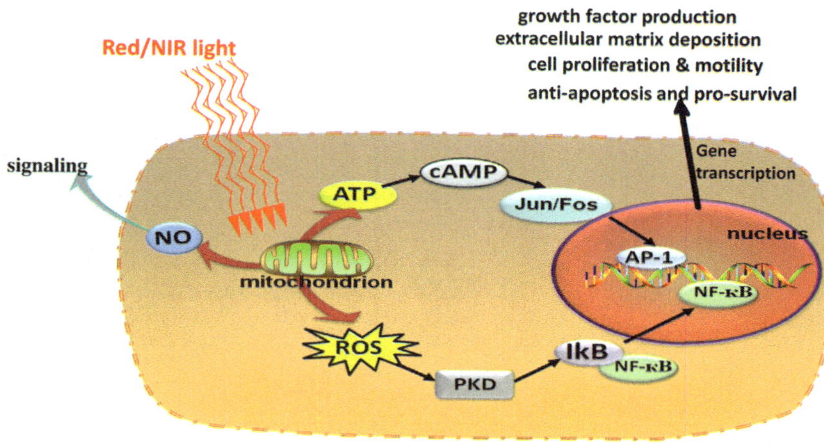


**Fig. 1 Mitochondria act as the central photoacceptor in PBM.** Signaling molecules (ROS, NO and AMP) are produced which can activate transcription factors such as Nf-kB or AP1.

selectively absorbs the light, which alters the tertiary conformation of the ion channel. It should be noted that TRPV channels are present in the mitochondria so the final results could be similar to those obtained by activation of the respiratory chain (Juárez-Contreras, Méndez-Reséndiz, Rosenbaum, González-Ramírez, & Morales-Lázaro, 2020). Fig. 2 illustrates the two possible mechanisms for activation of TRPV channels by blue/green or NIR light.

## 2.2 Cellular mechanisms

After the primary photon absorption event occurs, whether the photons are absorbed by CCO, or by TRP ion channels a series of secondary events occurs. One of these events is the generation of reactive oxygen species (ROS), which are thought to be produced inside the mitochondria due to an increase in electron transport, and a rise in the mitochondrial membrane potential above the baseline levels (Suski et al., 2012). It should be noted that mitochondrial ROS can be produced when MMP is raised above normal, and also when ROS is reduced below normal. It is thought that the ROS produced when MMP is lowered (mitochondrial dysfunction) are more damaging than the ROS produced when MMP is raised (mitochondrial stimulation). Nitric oxide is produced after PBM (Hamblin, 2008), possibly by photodissociation from CCO where it inhibits oxygen consumption and electron transport (Lane, 2006). There have also been

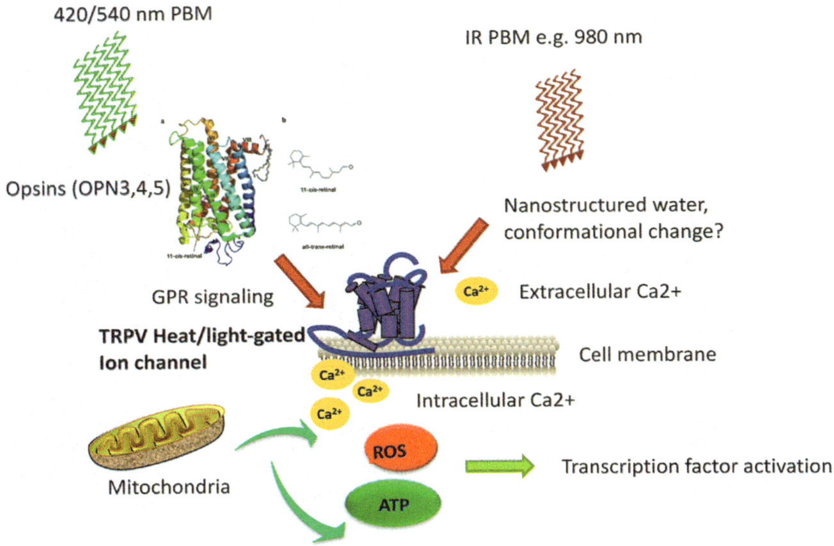


**Fig. 2 Role of opsins and TRPV ion channels in PBM.** Blue/green light may activate opsins via photoisomerization of retinal to carry out G-protein coupled receptor signaling to modulate TRPV ion channels. On the other hand, NIR light may be selectively absorbed by nanostructured water clusters to directly modulate the tertiary structure of TRPV channels.

additional hypotheses put forward to explain the increase in NO generation after PBM (Kashiwagi et al., 2023). NO could be photodissociated from nitrosylated hemoglobin (Vladimirov, Borisenko, Boriskina, Kazarinov, & Osipov, 2000) or from nitrosothiols (Keszler, Lindemer, Hogg, Weihrauch, & Lohr, 2018). Light could stimulate CCO to act as a nitrite reductase enzyme especially in hypoxic conditions (Poyton & Ball, 2011), or could stimulate the phosphorylation of endothelial nitric oxide synthase to increase its activity (Yokomizo et al., 2022).

Cyclic adenosine monophosphate (cAMP) (Gao & Xing, 2009) and intracellular calcium are increased by PBM (Alexandratou, Yova, Handris, Kletsas, & Loukas, 2002). Many of these substances increased by PBM (ROS, NO, cAMP, Ca2+) can act as secondary mediators in the signaling pathways triggered by PBM. These signaling pathways can induce activation of cellular transcription factors, that can in turn go on to upregulate or downregulate the expression levels of a large number of genes. One of the best-known transcription factors is NF-kB (nuclear factor kappa B) that can regulate expression of over one hundred genes including proteins with antioxidant, anti-apoptotic, pro-proliferation, and pro-migration functions. PBM (810 nm 3 J/cm$^2$) was shown to activate NF-kB in mouse embryonic fibroblasts via ROS production (Chen, Arany, et al., 2011). Since NF-kB is known to be a pro-inflammatory transcription factor, it might be thought that PBM would be pro-inflammatory. However it was also shown that NF-KB was decreased in already activated inflammatory dendritic cells (treated with Toll-like receptor ligands) when PBM (810 nm 3 J/cm$^2$) was applied (Chen, Huang, Sharma, & Hamblin, 2011).

## 2.3 Tissue mechanisms

The changes in expression levels of proteins involved in antioxidant and redox-regulation, anti-apoptotic and pro-survival, cellular proliferation, etc mean that distinct changes in tissue homeostasis, involving the promotion of healing and regeneration can be expected after PBM. For instance, structural proteins such as collagen are newly synthesized in order to repair tissue damage (Tatmatsu-Rocha et al., 2016). Cells at risk of dying in tissue that has been subjected to ischemic or other insults are protected from apoptosis (Sussai et al., 2010). Stem cells are activated to leave their niche, proliferate and differentiate (Oron & Oron, 2016; Zhang, Dong, Li, & Wu, 2016). Pain and inflammation are reduced (Chow, Johnson, Lopes-Martins, & Bjordal, 2009). Blood flow is increased (Samoilova, Zhevago, Petrishchev, & Zimin, 2008) (possibly as a result of the release of NO (Mitchell & Mack, 2013)), which also stimulates lymphatic drainage thereby reducing edema (Dirican et al., 2011).

## 2.4 Brain specific mechanisms

In addition to the foregoing, there are some PBM tissue mechanisms that are specific to the brain. One of the most important is an increase in cerebral blood flow often reported after transcranial photobiomodualtion (tPBM) (Salgado, Zangaro, Parreira, & Kerppers, 2015), leading to increased tissue oxygenation, and more oxidized CCO as measured by NIR spectroscopy (Rojas & Gonzalez-Lima, 2013). tPBM has been shown to reduce activated microglia in the brains of TBI mice as measured by IBA1 (ionized calcium-binding adapter molecule-1) expression thus demonstrating reduced neuroinflammation (Khuman et al., 2012). A systematic review confirmed this reduction of neuroinflammation by tPBM in laboratory studies in animal models (Cardoso, Salehpour, Coimbra, Gonzalez-Lima, & Gomes da Silva, 2022). tPBM has been shown to increase neurogenesis (formation of new brain cells derived from

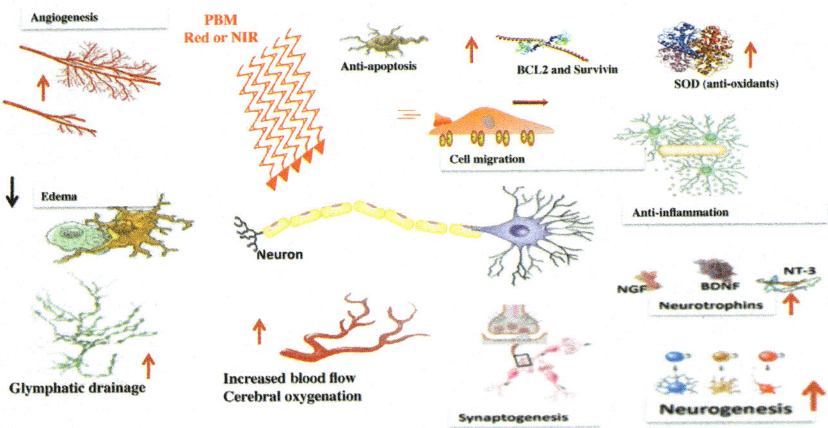


**Fig. 3 Brain specific mechanisms of PBM.** A wide range of biological processes have been shown to occur after PBM which may have major effects on the brain and its functions. These include increased blood flow, oxygenation and mitochondrial activity. Increased NO may decrease swelling and increase glymphatic drainage to remove toxins. Anti-apoptotic and pro-survival proteins are upregulated along with antioxidant enzymes. Neuroinflannation is reduced by modulation of microglial activation state, Brain repair is stimulated via increased angiogenesis, neurogenesis and synaptogenesis.

neuroprogenitor cells) (Xuan, Vatansever, Huang, & Hamblin, 2014), and synaptogenesis (formation of new connections between existing brain cells) (Xuan, Agrawal, Huang, Gupta, & Hamblin, 2015). Both of these mechanisms were observed in experimental studies carried out in TBI mice. The glymphatic drainage system is designed to drain away waste metabolites and amyloid-beta during the hours of sleep (Agaiby, Ghali, Wilson, & Dyson, 2000). Disturbances in glymphatic drainage may be implicated in various brain disorders (Buccellato, D'Anca, Serpente, Arighi, & Galimberti, 2022). The application of tPBM has been shown to increase glymphatic drainage in various animal models (Salehpour, Khademi, Bragin, & DiDuro, 2022).

Fig. 3 shows a graphical representation of some of these brain-specific tissue mechanisms by PBMT.

## 3. Transcranial photobiomodulation

Transcranial PBM is a growing approach to many different brain disorders that may be classified as sudden onset (stroke, TBI, global ischemia), neurodegenerative (Alzheimer's, Parkinson's, dementia), or psychiatric (depression, anxiety, posttraumatic stress disorder) (Hamblin, 2016; Hennessy & Hamblin, 2016; Thunshelle & Hamblin, 2016). In the following section some issues concerning where the light should be delivered, and the effects of PBM on uninjured mice and humans are addressed.

## 3.1 Light penetration

Several laboratories working in the field of tissue optics, have investigated the penetration of light of different wavelengths though the scalp and the skull, and to what depths into the parenchyma of the brain this light can penetrate. Answering the question "can light shone on the head sufficiently penetrate to reach the brain?" is difficult. The main reason is that at present it is unclear exactly what threshold of power density is necessary (expressed in $mW/cm^2$) at some depth inside the brain to have a biological effect. There clearly must be a minimum value below which, the light can be delivered for an infinite time without having any effect, but whether this threshold is in the region of $\mu W/cm^2$ or $mW/cm^2$ is unknown at present.

Haeussinger et al. (2011) estimated that the mean penetration depth (5 % remaining intensity) of NIR light through the scalp and skull was 23.6 $\pm$ 0:7 mm. Other studies have found comparable results with some variations depending on the precise location on the head and the precise wavelength studied (Okada & Delpy, 2003; Strangman, Zhang, & Li, 2014).

Jagdeo et al. (2012) used human cadaver heads (skull with intact soft tissue) to measure penetration of 830 nm light, and found that penetration depended on the anatomical region of the skull (0.9 % at the temporal region, 2.1 % at the frontal region, and 11.7 % at the occipital region). Tedford et al. (2015) also used human cadaver heads to compare penetration of 660 nm, 808 nm, and 940 nm light. They found that 808 nm light penetrated best, and could reach a depth in the brain of 40–50 mm. Lapchak et al. compared the transmission of 810 nm light through the skulls (no soft tissue) of four different species, and found the mouse skull transmitted 40 %, while for rat it was 21 %, for rabbit it was 11.3 and for the human skull it was only 4.2 % (Lapchak et al., 2015). Pitzschke and colleagues compared penetration of 670 nm and 810 nm light into the brain when delivered by a transcranial or a transphenoidal approach, and found that the best combination was 810 nm delivered transphenoidally (Pitzschke et al., 2015). Yaroslavsky et al. examined light penetration of different wavelengths through different parts of the brain tissue (white brain matter, gray brain matter, cerebellum, and brainstem tissues, pons, thalamus). The best penetration was found with wavelengths between 1000 and 1100 nm (Yaroslavsky et al., 2002). Henderson and Morries found that between 0.45 % and 2.90 % of 810 nm or 980 nm light penetrated through 3 cm of scalp, skull and brain tissue in ex vivo lamb heads (Henderson & Morries, 2015a).

## 3.2 Local vs systemic effects of light

It is possible that the beneficial effects of PBM on the brain cannot be entirely explained by penetration of light through the scalp and skull into the brain itself, at a sufficient intensity to have an effect on the brain cells. The surface power density that can be safely applied to the head, is limited by heating of the skin. Perceptible heating of the skin starts to be felt when the power density is over about 500 mW/cm$^2$, and can become severe at 1 W/cm$^2$.

There has been one study that explicitly addressed whether direct transcranial PBM or indirect PBM is best for the brain. In a study of PBM for Parkinson's disease in a mouse model, Mitrofanis and colleagues compared the direct delivery of light to the mouse head, and they also covered up the head with aluminum foil so that the light was delivered to the remainder of the mouse body. They found that there was a highly beneficial effect on brain histology with light delivered to the head, but nevertheless there was also a statistically significant although less pronounced benefit (referred to as an "abscopal effect") when the head was shielded from light. Moreover Oron and co-workers (Farfara et al., 2015) have shown that delivering NIR light to the mouse tibia (using either surface illumination or a fiber optic) resulted in improvements in memory and spatial learning in a transgenic mouse model of Alzheimer's disease. They proposed the mechanism involved PBM stimulating c-kit-positive mesenchymal stem cells (MSCs) that were normally resident in autologous bone marrow. These MSCs were proposed to be able to infiltrate the brain, and clear β-amyloid plaques (Oron & Oron, 2016). It should be noted in general that the calvarial bone marrow of the skull contains substantial numbers of stem cells (Iwashita, Tada, Zhan, Tanaka, & Hongo, 2003).

## 3.3 PBM for brain function in uninjured animals

Several laboratories have reported that shining light onto the head of uninjured healthy mice or rats can improve various cognitive and emotional parameters. The first study reported that exposure of the middle aged (12 months) CD1 female mice to 1072 nm LED arrays

(Michalikova, Ennaceur, van Rensburg, & Chazot, 2008) produced improved performance in a 3D maze compared to sham treated age-matched controls. Gonzalez-Lima and coworkers (Gonzalez-Lima & Barrett, 2014) showed that transcranial PBM (9 mW/cm$^2$ with a 660 nm LED array) delivered to rats induced dose-dependent increases in oxygen consumption (5 % after 1 J/cm$^2$ and 16 % after 5 J/cm$^2$) [113]. They also found that tPBM reduced fear renewal and prevented the reemergence of the extinguished conditioned fear-responses (Rojas, Bruchey, & Gonzalez-Lima, 2012).

## 3.4 PBM for enhancement of brain function in uninjured human volunteers

Gonzalez-Lima et al. delivered transcranial PBM (1064 nm laser, 60 J/cm$^2$ at 250 mW/cm$^2$) to the forehead in uninjured human volunteers in a placebo-controlled, randomized study. The goal was to improve performance of cognitive tasks related to the prefrontal cortex, including a psychomotor vigilance task (PVT), a delayed match-to-sample (DMS) memory task, and improved mood as measured by the positive and negative affect schedule (PANAS-X) (Barrett & Gonzalez-Lima, 2013). Subsequent studies in uninjured humans showed that tPBM with 1064 nm laser could improve performance in the Wisconsin Card Sorting Task (considered the gold standard test for executive function) (Blanco, Maddox, & Gonzalez-Lima, 2015). They also showed that tPBM to the right forehead (but not the left forehead) could improve attention bias modification (ABM) in humans with depression (Disner, Beevers, & Gonzalez-Lima, 2016).

Salgado et al. applied transcranial LED to enhance cerebral blood flow in healthy elderly women, as measured by transcranial Doppler ultrasound (TCD) of the right and left middle cerebral artery and basilar artery. Twenty-five non-institutionalized elderly women (mean age 72 years), with cognitive status $> 24$, were assessed using TCD before and after transcranial LED therapy. tPBM (627 nm, 70 mW/cm$^2$, 10 J/cm$^2$) was performed at four points of the frontal and parietal region for 30 s each twice a week for 4 weeks. There was a significant increase in the systolic and diastolic velocity of the left middle cerebral artery (25 % and 30 %, respectively) and the basilar artery (up to 17 % and 25 %), as well as a decrease in the pulsatility index and resistance index values of the three cerebral arteries analyzed (Salgado et al., 2015).

# 4. Animal studies of PBM in acute TBI models

## 4.1 Studies from the Oron laboratory

Oron's group was the first (Oron et al., 2007) to demonstrate that a single exposure of the head of a mouse a few hours after creation of a TBI lesion, using a NIR laser (808 nm) could improve neurological performance and reduce the size of the brain lesion. A weight-drop device was used to induce a closed-head TBI in the mice. An 808 nm diode laser with two energy densities calculated at the surface of the brain (1.2–2.4 J/cm$^2$ delivered by 2 min of irradiation with 200 mW laser power to the scalp) was delivered to the head 4 h after TBI was induced. Neurobehavioral function was assessed by the neurological severity score (NSS). There was no significant difference between the control and laser-treated group in NSS between the power densities (10 vs 20 mW/cm$^2$), and no significant difference at early time points (24 and 48 h) post TBI. However, there was a significant improvement (27 % lower NSS score) in the PBM group at times between 5 days and 4 weeks. The laser treated group also showed a smaller loss of cortical tissue than the sham group.

In another study Oron et al. (2012) they varied the pulse parameters (CW, 100 Hz, or 600 Hz) and tested whether the tPBM was equally effective when delivered at 4, 6, or 8 h post-TBI. They first established that a calculated dose to the cortical surface of 1.2 J/cm$^2$ of 808 nm laser at 200 mW applied to the head, was more effective when delivered at 6 h post TBI than at 8 h. They then selected an even shorter time post-TBI (4 h) and compared CW with 100 Hz and 600 Hz. At 56 days, more mice in the 100 Hz group (compared to the CW and 600 Hz groups) had fully recovered. The 600 Hz group had lower NSS scores than the CW and 100 Hz groups up to 20 days. Magnetic resonance imaging (MRI) analysis demonstrated significantly smaller lesion volumes in PBM-treated mice compared to controls.

## 4.2 Studies from the Hamblin laboratory

Wu et al. (2012) first explored the effect of varying the laser wavelengths of PBM had on closed-head TBI in mice. Mice were randomly assigned to a PBM treatment group with a particular wavelength, or to a sham treatment group as a control. Closed-head injury (CHI) was induced via a weight-drop apparatus. To analyze the severity of the TBI, the neurological severity score (NSS) was measured and recorded. The injured mice were then treated with varying wavelengths of laser light (665, 730, 810 or 980 nm) at an energy density of 36 J/cm$^2$ directed onto the scalp at 4 h post-TBI. The 665 nm and 810 nm laser groups showed significant improvement in NSS when compared to the control group between days 5–28. By contrast, the 730 nm and 980 nm laser groups did not show any significant improvement in NSS (Wu et al., 2012). The tissue chromophore cytochrome c oxidase (CCO) is proposed to be responsible for the underlying photon absorption process that underlies many PBM effects. CCO has absorption bands around 665 nm and 810 nm while it has a low absorption region at the wavelength of 730 nm (Karu, Pyatibrat, & Afanasyeva, 2005). It should be noted that this particular study Wu et al. (2012) found that the 980 nm did not produce the same positive effects as the 665 nm and 810 nm wavelengths did; nevertheless previous studies did find that the 980 nm wavelength was an active one for PBM (Anders et al., 2014). Wu et al. suggested that these contradictory results may be due to differences in the energy density, irradiance etc. between the other studies and the Wu study (Wu et al., 2012). In particular a much lower dose of 980 nm might have been effective had it been tested (Wang, Huang, Wang, Lyu, & Hamblin, 2016).

Ando et al. (2011) next used the 810 nm wavelength produced by a Ga-Al-As diode laser delivered with the parameters used in the Wu study, and varied the pulse modes of the laser. These modes consisted of either pulsed wave at 10 Hz or at 100 Hz (50 % duty cycle) or continuous wave laser. They used a different mouse model of TBI induced with a controlled cortical impact device directly inflicting a lesion onto the cortex via an open craniotomy. A single treatment with a power density of 50 mW/m$^2$ and an energy density of 36 J/cm$^2$ (duration of 12 min) was given via tPBM to the closed head in mice at 4 h post CCI. At 48 h to 28 days post TBI, all laser treated groups had significant decreases in the measured neurological severity score (NSS) when compared to the controls. Although all PBM treated groups had similar NSS improvement rates up to day 7, the PW 10 Hz group began to show even greater improvement beyond this time point. At day 28, the forced swim test for depression and anxiety was used and showed a significant decrease in the immobility time for the PW 10 Hz group. In the tail suspension test, which measures depression and anxiety, there was also a significant decrease in the immobility time at day 28, and also at day 1, in the PW 10 Hz group.

Studies using immunofluorescence staining of sections cut from mouse brains showed that tPBM increased neuroprogenitor cells (incorporating

BrdU) in the dentate gyrus (DG) of the hippocampus and the subventricular zone (SVZ) at 7 days after the treatment (Xuan et al., 2014). The neurotrophin known as brain derived neurotrophic factor (BDNF) was also increased in the DG and SVZ at 7 days, while the protein marker (synapsin-1) for synaptogenesis and neuroplasticity was increased in the cortex at 28 days but not in the DG, SVZ or in any location at 7 days (Xuan et al., 2015). Learning and memory as measured by the Morris water maze was also improved by tPBM (Xuan et al., 2014).

We had previously showed that PBM (810 nm, CW, 18 J/cm$^2$, 25 mW/cm$^2$) delivered to the mouse head daily for 3-days after a controlled cortical impact TBI gave a significant improvement in neurological/cognitive function (Xuan et al., 2013). However the same parameters delivered daily for 14 consecutive days gave significantly less benefit. This biphasic dose response intrigued us, and we decided to follow the mice that received 3X or 14X laser treatments out to 56-days post-TBI (Xuan, Huang, & Hamblin, 2016). We found the 14X group showed worse neurological function than the no-treatment TBI group at 2-weeks, but started to improve steadily during the next 6-weeks, and by 56- days were significantly better than the no-treatment TB mice, but still worse than the 3X mice. A marker of activated glial cells (GFAP) was significantly increased in the brain regions (compared to both untreated TBI and 3X groups) at 4-weeks in the 14X group, but the GFAP had fallen to low levels in both 3X and 14X groups by 8-weeks. We conclude that an excessive number of laser-treatments delivered to mice can temporarily inhibit the process of brain repair stimulated by tPBM, but then the inhibitory effect ceased, and brain repair could resume. The mechanism may be attributed to a temporary induction of reactive gliosis which delayed the PBM stimulation of brain repair.

## 4.3 Studies from the Wu laboratory

Zhang et al. (2014) first showed that secondary brain injury after an acute TBI occurred to a worse degree in mice that had been genetically engineered to lack "Immediate Early Response" gene X-1 (IEX-1). When these mice were exposed to a gentle head impact (thought to closely resemble mild TBI in humans) they had a worse NSS than uninjured mice receiving the same TBI. Exposure of IEX-1 knockout mice to PBM (810 nm,150 mW/cm$^2$, 4 min, and 36 J/cm$^2$) delivered at 4 h post injury, restored the NSS to almost baseline levels, suppressed proinflammatory cytokine expression of interleukin (IL)-I$\beta$ and IL-6, but increased TNF-$\alpha$.

The original lack of IEX-1 in the mice decreased ATP production, but exposing the injured brain to PBM restored ATP production back to near normal levels.

Dong et al. (2015) asked whether the beneficial effects of PBM on TBI in mice could be enhanced by combining PBM with administration of metabolic substrates such as pyruvate and/or lactate. The goal was to even further improve mitochondrial function in the brain. This combinatorial treatment was able to reverse memory and learning deficits in TBI injured mice back to normal levels as well as leaving the hippocampal region completely protected from tissue loss; a stark contrast to control TBI mice that exhibited severe tissue loss from secondary brain injury.

## 4.4 Studies from the Whalen laboratory

Khuman et al. (2012) delivered PBM (800 nm) either directly to the injured brain tissue (through the craniotomy) or transcranially in mice beginning 60–80 min after CCI TBI. At a dose of 60 J/cm$^2$ (500 mW/cm$^2$) the mice showed increased performance in the Morris water maze (latency to the hidden platform, $p < 0.05$, and probe trial, $p < 0.01$) compared to non-treated controls. When PBM was delivered via open craniotomy there was reduced microgliosis at 48 h (IbA-1 + cells, $p < 0.05$). Little or no effect of tPBM on post-injury cognitive function was observed using lower or higher doses, a 4-h administration time point or 60 J/cm$^2$ at 7-days post-TBI.

## 4.5 Studies from the Whelan laboratory

Quirk et al. (2012) studied Sprague-Dawley rats who had received a severe CCI TBI and were divided into three groups: real TBI, sham surgery, and anesthetization only. Each group received either real or sham PBM consisting of 670 nm LED treatments of 15 J/cm$^2$, 50 mW/cm$^2$, 5 min, given two times per day for 3 days (chemical analysis) or 10 days (behavioral analysis using a TruScan nose-poke device). Significant differences in task entries, repeat entries, and task errors were seen in the TBI rats treated with PBM vs untreated TBI mice, and in sham surgery mice treated with PBM vs untreated sham surgery mice. A statistically significant decrease was found in the pro-apoptotic marker Bax, and increases in the anti-apoptotic marker Bcl-2 and reduced glutathione (GSH) levels in tPBM TBI mice.

## 4.6 Studies from the Marques laboratory

Moreira et al. (2009) used a different model of TBI. Wistar rats received a craniotomy and a copper probe cooled in liquid nitrogen was applied to the surface of the brain to create a standardized cryogenic injury. They treated the rats with either a 780 nm or 660 nm laser at one of two different doses ($3\,J/cm^2$ or $5\,J/cm^2$) twice (once immediately after the injury and again 3 h later). Rats were sacrificed 6 h and 24 h after the injury. The 780 nm laser was better at reducing levels of pro-inflammatory cytokines (TNFα, IL1β, IL6) particularly at early time points (Moreira et al., 2009). In a follow-up study using $3\,J/cm^2$ (Moreira et al., 2011) these workers reported on the healing of the injuries in these rats at time points 6 h, 1, 7 and 14 days after the last irradiation. Cryogenic injury created focal lesions in the cortex characterized by necrosis, edema, hemorrhage and inflammatory infiltrate. The most striking findings were: PBM-treated lesions showed less tissue loss than control lesions at 6 h. During the first 24 h the amount of viable neurons was significantly higher in the PBM groups. PBM reduced the amount of GFAP (glial fibrillary acidic protein, a marker of astrogliosis) and the numbers of leukocytes and lymphocytes, thus demonstrating its anti-inflammatory effect.

## 4.7 Studies from the Ahmed laboratory

Stevens et al. (2024) investigated the use of tPBM in a rat model of mild TBI cause by a closed head weight drop technique. The PBM used lasers with wavelengths of 660 nm, 810 nm, or a combination of both. PBM was delivered on days 0, 1, and 2 starting at 10 min after the injury for 2 min per session (1 min per hemisphere). The power density was set at $20\,mW/cm^2$ calculated at the level of the cortex (2.4 J). For each hemisphere, the laser probe was positioned for 10 s on one of six adjacent points to avoid excessive heating in the skin. Neurological function was assessed using the novel object recognition (NOR) and beam balance tests, whilst immunohistochemistry was used to assess neuropathology in brain sections. They found that PBM at 810, 660 nm, and 810/660 nm all significantly improved NOR and beam balance performance, with 810 nm PBM having the greatest effect. Immunohistochemistry showed significantly reductions in CD11b+ microglia and GFAP+ astrocytes at 3 days post-injury. They also found reduced cortical localization of the pro-apoptotic marker, cleaved caspase-3, and modest reductions in extracellular matrix deposition after PBM treatment, limited to the choroid plexus and peri-ventricular areas.

## 4.8 Studies from the Yin laboratory

Bartelt et al. (2025) reasoned that a closed head TBI is often treated by a decompressive craniectomy surgery (DC) in order to reduce the intracranial pressure. The removal of a portion of the skull would be expected to increase the penetration of red/NIR light into the brain. They tested this hypothesis in a CHI model in Long-Evans rats using 660 nm and 850 nm PBMT. The surgical procedure increased penetration of 660 nm from 3.96 $mW/cm^2$–70.5 $mW/cm^2$ for 660 nm PBM and from 4.96 $mW/cm^2$–82 $mW/cm^2$ for 850 nm PBM. Both wavelengths led to significant improvements in a beam walk test for measuring motor function.

## 4.9 Studies on repeated TBI in animal models

It is known that repeated blows to the head in humans can have a cumulative effect, even if each impact itself is not particularly severe (Prins, Alexander, Giza, & Hovda, 2013). Repeated head trauma can lead to the development of "a vulnerable brain" (Povlishock, 2013). It Is also a major causative factor in the development of chronic traumatic encephalopathy (CTE) (Stern et al., 2011). There have been attempts to see if PBM can also have beneficial effects on repeated TBI in animal models. Yang et al. (2019) tested PBM in a repeated closed head injury (rCHI) rat model. Rats were administered a total of three closed head injuries, with each injury separated by 5 days. PBM treatment to the head was initiated 2 h after the first injury and administered daily for a total of 15 days. The PBM-treated rats showed a significant improvement in motor ability, anxiety and cognitive deficits compared to CHI group. Analysis of brain sections showed an increase of synaptic proteins and surviving neurons, along with a reduction in reactive gliosis and markers of neuronal damage. Wu et al. (2025) studied a similar rat model of repeated CHI by using a weight drop technique on days 0, 5, and 10. In this case they delivered remote PBM by shining a 808 nm laser on the chest of the rat to irradiate the lungs. They started PBM treatment immediately after the first injury, and repeated PBM each day for 20 days. The laser spot was 2 $cm^2$ with an irradiance of 350 $mW/cm^2$ for 1 min on each side of the chest. Behavioral assays showed that remote PBM produced significant improvements in the three-chamber social interaction test, wire grip test, forced swim test, open field test, and elevated plus maze. Immunofluorescence assays showed that PBM produced less neuronal damage, apoptosis, and degeneration induced by rCHI. PBM mitigated hyperactivation of cortical microglia and astrocytes and significantly reduced excessive synaptic phagocytosis by these cells. Analysis of lung microbiome changes revealed an enrichment of lipopolysaccharide biosynthesis pathways, suggesting a potential link to neuroprotection.

## 5. Clinical studies in patients with acute TBI

In 2020 the results of the first prospective, randomized, interventional clinical trial of transcranial PBMT for acute (within 72 h) moderate TBI in humans, conducted in the emergency department of the Massachusetts General Hospital (MGH) in Boston were published (Figueiro Longo et al., 2020). Between 2015 and 2019 a total of 4216 men and women aged 18–79 years with acute, blunt TBI admitted to MGH were screened to recruit moderate TBI patients within 72 h of a head injury. A total of 68 patients were enrolled in a single-center, prospective, double-blind, placebo-controlled, parallel-group trial in which patients with moderate TBI were randomly assigned to PBMT or sham treatment. Treatment was divided into 3 sessions of 20 min' duration with at least 12-hour intervals between treatments. The custom-made helmet had 18 clusters of 20 each 810 nm LEDS providing a fluence of 43 $J/cm^2$ at 0.036 $W/cm^2$ for 20 min to the scalp. The sham helmet had no active LEDs and neither participants nor staff could tell the difference between real and sham helmets. Following the PBMT, a baseline brain MRI scan was performed to evaluate the acute trauma. Subsequently two follow-up scans were carried out, one at 2–3 weeks after the TBI and another at 3 months follow up. Clinical assessment was performed with the Rivermead Post-Concussion Symptoms Questionnaire (RPQ) evaluating headache, dizziness, nausea, and vomiting. RPQ also evaluated noise sensitivity, sleep disturbance, fatigue, irritability, depressed mood, forgetfulness, poor concentration, longer thinking time, blurred vision, light sensitivity, double vision, and restlessness. The RPQ was carried out at baseline, 14 days, 3 months, and 6 months after the trauma. Out of the 68 patients enrolled, 43 patients (21 female, 22 male; mean age, 50.5 years) complated at least 1 MRI scan, and 40 patients completed 2 MRI scans (18 in the PBMT group and 22 in the sham group. No adverse events referable to PBMT were reported. In the MRI scans, radial diffusivity ($P < .001$), mean diffusivity ($P = .03$), and fractional anisotropy ($P < .001$) showed significant time and treatment interactions at the 3-month time point. The PBMT group had lower RPQ scores than the sham group, but this effect did not reach statistical significance.

A secondary analysis of the data from this trial was recently published (Chan et al., 2024). Functional MRI was used to assess the effect of PBMT on whole-brain resting-state functional connectivity (RSFC) in patients with moderate TBI at acute (within 1 week), subacute (2–3 weeks), and late-subacute (3 months) recovery phases. RSFC was evaluated in 17 PBMT-treated and 21 sham-treated participants. Seven brain region pairs exhibited a greater change in connectivity after PBMT compared to sham between the acute and subacute phases (P value range = .010–.047). Thirteen different brain region pairs showed an increase in connectivity in sham-treated participants between the subacute and late-subacute phases (P value range = .020–.047). They concluded that acute-phase PBMT may have an effect on resting-state neuronal circuits in the early recovery phase of moderate TBI.

## 6. Clinical studies in patients with chronic TBI

### 6.1 Naeser studies

In 2011 Naeser, Saltmarche et al., published the first report describing two chronic, TBI cases treated with tPBM (Naeser, Saltmarche, Krengel, Hamblin, & Knight, 2011). A 500 mW CW LED source (660 nm red and 870 nm NIR LEDs) with a power density of 22.2 mW/cm$^2$ (area of 22.48 cm$^2$), was applied all over the head, for 10 min at each placement location (13.3 J/cm$^2$). In the first case study the patient reported that she could concentrate on tasks for a longer period of time (the time able to work at a computer increased from 20 min to 3 h). She had a better ability to remember what she read, decreased sensitivity when receiving haircuts in the spots where PBM was applied, and improved mathematical skills after undergoing PBM. The second patient had statistically significant improvements compared to prior neuropsychological tests after 9 months of treatment. The patient had a 2 standard deviation (SD) increase on tests of inhibition and inhibition accuracy (9th percentile to 63rd percentile on the Stroop test for executive function and a 1 SD increase on the Wechsler Memory scale test for the logical memory test (83rd percentile to 99th percentile) (Naeser et al., 2011).

Naeser et al. then went on to report a case series containing a further eleven patients (Naeser et al., 2014). This was an open protocol study that examined whether scalp application of red and NIR LED could improve cognition in patients with chronic, mild TBI (mTBI). This study enrolled 11 participants ranging in age from 26 to 62 years (6 males, 5 females) who suffered from persistent cognitive dysfunction after mTBI. The injuries in the participants had been caused by motor vehicle accidents, sports related events and for one participant, an improvised explosive device (IED) blast. tPBM consisted of 18 sessions (Monday, Wednesday, and Friday for

6 weeks) and was started anywhere from 10 months to 8 years post-TBI. A total of 11 LED cluster heads (5.25 cm in diameter, 500 mW, 22.2 $mW/cm^2$, 13 $J/cm^2$) were applied for 10 min per set (5 or 6 LED placements per set, Set A and then Set B, in each session). Neuropsychological testing was performed pre-LED application and 1 week, 1 month and 2 months after the final treatment. They found that there was a significant positive linear trend for the Stroop Test for executive function, in trial 3 inhibition ($p = 0.004$); Stroop, trial 4 inhibition switching ($p = 0.003$); California Verbal Learning Test (CVLT)-II, total trials 1–5 ($p = 0.003$); CVLT-II, long delay free recall ($p = 0.006$). Improved sleep and fewer post-traumatic stress disorder (PTSD) symptoms, if present beforehand, were observed after treatment. Participants and family members also reported better social function and a better ability to perform interpersonal and occupational activities. Although these results were significant, the authors suggested that further placebo-controlled studies would be needed to ensure the reliability of this approach (Naeser et al., 2014).

Naeser has proposed Naeser et al. (2014, 2016) that specific scalp placements of the LED cluster heads may affect specific cortical nodes in the intrinsic networks of the brain, such as the default mode network (DMN), the salience network (SN), and the central executive network (CEN). These intrinsic brain networks are often dysregulated after TBI (Sharp, Scott, & Leech, 2014). Naeser proposed that the specific areas of the head that should receive light, in order to target cortical nodes in these networks were as follows:

1. For the DMN, placement of the LED cluster head on the midline of face, centered on the upper forehead and the front hairline, targeted the left and right mesial prefrontal cortex; and on a midline, scalp location half-way between the occipital protuberance and the vertex of the head, targeted the precuneus; and on left and right LED placements superior to the tip of each ear and posterior to each ear, targeted the inferior parietal cortex/angular gyrus areas.
2. For the SN, placement of LED cluster heads on the left and right temple areas, to target the anterior insula (but due to depth of insula, unknown if the photons reached the target); midline of the vertex of the head, to target the left and right presupplementary motor areas; and the LED cluster head placed on the midline of face, centered on the upper forehead and the front hairline, also targeted the left and right dorsal anterior cingulate cortex.

3. For the CEN, left and right scalp LED placements immediately posterior to the front hairline (on a line directly superior from the pupils of the eyes), targeted the dorso-lateral prefrontal cortex areas; and the left and right LED placements superior to the tip of each ear and posterior to each ear, also targeted the posterolateral inferior parietal cortex/angular gyrus areas (also treated as part of the DMN).

Further studies from Naeser and colleagues (Naeser et al., 2016) tested an intranasal LED (iLED) device. Two small iLEDs (one red and the other NIR) were clipped into each nostril and used at the same time for 25 min. The parameters were as follows: red, 633 nm, 8 mW CW, 1 $cm^2$, energy density 12 J/$cm^2$ (25 min); NIR 810 nm, 14.2 mW, pulsed 10 Hz, 1 $cm^2$, 21.3 J/$cm^2$. The first mTBI participant (24-year old female) who had sustained four sports-related concussions (two during snowboarding and two during field hockey), received iLED PBM three times per week for 6 weeks. Significant improvements were observed in tasks measuring executive function and verbal memory as well as attention and verbal fluency. At 1 week after the 18th iLED treatment, the average total time asleep had increased by 61 min per night and her sleep efficiency (total sleep time divided by total time in bed) had increased by 11 %. At 12 weeks after the last iLED treatment, she was able to discontinue all sleep medications that she had previously been using. The second, mTBI participant who received the intranasal only, LED treatment series is a 49 Yr. M (non-Veteran) who sustained mTBI in a MVA, 30 years prior to receiving the intranasal LED treatment series. He showed significant improvement on the Controlled Oral Word Association-FAS Test post- the iLED treatment series, improving by + 1.3 SD and + 1.5 SD at 1 and 2 months post- the 18th iLED treatment. His sleep data indicated he was already a good sleeper, at entry.

Bogdanova and Naeser reported Bogdanova et al. (2014) a case report of two patients (1 female) with moderate TBI (medical records and clinical evaluation) and persistent cognitive dysfunction (as measured by neuropsychological tests of executive function and memory). Patients received 18 sessions of transcranial LED therapy (3x/week for 6 weeks) using the mixed red/NIR cluster described above (Naeser et al., 2011). Standardized neuropsychological tests for executive function, memory, depression, PTSD and sleep measures (PSQI, actigraphy) were administered to participants pre-(T1), mid-(T2), and one week (T3) post-PBM treatment. Both PBM treated cases (P1 and P2) showed marked improvement in sleep (actigraphy total sleep) 1 week post-LED treatment (T3), as compared to pre-treatment (T1). P1 also improved in executive function, verbal memory, and sleep efficiency; while P2 significantly improved on measures of PTSD (PCL-M) and depression. No adverse events were reported.

Naeser et al. (2023) reported on four ex-NFL football players suffering from possible chronic traumatic encephalopathy (CTE), who had played for 11–16 years and then reported worsening cognitive performance accompanied by disturbances in behavior and mood. The four patients received in-office, red/near-infrared tPBM to the scalp, 3x/week for 6 weeks. The three younger cases (ages 55, 57, 65) improved 2 SD ($p < 0.05$) on three to six neuropsychological tests/subtests at 1 week or 1 month post-tPBM, compared to pre-treatment, while the older case (age 74) improved by 1.5 SD on three tests. There was significant improvement at 1 month in post-traumatic stress disorder (PTSD), depression, pain, and sleep. One case discontinued narcotic pain medication and had reduced tinnitus. After 2 months two of the cases regressed. Then, home tPBM was applied to cortical nodes and DMN for 12 weeks and significant improvements were observed again. Functional MRI showed changes in resting-state functional-connectivity in the salience, central executive, and default mode networks, while MRI spectroscopy, showed increases in n-acetyl-aspartate (NAA), oxygen consumption, and mitochondrial activity in the anterior cingulate cortex.

## 6.2 Studies from Henderson and Morries

Henderson and Morries (2015b) used a high-power NIR laser (10–15 W at 810 and 980 nm) and applied it to the head to treat a patient with moderate TBI. The patient received 20 NIR applications over a 2-month period. They carried out anatomical magnetic resonance imaging (MRI) and perfusion single-photon emission computed tomography (SPECT). The patient showed decreased depression, anxiety, headache, and insomnia, whereas cognition and quality of life improved, accompanied by changes in the SPECT imaging.

They next reported Morries, Cassano, and Henderson (2015) a series of ten patients with chronic TBI (average time since injury 9.3 years) where each patient received ten treatments over the course of 2 months using a high-power NIR laser (13.2 W/0.89 $cm^2$ equivalent to 14.6 W/$cm^2$ at 810 nm; or 9 W/0.89 $cm^2$ equivalent to 10.11 W/$cm^2$ at 980 nm). A continuous sweeping motion over the forehead was utilized to minimize skin heating and cover a larger area. Skin temperature increased no more than 3 °C. Overall symptoms of headache, sleep disturbance, cognition, mood dysregulation, anxiety, and irritability improved. Symptoms were monitored by depression scales and a novel patient diary system specifically designed for this study. These authors have proposed that high power lasers are preferable for tPBM treatments because the photons can better reach the brain (Henderson & Morries, 2015a).

## 6.3 Case series from Hipskind

Hipskind et al. (2018) treated 12 symptomatic military Veterans diagnosed with chronic TBI > 18 months post-trauma with pulsed transcranial PBMT (tPBMT). They used two neoprene therapy pads containing 220 infrared and 180 red LEDs, generating a power output of 3.3 W and an average power density of 6.4 mW/cm$^2$ for 20 min, 3X/week for 6 weeks. They employed standardized neuropsychological test scores to measure cognitive function and single photon emission computed tomography (SPECT) to measure regional cerebral blood flow (rCBF) tPBMT significantly improved neuropsychological scores in 6 out of 15 subscales (40.0 %; $p < 0.05$). SPECT analysis showed an increase in rCBF in 8 of 12 (66.7 %) study participants.

## 6.4 Case series from Rindner

Rindner et al. (2022) treated eleven patients who had been diagnosed with TBI after full neurological examination and MRI evaluation. The time elapsing since the trauma varied from 9 days to >10 years. They used a 1064 nm laser with a spot size of 16 cm$^2$ and a power density of 250 mW/cm$^2$ applied for 10 min to each hemisphere at Brodmann area 10. The delivered fluence was 150 J/cm$^2$ per site. Eight treatments were given over 8 weeks. All 9 participants who completed the protocol reported clinically significant improvements in the self-assessed GRC scale. In neurocognitive testing (executive function and attention) 8 out 11 improved by at least 1 point, while 3 remained the same. Eight out of 11 patients improved by at least 3 points on the depression scale and 6 out of 11 on the anxiety scale.

## 6.5 Pilot clinical trial from Liebel

Liebel et al. reported a pilot clinical trial as a conference abstract (Liebel et al., 2022). They treated 49 former male and female athletes (mean age = 45.9 years) with a history of repetitive head impacts. They used a 810 nm LED device delivering 100 mW/cm$^2$ at 40 Hz. Participants received tPBM treatment once a week for 8 weeks. Following tPBM treatment, study participants demonstrated statistically significant reductions in self-reported depression ($p < 0.001$), posttraumatic stress ($p < 0.001$), and adjustment (0.001) symptoms compared to pre-treatment levels. Sleep quality ($p < 0.001$), simple reaction time ($p < 0.001$), dominant hand grip strength ($p = 0.013$), and nondominant hand grip strength ($p < 0.001$) all improved following tPBM treatment.

## 6.6 Case study from Nawashiro

Nawashiro et al. (2012) treated a single patient who had suffered a severe TBI. The patient survived but was left in a persistent vegetative state for 8 months after the accident. He showed no spontaneous movement of limbs and a CT scan of the head 8 months after the accident showed a focal low-density area in the right frontal lobe. The device had 23 individual 850 nm LEDs (13 mW each; total power 299 mW, total area 57 $cm^2$). A treatment time of 30 min per session delivered 20.5 $J/cm^2$ over the left and right forehead areas repeated twice daily (6 h apart), for 73 days. Five days after beginning the PBM (after 10 treatments), the patient began to spontaneously move his left arm and hand, which had not occurred during the previous 8 months. Single-photon emission computed tomography with N-isopropyl-[123I]p-iodoamphetamine (IMP-SPECT) was performed twice. The IMP-SPECT scans showed a focal increase (20 % higher) in cerebral blood flow in the uninjured left anterior frontal lobe 30 min after the last (146th) PBM treatment, compared to before PBM began.

## 6.7 Case study from Chao

Chao et al. reported a single case of a 23-year professional hockey player with a history of multiple concussions, presumed to have caused his symptoms of headache, mild anxiety, and difficulty concentrating (Chao, Barlow, Karimpoor, & Lim, 2020). The subject treated himself at home every two days with a commercially available (Vielight Neuro), PBM device that used light-emitting diodes (LEDs) to emit 810-nm light pulsed at 10 or 40 Hz delivered by an intranasal and four transcranial modules that targeted the default mode network (DMN) with a maximum power density of 100 $mW/cm^2$ and delivered a total of 240 J per 20-min treatment session. MRI with arterial spin labeling, echo-planar imaging, and resting-state functional MRI were carried out at baseline and after 8 weeks of PBMT. Scores related to verbal learning and memory, executive function, verbal fluency, attention, processing, speed word reading all improved after 8 weeks of PBMT. Neuroimaging showed an increase in brain volume, improved functional connectivity, and increased cerebral perfusion after 8 weeks of PBMT.

### 6.8 Case series from Hesse

Hesse et al. reported a case series of five TBI patients treated with PBMT using a 785 nm laser, continuous wave and10 mW/cm$^2$ power density over 10 weeks. The patients included four chronic TBI cases in a state of unresponsive wakefulness or minimal consciousness and one subacute patient with akinetic mutism. The treatment protocol improved alertness and awareness in the patients, but epileptic fits were considered to be potential side effects.

## 7. Conclusion and future prospects

One of the most important questions to be answered when contemplating clinical treatment of TBI patients with tPBM, is what is the best time to administer the treatment? Most of the available reports of studies using PBM in laboratory animal models of TBI and stroke, and also in patients treated for stroke, have been in the acute phase where the overall goal of the intervention can be best described as neuroprotection. Not only that but there are several reports (Lapchak, Salgado, Chao, & Zivin, 2007; Oron et al., 2012) that PBM for both TBI and stroke is most effective when it is delivered as soon as possible after the actual event (head impact or ischemic stroke). The protocols for the series of NEST clinical trials specified that patients should be treated with PBM within 24 h of the stroke occurring. By contrast, only one trial has treated patients suffering from acute TBI with tPBMT, and although significant improvements in MRI markers were observed, the improvement in clinical symptoms was not significant (Figueiro Longo et al., 2020). It is possible that better results would have been achieved if the treatment could have been started at 4–6 h post-TBI, and continued for a limited number of times after the injury; perhaps once a day for 7 days?

The majority of the clinical trials of PBM for patients with TBI, that have so far been carried out, have been with chronic TBI patients, with varying periods of time having elapsed after the original head injury, sometimes as long as 10 years. The dosimetry and optimum delivery apparatus of tPBM is still uncertain. Although there is some consensus that wavelengths in the region of 800–900 nm will penetrate the scalp and skull, other workers have used longer NIR wavelengths, 980 nm, 1064 nm, or 1267 nm. Pulsing or CW is another unresolved question, although a consensus may be emerging that 10 Hz (alpha) or 40 Hz (gamma) may be the best. The exact locations on the head that should receive the light are still unknown. Naeser has proposed (Naeser et al., 2016) some interesting considerations regarding the scalp placements of the tLEDS, and their effect on various intrinsic cortical networks of the brain. Targeted LED placements could promote better neuromodulation (activation/deactivation) in specific cortical nodes. It is possible that communication between nodes within one single network, and/or across networks could be improved. Moreover preliminary data indicate that intranasal, red plus near-infrared LEDs can also benefit TBI patients, although the degree to which light incident on the nasal mucosa, and possibly delivered transsphenoidally (Pitzschke et al., 2015) can penetrate directly into the brain, remains to be determined.

An advantage of intranasal and/or transcranial LED PBM therapy is that it can be performed in the home, for long-term use (Naeser et al., 2011). A pilot trial with 5 chronic, mild to moderately-severe dementia cases recently showed significant improvement on the Mini-Mental State Examination ($p < 0.003$), and on the Alzheimer's Disease Assessment Scale-Cognitive subscale ($p < 0.023$) after 12 weeks of daily, at-home, intranasal, near-infrared LED PBM treatments (810 nm, pulsed at 10 Hz), and once-a-week in-office, tLED treatments applied to the cortical nodes of the Default Mode Network (Saltmarche, Naeser, Ho, Hamblin, & Lim, 2017). Anecdotally, there was also improved sleep, fewer angry outbursts, and less wandering. When all LED treatments were withdrawn after 12 weeks of active LED PBM treatment, there was precipitous decline in cognition and behavior. Thus, at-home, long-term use of iLED plus tLED PBM offers a potential therapy to mitigate the sequelae of Alzheimer's disease and possibly other neurodegenerative disorders, as well as TBI and stroke.

One highly distressing aspect of TBI symptomatology that has not so far been addressed by PBM, is that of post-traumatic epilepsy (PTE). TBI is the most significant cause of symptomatic epilepsy in people from 15 to 24 years of age. The frontal and temporal lobes are the most frequently affected regions, but imaging (MRI) often fails to show the precise cause. During PTE seizures there is an abnormal electrical discharge in the brain, with staring and unresponsiveness, stiffening or shaking of the body, legs, arms or head; strange sounds, tastes, visual images, feelings or smells; inability to speak or understand, etc (Cotter, Kelso, & Neligan, 2017). Epilepsy has traditionally been considered to be a contra-indication for PBMT (Navratil & Kymplova, 2002). However the knowledge that has recently been gained concerning the beneficial effects of PBMT on the damaged brain, suggests that this view may need to be critically revisited (Hamblin, 2023).

Moreover the use of tPBM to treat a wide range of brain disorders is steadily increasing (Hamblin, 2016, 2024). These disorders include Parkinson's disease (Purushothuman, Nandasena, Johnstone, Stone, & Mitrofanis, 2013), Alzheimer's disease/dementia (Salehpour, Khademi, & Hamblin, 2021), depression (Salehpour & Rasta, 2017), anxiety (Cassano et al., 2022), post-traumatic stress disorder (Li et al., 2021), autism spectrum disorder (Hamilton, Liebert, Pang, Magistretti, & Mitrofanis, 2022) and so on. These disorders are discussed at length in other chapters in this volume.

## References


Agaiby, A. D., Ghali, L. R., Wilson, R., & Dyson, M. (2000). Laser modulation of angiogenic factor production by T-lymphocytes. *Lasers in Surgery and Medicine, 26*(4), 357–363. https://doi.org/10.1002/(SICI)1096-9101(2000)26:4<357::AID-LSM3>3.0.CO;2-O [pii].

Aimbire, F., Albertini, R., Pacheco, M. T., Castro-Faria-Neto, H. C., Leonardo, P. S., Iversen, V. V., ... Bjordal, J. M. (2006). Low-level laser therapy induces dose-dependent reduction of TNFalpha levels in acute inflammation. *Photomedicine and Laser Surgery, 24*(1), 33–37.

Alexandratou, E., Yova, D., Handris, P., Kletsas, D., & Loukas, S. (2002). Human fibroblast alterations induced by low power laser irradiation at the single cell level using confocal microscopy. *Photochemical & Photobiological Sciences: Official Journal of the European Photochemistry Association and the European Society for Photobiology, 1*(8), 547–552.

Anders, J. J., Moges, H., Wu, X., Erbele, I. D., Alberico, S. L., Saidu, E. K., ... Pryor, B. A. (2014). In vitro and in vivo optimization of infrared laser treatment for injured peripheral nerves. *Lasers in Surgery and Medicine, 46*(1), 34–45. https://doi.org/10.1002/lsm.22212.

Ando, T., Xuan, W., Xu, T., Dai, T., Sharma, S. K., Kharkwal, G. B., ... Hamblin, M. R. (2011). Comparison of therapeutic effects between pulsed and continuous wave 810-nm wavelength laser irradiation for traumatic brain injury in mice. *PLoS One, 6*(10), e26212–e26220.

Barrett, D. W., & Gonzalez-Lima, F. (2013). Transcranial infrared laser stimulation produces beneficial cognitive and emotional effects in humans. *Neuroscience, 230*, 13–23. https://doi.org/10.1016/j.neuroscience.2012.11.016.

Bartelt, D. C., Osborn, S. G., Sowers, L. P., Bassuk, A., Howard, M. A., Andrews, B. T., & Yin, T. C. (2025). Photobiomodulation promotes motor recovery following decompressive craniectomy for traumatic brain injury. *Photobiomodulation, Photomedicine, and Laser Surgery, 43*(9), 394–399. https://doi.org/10.1089/photob.2025.0021.

Blanco, N. J., Maddox, W. T., & Gonzalez-Lima, F. (2015). Improving executive function using transcranial infrared laser stimulation. *Journal of Neuropsychology, 11*(1), 14–25. https://doi.org/10.1111/jnp.12074.

Blennow, K., Brody, D. L., Kochanek, P. M., Levin, H., McKee, A., Ribbers, G. M., ... Zetterberg, H. (2016). Traumatic brain injuries. *Nature Reviews Disease Primers, 2*, 16084. https://doi.org/10.1038/nrdp.2016.84.

Bogdanova, Y., Martin, P. I., Ho, M. D., Krengel, M. H., Ho, V. T., Yee, M. K., ... Naeser, M. A. (2014). LED therapy improves sleep and cognition in chronic moderate TBI: Pilot case studies. *Archives Phys Med Rehab, 95*(10), e77.

Buccellato, F. R., D'Anca, M., Serpente, M., Arighi, A., & Galimberti, D. (2022). The role of glymphatic system in Alzheimer's and Parkinson's disease pathogenesis. *Biomedicines, 10*(9), 2261.

Cardoso, F. d S., Salehpour, F., Coimbra, N. C., Gonzalez-Lima, F., & Gomes da Silva, S. (2022). Photobiomodulation for the treatment of neuroinflammation: A systematic review of controlled laboratory animal studies. *Frontiers in Neuroscience, 16*, 1006031.

Cassano, P., Norton, R., Caldieraro, M. A., Vahedifard, F., Vizcaino, F., McEachern, K. M., & Iosifescu, D. (2022). *Tolerability and safety of transcranial photobiomodulation for mood and anxiety disorders*. Paper presented at the Photonics.

Caterina, M. J., & Pang, Z. (2016). TRP channels in skin biology and pathophysiology. *Pharmaceuticals (Basel), 9*(4), https://doi.org/10.3390/ph9040077.

Chan, S.-t, Mercaldo, N., Figueiro Longo, M. G., Welt, J., Avesta, A., Lee, J., ... Gupta, R. (2024). Effects of low-level light therapy on resting-state connectivity following moderate traumatic brain injury: Secondary analyses of a double-blinded placebo-controlled study. *Radiology, 311*(2), https://doi.org/10.1148/radiol.230999.

Chao, L. L., Barlow, C., Karimpoor, M., & Lim, L. (2020). Changes in brain function and structure after self-administered home photobiomodulation treatment in a concussion case. *Frontiers in Neurology, 11*. https://doi.org/10.3389/fneur.2020.00952.

Chen, A. C., Arany, P. R., Huang, Y. Y., Tomkinson, E. M., Sharma, S. K., Kharkwal, G. B., ... Hamblin, M. R. (2011). Low-level laser therapy activates NF-kB via generation of reactive oxygen species in mouse embryonic fibroblasts. *PLoS One, 6*(7), e22453. https://doi.org/10.1371/journal.pone.0022453.

Chen, A. C., Huang, Y. Y., Sharma, S. K., & Hamblin, M. R. (2011). Effects of 810-nm laser on murine bone-marrow-derived dendritic cells. *Photomedicine and Laser Surgery, 29*(6), 383–389. https://doi.org/10.1089/pho.2010.2837.

Chow, R. T., Johnson, M. I., Lopes-Martins, R. A., & Bjordal, J. M. (2009). Efficacy of low-level laser therapy in the management of neck pain: A systematic review and meta-analysis of randomised placebo or active-treatment controlled trials. *Lancet, 374*(9705), 1897–1908. https://doi.org/10.1016/S0140-6736(09)61522-1.

Cotter, D., Kelso, A., & Neligan, A. (2017). Genetic biomarkers of posttraumatic epilepsy: A systematic review. *Seizure: the Journal of the British Epilepsy Association, 46*, 53–58. https://doi.org/10.1016/j.seizure.2017.02.002.

de Freitas, L. F., & Hamblin, M. R. (2016). Proposed mechanisms of photobiomodulation or low-level light therapy. *Journal of Selected Topics in Quantum Electronics, 22*(3), 7000417–7000434.

Dirican, A., Andacoglu, O., Johnson, R., McGuire, K., Mager, L., & Soran, A. (2011). The short-term effects of low-level laser therapy in the management of breast-cancer-related lymphedema. *Supportive Care in Cancer: Official Journal of the Multinational Association of Supportive Care in Cancer, 19*(5), 685–690. https://doi.org/10.1007/s00520-010-0888-8.

Disner, S. G., Beevers, C. G., & Gonzalez-Lima, F. (2016). Transcranial laser stimulation as neuroenhancement for attention bias modification in adults with elevated depression symptoms. *Brain Stimulation, 9*(5), 780–787. https://doi.org/10.1016/j.brs.2016.05.009.

Dong, T., Zhang, Q., Hamblin, M. R., & Wu, M. X. (2015). Low-level light in combination with metabolic modulators for effective therapy of injured brain. *Journal of Cerebral Blood Flow and Metabolism: Official Journal of the International Society of Cerebral Blood Flow and Metabolism*. https://doi.org/10.1038/jcbfm.2015.87.

Farfara, D., Tuby, H., Trudler, D., Doron-Mandel, E., Maltz, L., Vassar, R. J., ... Oron, U. (2015). Low-level laser therapy ameliorates disease progression in a mouse model of Alzheimer's disease. *Journal of Molecular Neuroscience: MN, 55*(2), 430–436. https://doi.org/10.1007/s12031-014-0354-z.

Faul, M., Xu, L., Wald, M. M., & Coronado, V. (2010). *Traumatic brain injury in the United States: Emergency department visits, hospitalizations and deaths, 2002-2006*. Atlanta, GA: Centers for Disease Control and Prevention, National Center for Injury Prevention and Control,.

Figueiro Longo, M. G., Tan, C. O., Chan, S.-t, Welt, J., Avesta, A., Ratai, E., ... Gupta, R. (2020). Effect of transcranial low-level light therapy vs sham therapy among patients with moderate traumatic brain injury. *JAMA Network Open, 3*(9), e2017337. https://doi.org/10.1001/jamanetworkopen.2020.17337.

Gam, A. N., Thorsen, H., & Lonnberg, F. (1993). The effect of low-level laser therapy on musculoskeletal pain: A meta-analysis. *Pain, 52*(1), 63–66.

Gao, X., & Xing, D. (2009). Molecular mechanisms of cell proliferation induced by low power laser irradiation. *Journal of Biomedical Science, 16*, 4. https://doi.org/10.1186/1423-0127-16-4.

Gonzalez-Lima, F., & Barrett, D. W. (2014). Augmentation of cognitive brain functions with transcranial lasers. *Frontiers in Systems Neuroscience, 8*, 36. https://doi.org/10.3389/fnsys.2014.00036.

Haeussinger, F. B., Heinzel, S., Hahn, T., Schecklmann, M., Ehlis, A. C., & Fallgatter, A. J. (2011). Simulation of near-infrared light absorption considering individual head and prefrontal cortex anatomy: Implications for optical neuroimaging. *PLoS One, 6*(10), e26377. https://doi.org/10.1371/journal.pone.0026377.

Hamblin, M. R. (2008). The role of nitric oxide in low level light therapy. In M. R. Hamblin, J. J. Anders, & R. W. Waynant (Eds.). *Mechanisms for Low-Light Therapy II (Vol. in press)*. Bellingham, WA: The International Society for Optical Engineering.

Hamblin, M. R. (2016). Shining light on the head: Photobiomodulation for brain disorders. *BBA Clincal, 6*, 113–124. https://doi.org/10.1016/j.bbacli.2016.09.002.

Hamblin, M. R. (2018). Photobiomodulation for traumatic brain injury and stroke. *Journal of Neuroscience Research, 96*(4), 731–743.

Hamblin, M. R. (2023). Could photobiomodulation therapy be helpful in epilepsy? *Photobiomodulation, Photomedicine, and Laser Surgery, 41*(2), 35–36.

Hamblin, M. R. (2024). Transcranial photobiomodulation for the brain: A wide range of clinical applications. *Neural Regeneration Research, 19*(3), 483–484.

Hamblin, M. R., Agrawal, T., & de Sousa, M. V. (2016). *Handbook of Low Level Laser Therapy*. New York: Jenny Stanford Publishing. https://doi.org/10.1201/9781315364827.

Hamilton, C., Liebert, A., Pang, V., Magistretti, P., & Mitrofanis, J. (2022). Lights on for autism: Exploring Photobiomodulation as an effective therapeutic option. *Neurology International, 14*(4), 884–893.

Henderson, T. A., & Morries, L. D. (2015a). Near-infrared photonic energy penetration: Can infrared phototherapy effectively reach the human brain? *Neuropsychiatric Disease and Treatment, 11*, 2191–2208. https://doi.org/10.2147/NDT.S78182.

Henderson, T. A., & Morries, L. D. (2015b). SPECT perfusion imaging demonstrates improvement of traumatic brain injury with transcranial near-infrared laser phototherapy. *Advances in Mind-Body Medicine, 29*(4), 27–33.

Hennessy, M., & Hamblin, M. R. (2016). Photobiomodulation and the brain: A new paradigm. *J Optics, 19*(1), 013003. https://doi.org/10.1088/2040-8986/19/1/013003.

Hipskind, S. G., Grover, F. L., Fort, T. R., Helffenstein, D., Burke, T. J., Quint, S. A., ... Hurtado, T. (2018). Pulsed transcranial red/near-infrared light therapy using light-emitting diodes improves cerebral blood flow and cognitive function in veterans with chronic traumatic brain injury: A case series. *Photomedicine and Laser Surgery*. https://doi.org/10.1089/pho.2018.4489.

Hopkins, J. T., McLoda, T. A., Seegmiller, J. G., & David Baxter, G. (2004). Low-level laser therapy facilitates superficial wound healing in humans: A triple-blind, sham-controlled study. *Journal of Athletic Training, 39*(3), 223–229.

Iwashita, T., Tada, T., Zhan, H., Tanaka, Y., & Hongo, K. (2003). Harvesting blood stem cells from cranial bone at craniotomy–A preliminary study. *Journal of Neuro-Oncology, 64*(3), 265–270.

Jagdeo, J. R., Adams, L. E., Brody, N. I., & Siegel, D. M. (2012). Transcranial red and near infrared light transmission in a cadaveric model. *PLoS One,* 7(10), e47460. https://doi.org/10.1371/journal.pone.0047460.

Juárez-Contreras, R., Méndez-Reséndiz, K. A., Rosenbaum, T., González-Ramírez, R., & Morales-Lázaro, S. L. (2020). TRPV1 channel: A noxious signal transducer that affects mitochondrial function. *International Journal of Molecular Sciences, 21*(23), 8882.

Kamins, J., & Giza, C. C. (2016). Concussion-mild traumatic brain injury: Recoverable injury with potential for serious sequelae. *Neurosurgery Clinics of North America, 27*(4), 441–452. https://doi.org/10.1016/j.nec.2016.05.005.

Karu, T. (1999). Primary and secondary mechanisms of action of visible to near-IR radiation on cells. *Journal of Photochemistry and Photobiology. B, Biology, 49*(1), 1–17.

Karu, T. I., & Kolyakov, S. F. (2005). Exact action spectra for cellular responses relevant to phototherapy. *Photomedicine and Laser Surgery, 23*(4), 355–361.

Karu, T. I., Pyatibrat, L. V., & Afanasyeva, N. I. (2004). A novel mitochondrial signaling pathway activated by visible-to-near infrared radiation. *Photochemistry and Photobiology, 80*(2), 366–372.

Karu, T. I., Pyatibrat, L. V., & Afanasyeva, N. I. (2005). Cellular effects of low power laser therapy can be mediated by nitric oxide. *Lasers in Surgery and Medicine, 36*(4), 307–314. https://doi.org/10.1002/lsm.20148.

Karu, T. I., Pyatibrat, L. V., & Kalendo, G. S. (2004). Photobiological modulation of cell attachment via cytochrome c oxidase. *Photochemical & Photobiological Sciences: Official Journal of the European Photochemistry Association and the European Society for Photobiology, 3*(2), 211–216.

Kashiwagi, S., Morita, A., Yokomizo, S., Ogawa, E., Komai, E., Huang, P. L., ... Atochin, D. N. (2023). Photobiomodulation and nitric oxide signaling. *Nitric Oxide: Biology and Chemistry / Official Journal of the Nitric Oxide Society, 130*, 58–68.

Keszler, A., Lindemer, B., Hogg, N., Weihrauch, D., & Lohr, N. L. (2018). Wavelength-dependence of vasodilation and NO release from S-nitrosothiols and dinitrosyl iron complexes by far red/near infrared light. *Archives of Biochemistry and Biophysics, 649*, 47–52.

Khuman, J., Zhang, J., Park, J., Carroll, J. D., Donahue, C., & Whalen, M. J. (2012). Low-level laser light therapy improves cognitive deficits and inhibits microglial activation after controlled cortical impact in mice. *Journal of Neurotrauma, 29*(2), 408–417. https://doi.org/10.1089/neu.2010.1745.

Kovacs, I. B., Mester, E., & Gorog, P. (1974). Stimulation of wound healing with laser beam in the rat. *Experientia, 30*(11), 1275–1276.

Lane, N. (2006). Cell biology: Power games. *Nature, 443*(7114), 901–903.

Lapchak, P. A., Boitano, P. D., Butte, P. V., Fisher, D. J., Holscher, T., Ley, E. J., ... Rajput, P. S. (2015). Transcranial near-infrared laser transmission (NILT) profiles (800 nm): Systematic comparison in four common research species. *PLoS One, 10*(6), e0127580. https://doi.org/10.1371/journal.pone.0127580.

Lapchak, P. A., Salgado, K. F., Chao, C. H., & Zivin, J. A. (2007). Transcranial near-infrared light therapy improves motor function following embolic strokes in rabbits: An extended therapeutic window study using continuous and pulse frequency delivery modes. *Neuroscience, 148*(4), 907–914. https://doi.org/10.1016/j.neuroscience.2007.07.002.

Li, Y., Dong, Y., Yang, L., Tucker, L., Zong, X., Brann, D., ... Zhang, Q. (2021). Photobiomodulation prevents PTSD-like memory impairments in rats. *Molecular Psychiatry, 26*(11), 6666–6679.

Liebel, S. W., Johnson, P. K., Lindsey, H. M., Russell, H. A., Hovenden, E. S., Velez, C., ... Tate, D. F. (2022). A-25 transcranial photobiomodulation treatment effects in former athletes with repetitive head hits. *Archives of Clinical Neuropsychology, 37*(5), 1066. https://doi.org/10.1093/arclin/acac32.25.

Lim, L. (2024). Traumatic brain injury recovery with photobiomodulation: Cellular mechanisms, clinical evidence, and future potential. *Cells, 13*(5), 385.

Loane, D. J., & Faden, A. I. (2010). Neuroprotection for traumatic brain injury: Translational challenges and emerging therapeutic strategies. *Trends in Pharmacological Sciences, 31*(12), 596–604. https://doi.org/10.1016/j.tips.2010.09.005.

McKee, A. C., Alosco, M. L., & Huber, B. R. (2016). Repetitive head impacts and chronic traumatic encephalopathy. *Neurosurgery Clinics of North America, 27*(4), 529–535. https://doi.org/10.1016/j.nec.2016.05.009.

Michalikova, S., Ennaceur, A., van Rensburg, R., & Chazot, P. L. (2008). Emotional responses and memory performance of middle-aged CD1 mice in a 3D maze: Effects of low infrared light. *Neurobiology of Learning and Memory, 89*(4), 480–488. https://doi.org/10.1016/j.nlm.2007.07.014.

Mitchell, U. H., & Mack, G. L. (2013). Low-level laser treatment with near-infrared light increases venous nitric oxide levels acutely: A single-blind, randomized clinical trial of efficacy. *American Journal of Physical Medicine & Rehabilitation / Association of Academic Physiatrists, 92*(2), 151–156. https://doi.org/10.1097/PHM.0b013e318269d70a.

Moreira, M. S., Velasco, I. T., Ferreira, L. S., Ariga, S. K., Abatepaulo, F., Grinberg, L. T., & Marques, M. M. (2011). Effect of laser phototherapy on wound healing following cerebral ischemia by cryogenic injury. *Journal of Photochemistry and Photobiology. B, Biology, 105*(3), 207–215. https://doi.org/10.1016/j.jphotobiol.2011.09.005.

Moreira, M. S., Velasco, I. T., Ferreira, L. S., Ariga, S. K., Barbeiro, D. F., Meneguzzo, D. T., ... Marques, M. M. (2009). Effect of phototherapy with low intensity laser on local and systemic immunomodulation following focal brain damage in rat. *Journal of Photochemistry and Photobiology. B, Biology, 97*(3), 145–151. https://doi.org/10.1016/j.jphotobiol.2009.09.002.

Morries, L. D., Cassano, P., & Henderson, T. A. (2015). Treatments for traumatic brain injury with emphasis on transcranial near-infrared laser phototherapy. *Neuropsychiatric Disease and Treatment, 11*, 2159–2175. https://doi.org/10.2147/NDT.S65809.

Naeser, M. A., & Hamblin, M. R. (2011). Potential for transcranial laser or LED therapy to treat stroke, traumatic brain injury, and neurodegenerative disease. *Photomedicine and Laser Surgery, 29*(7), 443–446. https://doi.org/10.1089/pho.2011.9908.

Naeser, M. A., & Hamblin, M. R. (2015). Traumatic brain injury: A major medical problem that could be treated using transcranial, red/near-infrared LED photobiomodulation. *Photomedicine and Laser Surgery*. https://doi.org/10.1089/pho.2015.3986.

Naeser, M. A., Martin, P. I., Ho, M. D., Krengel, M. H., Bogdanova, Y., Knight, J. A., ... Koo, B. (2023). Transcranial photobiomodulation treatment: Significant improvements in four ex-football players with possible chronic traumatic encephalopathy. *Journal of Alzheimer's Disease Reports,* 7(1), 77–105. https://doi.org/10.3233/adr-220022.

Naeser, M. A., Martin, P. I., Ho, M. D., Krengel, M. H., Bogdanova, Y., Knight, J. A., ... Koo, B. B. (2016). Transcranial, red/near-infrared light-emitting diode therapy to improve cognition in chronic traumatic brain injury. *Photomedicine and Laser Surgery, 34*(12), 610–626. https://doi.org/10.1089/pho.2015.4037.

Naeser, M. A., Saltmarche, A., Krengel, M. H., Hamblin, M. R., & Knight, J. A. (2011). Improved cognitive function after transcranial, light-emitting diode treatments in chronic, traumatic brain injury: Two case reports. *Photomedicine and Laser Surgery, 29*(5), 351–358. https://doi.org/10.1089/pho.2010.2814.

Naeser, M. A., Zafonte, R., Krengel, M. H., Martin, P. I., Frazier, J., Hamblin, M., ... Baker, E. H. (2014). Significant improvements on cognitive performance post- transcranial, red/near-infrared LED treatments in chronic, mild TBI: Open-protocol study. *Journal of Neurotrauma*. https://doi.org/10.1089/neu.2013.3244.

Navratil, L., & Kymplova, J. (2002). Contraindications in noninvasive laser therapy: truth and fiction. *Journal of Clinical Laser Medicine & Surgery, 20*(6), 341–343. https://doi.org/10.1089/104454702320901134.

Nawashiro, H., Wada, K., Nakai, K., & Sato, S. (2012). Focal increase in cerebral blood flow after treatment with near-infrared light to the forehead in a patient in a persistent vegetative state. *Photomedicine and Laser Surgery, 30*(4), 231–233. https://doi.org/10.1089/pho.2011.3044.

Okada, E., & Delpy, D. T. (2003). Near-infrared light propagation in an adult head model. II. Effect of superficial tissue thickness on the sensitivity of the near-infrared spectroscopy signal. *Applied Optics, 42*(16), 2915–2922.

Oron, A., & Oron, U. (2016). Low-level laser therapy to the bone marrow ameliorates neurodegenerative disease progression in a mouse model of Alzheimer's disease: A minireview. *Photomedicine and Laser Surgery*. https://doi.org/10.1089/pho.2015.4072.

Oron, A., Oron, U., Streeter, J., de, Taboada, L., Alexandrovich, A., ... Shohami, E. (2007). low-level laser therapy applied transcranially to mice following traumatic brain injury significantly reduces long-term neurological deficits. *Journal of Neurotrauma, 24*(4), 651–656. https://doi.org/10.1089/neu.2006.0198.

Oron, A., Oron, U., Streeter, J., De Taboada, L., Alexandrovich, A., Trembovler, V., & Shohami, E. (2012). Near infrared transcranial laser therapy applied at various modes to mice following traumatic brain injury significantly reduces long-term neurological deficits. *Journal of Neurotrauma, 29*(2), 401–407. https://doi.org/10.1089/neu.2011.2062.

Passarella, S., Casamassima, E., Molinari, S., Pastore, D., Quagliariello, E., Catalano, I. M., & Cingolani, A. (1984). Increase of proton electrochemical potential and ATP synthesis in rat liver mitochondria irradiated in vitro by helium-neon laser. *FEBS Letters, 175*(1), 95–99.

Pitzschke, A., Lovisa, B., Seydoux, O., Zellweger, M., Pfleiderer, M., Tardy, Y., & Wagnieres, G. (2015). Red and NIR light dosimetry in the human deep brain. *Physics in Medicine and Biology, 60*(7), 2921–2937. https://doi.org/10.1088/0031-9155/60/7/2921.

Povlishock, J. T. (2013). The window of risk in repeated head injury. *Journal of Neurotrauma, 30*(1), 1. https://doi.org/10.1089/neu.2013.9942.

Poyton, R. O., & Ball, K. A. (2011). Therapeutic photobiomodulation: Nitric oxide and a novel function of mitochondrial cytochrome c oxidase. *Discovery Medicine, 11*(57), 154–159.

Prins, M. L., Alexander, D., Giza, C. C., & Hovda, D. A. (2013). Repeated mild traumatic brain injury: Mechanisms of cerebral vulnerability. *Journal of Neurotrauma, 30*(1), 30–38. https://doi.org/10.1089/neu.2012.2399.

Purushothuman, S., Nandasena, C., Johnstone, D. M., Stone, J., & Mitrofanis, J. (2013). The impact of near-infrared light on dopaminergic cell survival in a transgenic mouse model of parkinsonism. *Brain Research, 1535*, 61–70. https://doi.org/10.1016/j.brainres.2013.08.047.

Quirk, B. J., Torbey, M., Buchmann, E., Verma, S., & Whelan, H. T. (2012). Near-infrared photobiomodulation in an animal model of traumatic brain injury: Improvements at the behavioral and biochemical levels. *Photomedicine and Laser Surgery, 30*(9), 523–529. https://doi.org/10.1089/pho.2012.3261.

Rindner, E. S., Haroon, J. M., Jordan, K. G., Mahdavi, K. D., Surya, J. R., Zielinski, M. A., ... Jordan, S. E. (2022). Transcranial infrared laser stimulation for the treatment of traumatic brain injury: A case series. *Journal of Lasers in Medical Sciences, 13*, e65. https://doi.org/10.34172/jlms.2022.65.

Rojas, J. C., Bruchey, A. K., & Gonzalez-Lima, F. (2012). Low-level light therapy improves cortical metabolic capacity and memory retention. *Journal of Alzheimer's Disease: JAD, 32*(3), 741–752. https://doi.org/10.3233/JAD-2012-120817.

Rojas, J. C., & Gonzalez-Lima, F. (2013). Neurological and psychological applications of transcranial lasers and LEDs. *Biochemical Pharmacology, 86*(4), 447–457. https://doi.org/10.1016/j.bcp.2013.06.012.

Safinia, C., Bershad, E. M., Clark, H. B., SantaCruz, K., Alakbarova, N., Suarez, J. I., & Divani, A. A. (2016). Chronic traumatic encephalopathy in athletes involved with high-impact sports. *Journal of Vascular and Interventional Radiology, 9*(2), 34–48.

Salehpour, F., Khademi, M., Bragin, D. E., & DiDuro, J. O. (2022). Photobiomodulation therapy and the glymphatic system: Promising applications for augmenting the brain lymphatic drainage system. *International Journal of Molecular Sciences, 23*(6), 2975.

Salehpour, F., Khademi, M., & Hamblin, M. R. (2021). Photobiomodulation therapy for dementia: A systematic review of pre-clinical and clinical studies. *Journal of Alzheimer's Disease, 83*(4), 1431–1452.

Salehpour, F., & Rasta, S. H. (2017). The potential of transcranial photobiomodulation therapy for treatment of major depressive disorder. *Reviews in the Neurosciences, 28*(4), 441–453.

Salgado, A. S., Zangaro, R. A., Parreira, R. B., & Kerppers, I. I. (2015). The effects of transcranial LED therapy (TCLT) on cerebral blood flow in the elderly women. *Lasers in Medical Science, 30*(1), 339–346. https://doi.org/10.1007/s10103-014-1669-2.

Saltmarche, A. E., Naeser, M. A., Ho, K. F., Hamblin, M. R., & Lim, L. (2017). Significant improvement in cognition in mild to moderately severe dementia cases treated with transcranial plus intranasal photobiomodulation: Case series report. *Photomedicine and Laser Surgery, 35*(8), 432–441. https://doi.org/10.1089/pho.2016.4227.

Samoilova, K. A., Zhevago, N. A., Petrishchev, N. N., & Zimin, A. A. (2008). Role of nitric oxide in the visible light-induced rapid increase of human skin microcirculation at the local and systemic levels: II. Healthy volunteers. *Photomedicine and Laser Surgery, 26*(5), 443–449. https://doi.org/10.1089/pho.2007.2205.

Sharma, S. K., Sardana, S., & Hamblin, M. R. (2023). Role of opsins and light or heat activated transient receptor potential ion channels in the mechanisms of photobiomodulation and infrared therapy. *Journal of Photochemistry and Photobiology, 13*, 100160.

Sharp, D. J., Scott, G., & Leech, R. (2014). Network dysfunction after traumatic brain injury. *Nature Reviews Neurology, 10*(3), 156–166. https://doi.org/10.1038/nrneurol.2014.15.

Stern, R. A., Riley, D. O., Daneshvar, D. H., Nowinski, C. J., Cantu, R. C., & McKee, A. C. (2011). Long-term consequences of repetitive brain trauma: Chronic traumatic encephalopathy. *Pm&R, 3*(10 Suppl 2), S460–S467. https://doi.org/10.1016/j.pmrj.2011.08.008.

Stevens, A. R., Hadis, M., Milward, M., Ahmed, Z., Belli, A., Palin, W., & Davies, D. J. (2023). Photobiomodulation in acute traumatic brain injury: A systematic review and meta-analysis. *Journal of Neurotrauma, 40*(3-4), 210–227.

Stevens, A. R., Hadis, M., Thareja, A., Anderson, F. G., Milward, M. R., Di Pietro, V., ... Ahmed, Z. (2024). Photobiomodulation improves functional recovery after mild traumatic brain injury. *Bioengineering & Translational Medicine, 10*(2), https://doi.org/10.1002/btm2.10727.

Strangman, G. E., Zhang, Q., & Li, Z. (2014). Scalp and skull influence on near infrared photon propagation in the Colin27 brain template. *Neuroimage, 85*(Pt 1), 136–149. https://doi.org/10.1016/j.neuroimage.2013.04.090.

Suski, J. M., Lebiedzinska, M., Bonora, M., Pinton, P., Duszynski, J., & Wieckowski, M. R. (2012). Relation between mitochondrial membrane potential and ROS formation. *Methods in Molecular Biology, 810*, 183–205. https://doi.org/10.1007/978-1-61779-382-0_12.

Sussai, D. A., Carvalho Pde, T., Dourado, D. M., Belchior, A. C., dos Reis, F. A., & Pereira, D. M. (2010). Low-level laser therapy attenuates creatine kinase levels and apoptosis during forced swimming in rats. *Lasers in Medical Science, 25*(1), 115–120. https://doi.org/10.1007/s10103-009-0697-9.

Tatmatsu-Rocha, J. C., Ferraresi, C., Hamblin, M. R., Damasceno Maia, F., do Nascimento, N. R., Driusso, P., & Parizotto, N. A. (2016). Low-level laser therapy (904nm) can increase collagen and reduce oxidative and nitrosative stress in diabetic wounded mouse skin. *Journal of Photochemistry and Photobiology. B, Biology, 164*, 96–102. https://doi.org/10.1016/j.jphotobiol.2016.09.017.

Tedford, C. E., DeLapp, S., Jacques, S., & Anders, J. (2015). Quantitative analysis of transcranial and intraparenchymal light penetration in human cadaver brain tissue. *Lasers in Surgery and Medicine, 47*(4), 312–322. https://doi.org/10.1002/lsm.22343.

Thunshelle, C., & Hamblin, M. R. (2016). Transcranial low-level laser (light) therapy for brain injury. *Photomedicine and Laser Surgery, 34*(12), 587–598. https://doi.org/10.1089/pho.2015.4051.

Vladimirov, Y., Borisenko, G., Boriskina, N., Kazarinov, K., & Osipov, A. (2000). NO–hemoglobin may be a light-sensitive source of nitric oxide both in solution and in red blood cells. *Journal of Photochemistry and Photobiology B: Biology, 59*(1-3), 115–122.

Wang, Y., Huang, Y. Y., Wang, Y., Lyu, P., & Hamblin, M. R. (2016). Photobiomodulation of human adipose-derived stem cells using 810nm and 980nm lasers operates via different mechanisms of action. *Biochimica et Biophysica Acta*. https://doi.org/10.1016/j.bbagen.2016.10.008.

Wu, C., Li, M., Chen, Z., Feng, S., Deng, Q., Duan, R., ... Yang, L. (2025). Remote photobiomodulation ameliorates behavioral and neuropathological outcomes in a rat model of repeated closed head injury. *Translational Psychiatry, 15*(1), https://doi.org/10.1038/s41398-025-03228-3.

Wu, Q., Xuan, W., Ando, T., Xu, T., Huang, L., Huang, Y. Y., ... Hamblin, M. R. (2012). Low-level laser therapy for closed-head traumatic brain injury in mice: Effect of different wavelengths. *Lasers in Surgery and Medicine, 44*, 218–226. https://doi.org/10.1002/lsm.22003.

Xuan, W., Agrawal, T., Huang, L., Gupta, G. K., & Hamblin, M. R. (2015). Low-level laser therapy for traumatic brain injury in mice increases brain derived neurotrophic factor (BDNF) and synaptogenesis. *Journal of Biophotonics, 8*(6), 502–511. https://doi.org/10.1002/jbio.201400069.

Xuan, W., Huang, L., & Hamblin, M. R. (2016). Repeated transcranial low-level laser therapy for traumatic brain injury in mice: Biphasic dose response and long-term treatment outcome. *Journal of Biophotonics, 9*(11-12), 1263–1272. https://doi.org/10.1002/jbio.201500336.

Xuan, W., Vatansever, F., Huang, L., & Hamblin, M. R. (2014). Transcranial low-level laser therapy enhances learning, memory, and neuroprogenitor cells after traumatic brain injury in mice. *Journal of Biomedical Optics, 19*(10), 108003. https://doi.org/10.1117/1.JBO.19.10.108003.

Xuan, W., Vatansever, F., Huang, L., Wu, Q., Xuan, Y., Dai, T., ... Hamblin, M. R. (2013). Transcranial low-level laser therapy improves neurological performance in traumatic brain injury in mice: Effect of treatment repetition regimen. *PLoS One, 8*(1), e53454. https://doi.org/10.1371/journal.pone.0053454.

Yang, B., Xu, J., Li, Y., Dong, Y., Li, Y., Tucker, L., ... Yan, X. (2019). Photobiomodulation therapy for repeated closed head injury in rats. *Journal of Biophotonics, 13*(2), https://doi.org/10.1002/jbio.201960117.

Yaroslavsky, A. N., Schulze, P. C., Yaroslavsky, I. V., Schober, R., Ulrich, F., & Schwarzmaier, H. J. (2002). Optical properties of selected native and coagulated human brain tissues in vitro in the visible and near infrared spectral range. *Physics in Medicine and Biology, 47*(12), 2059–2073.

Yokomizo, S., Roessing, M., Morita, A., Kopp, T., Ogawa, E., Katagiri, W., ... Kashiwagi, S. (2022). Near-infrared II photobiomodulation augments nitric oxide bioavailability via phosphorylation of endothelial nitric oxide synthase. *The FASEB Journal, 36*(9), e22490.

Zecha, J. A., Raber-Durlacher, J. E., Nair, R. G., Epstein, J. B., Elad, S., Hamblin, M. R., ... Bensadoun, R. J. (2016). Low-level laser therapy/photobiomodulation in the management of side effects of chemoradiation therapy in head and neck cancer: Part 2: Proposed applications and treatment protocols. *Supportive Care in Cancer: Official Journal of the Multinational Association of Supportive Care in Cancer*. https://doi.org/10.1007/s00520-016-3153-y.

Zecha, J. A., Raber-Durlacher, J. E., Nair, R. G., Epstein, J. B., Sonis, S. T., Elad, S., ... Bensadoun, R. J. (2016). Low level laser therapy/photobiomodulation in the management of side effects of chemoradiation therapy in head and neck cancer: part 1: Mechanisms of action, dosimetric, and safety considerations. *Supportive Care in Cancer: Official Journal of the Multinational Association of Supportive Care in Cancer*. https://doi.org/10.1007/s00520-016-3152-z.

Zhang, Q., Dong, T., Li, P., & Wu, M. X. (2016). Noninvasive low-level laser therapy for thrombocytopenia. *Science Translational Medicine, 8*(349), 349ra101. https://doi.org/10.1126/scitranslmed.aaf4964.

Zhang, Q., Zhou, C., Hamblin, M. R., & Wu, M. X. (2014). Low-level laser therapy effectively prevents secondary brain injury induced by immediate early responsive gene X-1 deficiency. *Journal of Cerebral Blood Flow and Metabolism: Official Journal of the International Society of Cerebral Blood Flow and Metabolism*. https://doi.org/10.1038/jcbfm.2014.95.

CHAPTER FIVE

# Cognitive enhancement through transcranial infrared laser stimulation: Insights from our research

**Francisco Gonzalez-Lima***
Departments of Psychology, Psychiatry and Behavioral Sciences, Texas Consortium in Behavioral Neuroscience, The University of Texas at Austin, Austin, TX, United States
*Corresponding author. e-mail address: gonzalezlima@utexas.edu

## Contents



*International Review of Neurobiology*, Volume 189
ISSN 0074-7742, https://doi.org/10.1016/bs.irn.2026.01.015

## Abstract

Photobiomodulation (PBM), particularly transcranial infrared laser stimulation (TILS), has emerged as a promising non-invasive method to enhance cognitive functions in healthy individuals and those with neuropsychiatric or neurodevelopmental disorders. Conventional pharmacotherapy often fails to improve cognition effectively and can produce adverse effects. In contrast, TILS of the prefrontal cortex modulates brain energy metabolism through mechanisms like mitochondrial upregulation, improved cerebral oxygenation, and enhanced neuroplasticity. This chapter synthesizes findings and insights from human studies led by our group to evaluate TILS mechanisms and efficacy in cognitive enhancement. Examples of our research include studies on healthy young and older adults, and patients with neurocognitive disorder, bipolar disorder, depression, ADHD and autism. The evidence highlights TILS as a safe and physiologically grounded approach with wide-ranging applications. To maximize its potential, future research should explore optimal dosing, individual differences in response, long-term outcomes, and the use of neuroimaging biomarkers for personalized treatment strategies.

## 1. Introduction

Cognitive enhancement refers to efforts to improve mental abilities such as memory, attention, learning and executive functions. Cognitive enhancement has long been a central goal in both neuroscience and psychology, particularly given the rising prevalence of neurodegenerative diseases, age-related cognitive decline, and neuropsychiatric conditions that impair executive function, memory, and attention. Historically, this effort began with natural substances like caffeine and ginseng, used in traditional medicine for their perceived cognitive-boosting properties. Practices like meditation and yoga were also used to enhance attention and emotional regulation. In the 20th century, pharmacological methods emerged with drugs like amphetamines, methylphenidate (Ritalin) and modafinil that improve focus and alertness (Repantis et al., 2010), and low-dose methylene blue that improves learning and memory (Auchter et al., 2014; Rojas et al., 2012). These drugs were originally used for medical conditions but they are also used off-label to enhance cognitive performance in healthy individuals. This raised ethical and safety concerns despite their effectiveness.

More recently, neurotechnological approaches such as transcranial direct current stimulation (tDCS), transcranial magnetic stimulation (TMS), and brain-computer interfaces have been developed to modulate brain activity and potentially enhance learning and memory. Though promising, these methods remain largely experimental. Indeed, interventions such as pharmacological agents, cognitive training, and electrical and magnetic brain stimulation techniques (e.g., tDCS and TMS) have demonstrated variable efficacy and often carry limitations, including side effects, high variability across individuals, and limited long-term benefits (Bikson et al., 2018; Cobb et al., 2021, 2025). Overall, cognitive enhancement has evolved from natural methods to high-tech interventions, reflecting ongoing scientific progress and shifting societal attitudes toward mental performance.

Pioneering work in the field of **photobiomodulation (PBM)** has illuminated the potential of **transcranial infrared laser stimulation (TILS)** as a non-invasive intervention to enhance cognitive functions in healthy and disease populations (recently reviewed in Gamboa et al., 2025; Wang et al., 2024). Cognitive decline is a hallmark of aging and of various neuropsychiatric disorders, including mood disorders, and aging-related cognitive impairment and Alzheimer's disease. Cognitive dysfunction is also present in neurodevelopmental disorders such autism spectrum disorder (ASD) and attention deficit hyperactivity disorder (ADHD). Conventional pharmacological treatments often yield limited cognitive efficacy and are accompanied by adverse effects.

In this context, our lab's research has focused on harnessing the potential of PBM, particularly TILS to the frontal cortex, to modulate brain metabolism and augment cognitive functions (Gonzalez-Lima, 2021). In this chapter we summarize our main findings and explain the conceptual insights arising from our group's human studies to elucidate the in vivo mechanisms, efficacy, and therapeutic applications of TILS in cognitive enhancement. We argue that underlying mechanisms of TILS—including mitochondrial upregulation, increased cerebral oxygenation, and neuroplasticity—create a broad neurotherapeutic landscape similar to that of low-dose methylene blue (Gonzalez-Lima & Auchter, 2015). To fully realize this potential, future PBM research should focus on dose-response relationships, individual variability in treatment response, long-term effects, and the integration of neuroimaging biomarkers into personalized treatment protocols.

## 2. Background on PBM and TILS

The pursuit of safe, effective, and non-invasive methods to enhance cognitive performance has intensified with the global rise in age-related cognitive decline, neuropsychiatric disorders, and interest in human performance optimization. A novel class of interventions collectively referred to as low-level light/laser therapy (LLLT) or PBM (Anders et al., 2015), has emerged as a promising alternative to pharmacology, tDCS and TMS that is grounded in fundamental principles of light and cellular bioenergetics (Rojas & Gonzalez-Lima, 2011).

## 2.1 What is photobiomodulation (PBM)?

PBM involves the application of low-level light, typically in the red to near-infrared (NIR) spectrum (600–1100 nm), to stimulate cellular function through the absorption of photons by mitochondrial chromophores. In the field of LLLT or PBM, the development of mechanistic models describing how red to NIR light modulates cellular bioenergetics has evolved over the past four decades, grounded in both in vitro and in vivo experimental findings. A previous comprehensive chapter on PBM has thoroughly documented its historical evolution, underlying principles, and a broad range of clinical and experimental applications (Rojas & Gonzalez-Lima, 2017).

The principal intracellular photoacceptor implicated in in vivo PBM is **cytochrome c oxidase (CCO)**, the terminal enzyme of the mitochondrial electron transport chain (Hayworth et al., 2010; Rojas et al., 2008; Rojas & Gonzalez-Lima, 2011). Frankowski et al. (2025) recently reviewed the multiple hypothesized mechanisms underlying the biological effects of PBM, including the photonic activation of mitochondrial CCO, modulation of transmembrane transporters and receptor signaling pathways, and the upregulation of transforming growth factor-beta 1 (TGF-β1). Furthermore, they evaluated the emerging therapeutic potential of PBM across a spectrum of pathological conditions, encompassing cardiovascular pathologies, retinal degenerative diseases, Parkinson's disease, and various forms of cognitive dysfunction.

The transcranial application of PBM to the brain, known as **transcranial photobiomodulation (tPBM)**, has enabled targeted neuromodulation without the invasiveness of surgical or electrical interventions. Transcranial absorption of NIR light by CCO facilitates an increase in mitochondrial membrane potential, ATP production, and a reduction in oxidative stress, thereby supporting neuronal metabolic demands, network connectivity and other functions important for brain activity (Mitrofanis & Henderson, 2020). These mechanisms are particularly relevant for cortical brain regions involved in higher-order cognition, which require substantial metabolic resources and network interactions (Gonzalez-Lima et al., 2014) (Fig. 1).

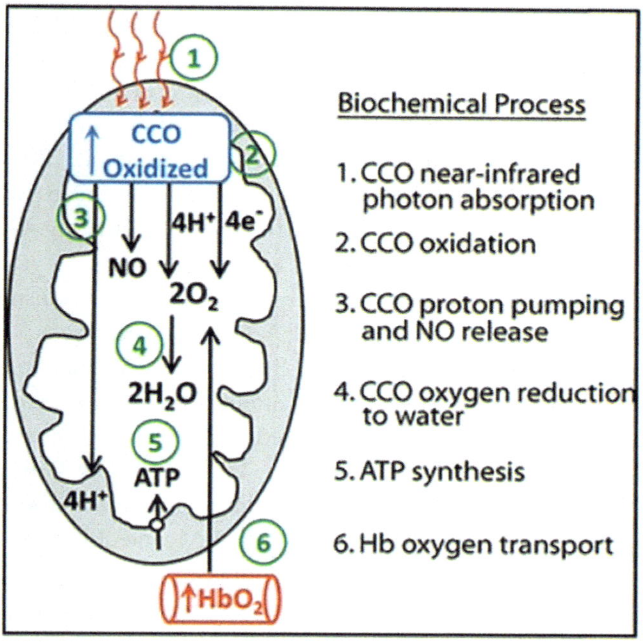


**Fig. 1** CCO photo-oxidation is the removal of electrons from CCO caused by the combined action of photons and oxygen. After photon absorption, CCO envelops oxygen, which is then reduced to water in a 6-steps process coupled with ATP and NO production in vivo. *Modified from Wang et al. (2017).*

## 2.2 The birth of TILS: A prefrontal-targeted tPBM method

Building on decades of animal research demonstrating PBM effects on wound healing and inflammation, the field advanced toward cognitive applications with the development of TILS, a technique pioneered and validated by researchers in our Texas group. TILS utilizes coherent NIR laser light at 1064 nm to penetrate the scalp and skull with higher spatial specificity and depth than broader band light sources like LEDs.

Crucially, TILS allows **targeted stimulation of the prefrontal cortex (PFC)**—a hub for executive control, working memory, and decision-making. By enhancing mitochondrial metabolism and promoting oxygen delivery in the PFC, TILS has been shown to improve performance on tasks involving attention, memory, and cognitive flexibility. In contrast to pharmacological interventions, TILS offers a **mechanism-based approach** that supports the brain endogenous energy metabolism and plasticity without exogenous chemicals or systemic side effects.

Specifically, TILS typically utilizes a wavelength of 1064 nm because it allows for less light scattering and deeper penetration than shorter wavelengths through the human scalp and skull (Huang et al., 2021; Tian et al., 2020; Wang et al., 2024), and enhanced CCO stimulation in cortical tissue (O'Donnell et al., 2023; Saucedo et al., 2021; Wang et al., 2017). To our knowledge, no other wavelength has been demonstrated to cause CCO photo-oxidation in the human brain in vivo. Up-regulating CCO in the human brain was of great interest to us because we have previously found that CCO plays a critical role in neuronal energy metabolism and Alzheimer's disease (Gonzalez-Lima, 1998; Valla et al., 2001).

The primary cortical target for TILS is the PFC, a region integral to working memory, cognitive flexibility, attentional control, and emotional regulation—all of which are compromised in various psychiatric and neurological disorders (Blanco, Maddox et al., 2017; Rojas & Gonzalez-Lima, 2013). A growing body of literature, including randomized controlled trials and pilot studies, supports the efficacy of tPBM and TILS in improving cognitive performance across a range of populations (Gamboa et al., 2025).

In healthy adults, TILS has been shown to enhance working memory, sustained attention, executive function, category learning and mood (Barrett & Gonzalez-Lima, 2013; Gonzalez-Lima, 2021; Holmes et al., 2019). In addition, clinical applications of TILS have demonstrated promising effects in populations with mild cognitive impairment (MCI), depression, and bipolar disorder (Barrett, Beevers et al., 2025; Barrett, Davis et al., 2025; Disner et al., 2016; Vargas et al., 2017). These effects are accompanied by physiological markers such as increased prefrontal oxygenation, improved cerebral blood flow, and alterations in functional brain connectivity, suggesting a multi-level impact of TILS on neural function.

Among the key contributors to this field, our research group has helped establish both the mechanistic underpinnings and clinical utility of TILS for cognitive enhancement. This work has provided some of the first evidence that tPBM targeting the PFC in humans can induce reliable enhancements in cognition, affect regulation, and neuroplasticity. By integrating animal models, neurophysiological measures, and rigorous clinical designs, we have positioned TILS as a mechanistically grounded and empirically validated cognitive intervention.

This chapter aims to systematically examine our human research studies that explore the effects of TILS on cognitive function in healthy people, with an additional focus on their application in specific clinical populations displaying mitochondrial dysfunction and PFC hypometabolism. We further discuss the physiological mechanisms implicated in these effects, including mitochondrial bioenergetics, neurovascular coupling, and large-scale neural network modulation.

## 3. Studies in healthy people

The discovery that TILS can enhance PFC functions—such as sustained attention, working memory, and affective regulation—represents a significant advancement in neuromodulation research. Utilizing low power density (0.25 W/cm$^2$) yet high energy density (60 J/cm$^2$) monochromatic light within the near-infrared spectrum (1064 nm CW), this non-invasive technique modulates neural activity in a non-thermal, non-destructive manner, offering promising neurocognitive enhancement potential. Clinical trials have demonstrated that TILS improves sustained attention, short-term memory, and emotional well-being in healthy individuals. Notably, participants exhibited faster reaction times and better performance in delayed match-to-sample memory tasks following TILS treatment. Additionally, self-reported measures of positive affect increased, suggesting mood enhancement as a secondary benefit. Importantly, no adverse effects have been found in any of our TILS studies, now exceeding 1000 participants. Our standard treatment targets the anterior PFC and involves the application of a 4 cm diameter laser beam to the right forehead (Fp2, 10–20 EEG coordinate system) in younger adults, or bilateral treatment (Fp1 and Fp2, 10–20 EEG coordinate system) for older adults (Fig. 2).

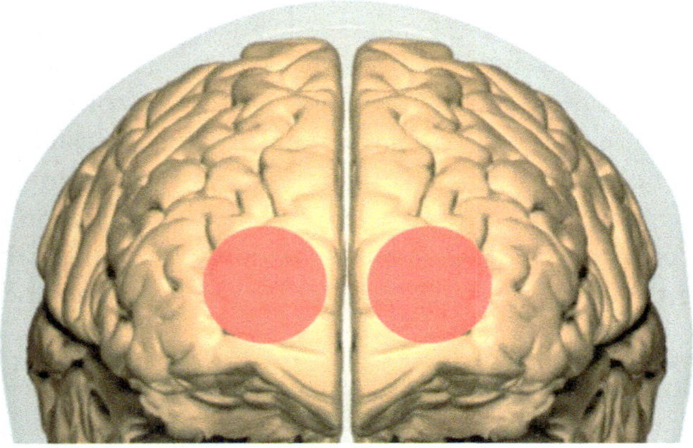

**Fig. 2** Brain schematic of bilateral stimulation sites (4 cm diameter) on the anterior PFC.

## 3.1 Cognitive and emotional benefits in healthy people

Barrett and Gonzalez-Lima (2013) conducted the first controlled study demonstrating that transcranial laser stimulation can enhance cognitive and emotional brain functions in humans. In this study, TILS was applied to the PFC of healthy volunteers. The results indicated significant improvements in sustained attention and short-term memory performance, as well as enhanced positive emotional states. These effects were attributed to increased mitochondrial activity and improved cerebral oxygenation, highlighting the potential of tPBM to enhance cognitive and emotional functions in healthy individuals. Similarly, the controlled study by Hwang et al. (2016) evaluated the combined effects of TILS and acute aerobic exercise on cognitive performance. The findings suggested that the combination was both safe and effective in enhancing cognitive function, further corroborating the safety and efficacy of TILS interventions.

## 3.2 Transcranial laser as neuroenhancement for attention bias modification

In a controlled study focusing on individuals with elevated depression symptoms, Disner et al. (2016) explored the potential of TILS to modify negative attention biases mediated by the right PFC. The findings demonstrated that TILS could serve as a neuroenhancement tool, effectively altering attention biases in the targeted individuals. This research underscored for the first time the therapeutic potential of TILS in addressing cognitive aspects associated with mood disorders.

## 3.3 Transcranial infrared laser stimulation improves executive function

In another controlled study, Blanco, Maddox et al. (2017) investigated the effects of TILS on executive function using the Wisconsin Card Sorting Test (WCST), a gold standard for assessing cognitive flexibility and PFC function. The results indicated that TILS enhanced performance on the WCST, suggesting PFC improvements in executive function. This study extended previous cognitive findings and provided the first evidence supporting the efficacy of TILS in enhancing executive abilities.

## 3.4 Transcranial infrared laser stimulation improves category learning

Blanco, Saucedo et al. (2017) conducted a randomized controlled trial to assess the impact of TILS on two distinct category learning systems: the

PFC-mediated rule-based system and the striatum-mediated information-integration system. Participants (n = 118) received either active or sham TILS targeting the lateral PFC. Results indicated that TILS significantly improved performance in the rule-based category learning task but had no effect on the information-integration task. These findings suggest that TILS enhances cognitive functions associated with the PFC, potentially offering a non-invasive method for cognitive enhancement in tasks reliant on explicit rule application.

## 3.5 Cognitive enhancement by TILS associated with cerebrovascular oxygenation

In this controlled study, we utilized functional near-infrared spectroscopy (fNIRS) to investigate the hemodynamic effects of TILS on the PFC during cognitive tasks involving working memory and sustained attention. The results revealed that TILS was associated with increased cerebrovascular

**Table 1** Summary of our TILS controlled studies of cognitive enhancement.

| |
|---|
| Barrett and Gonzalez-Lima (2013), Wang et al. (2017)<br>**Findings:** TILS improved attention and memory performance in healthy adults.<br>**Mechanisms:** Increased CCO concentration and enhanced cerebral oxygenation. |
| Hwang et al. (2016), Blanco, Maddox et al. (2017), Blanco, Saucedo et al. (2017), Urquhart et al. (2020)<br>**Findings:** TILS improved executive function, cognitive flexibility and category learning, and synergizes with cognitive-enhancing aerobic exercise.<br>**Mechanisms:** Modulation of neural network connectivity, circulation and PFC-based brain systems. |
| Vargas et al. (2017), Holmes et al. (2019)<br>**Findings:** TILS increased HbO levels in the PFC in resting conditions and during cognitive tasks in young and older healthy adults.<br>**Mechanisms:** Enhanced cerebral blood flow and oxygen delivery in frontal lobe. |
| Wang et al. (2017, 2019, 2024)<br>**Findings:** TILS increased EEG rhythms associated with cognitive processing independently of thermal effects.<br>**Mechanisms:** Enhanced power density of alpha and beta brain waves. |
| Saucedo et al. (2021), O'Donnell et al. (2022, 2023), Barrett, Beevers, and Gonzalez-Lima (2025), Barrett, Davis et al. (2025)<br>**Findings:** TILS improved cognitive performance in young and older adults with and without neuropsychiatric disorders.<br>**Mechanisms:** Increased CCO photo-oxidation in PFC and modulated neural network connectivity. |

oxygenation, which correlated with enhanced cognitive performance. This finding provides insight into the physiological mechanisms underlying the cognitive benefits of TILS (Holmes et al., 2019).

### 3.6 Beneficial neurocognitive effects of transcranial laser in older adults

In a subsequent controlled study, we compared the metabolic responses of younger and older adults to TILS, and found that older adults show a higher dose-dependent increase in mitochondrial function as measured by CCO photo-oxidation (Saucedo et al., 2021). These findings support the use of TILS for prevention of age-related decline in mitochondrial respiration and cognitive function. They justify further research on its therapeutic potential for diseases common in old age (Table 1).

## 4. Clinical applications in neuropsychiatric populations

### 4.1 Applications in depression and anxiety disorders

Our CCO histochemical studies in animal models of depression showing PFC hypometabolism (Shumake, Poremba et al., 2000; Shumake, Conejo-Jimenez et al., 2004), suggested that TILS to the PFC may upregulate CCO and thereby reduce depression and anxiety symptoms in humans. Inspired by promising results in a pilot study using LEDs by Schiffer et al. (2009), subsequent studies explored effects on cognitive and emotional functions in individuals with depression and anxiety disorders. Our study by Disner et al. (2016) examined the efficacy of TILS in enhancing attention and emotional regulation in adults with elevated depressive symptoms. The results indicated that TILS to the right PFC improved attention bias modification, suggesting its potential as an adjunctive treatment for emotional dysregulation in depression (Disner et al., 2016).

A different group led by Cassano et al. (2018) conducted a randomized controlled trial to evaluate the effects of tPBM using LEDs on emotional symptoms in patients with major depressive disorder (MDD). They found that tPBM led to improvements in mood, supporting its role as a non-invasive therapeutic option for depression. Additional MDD studies by this group and others have been reviewed by Vieira et al. (2023).

Zaizar et al. (2021) conducted a randomized controlled trial to evaluate TILS effects on anxiety disorders. Participants with elevated pathological fear in domains such as enclosed spaces, contamination, public speaking, or anxiety-related bodily sensations were recruited and randomized into four treatment arms: (1) Exposure + TILS, (2) Exposure + sham TILS, (3) TILS alone, and (4) Sham TILS alone. The study indicated that TILS administered immediately after exposure therapy did not significantly enhance fear extinction retention compared to sham TILS. However, TILS alone could significantly reduce pathological fear.

Barrett, Beevers et al. (2025) conducted a randomized controlled trial to evaluate the combined effects of internet-based Cognitive Behavioral Therapy (CBT) and TILS on individuals with MDD. The study aimed to determine whether TILS could enhance the antidepressant effects of CBT, specifically the Deprexis program. The study involved participants diagnosed with MDD, aged 18–45, who were provided access to Deprexis for 12 weeks. After an initial 2-week period, participants who showed at least a 10 % improvement in depressive symptoms were randomly assigned to receive either TILS or a sham treatment. TILS was administered once a week for 4 weeks to the right forehead, targeting the dorsolateral PFC. The primary outcome measure was the Quick Inventory of Depressive Symptomatology–Self-Report (QIDS-SR). The results indicated that participants receiving both Deprexis and TILS experienced a 56 % reduction in QIDS-SR scores, compared to a 43 % reduction in the sham group. This represents an additional 30 % improvement attributable to TILS. Importantly, no adverse effects were reported, suggesting that TILS is a safe adjunct to CBT.

We hypothesize that TILS enhances PFC function by increasing mitochondrial CCO activity and cerebral blood flow, thereby improving cognitive and emotional regulation in depression and anxiety disorders. For instance, this PFC augmentation may facilitate the application of adaptive coping strategies learned during CBT, supporting the efficacy of combining TILS with CBT for treating MDD. The findings suggest that TILS may serve as a promising adjunct neurotherapy, potentially improving outcomes for individuals undergoing cognitive therapies.

## 4.2 Efficacy in bipolar disorder

Bipolar disorder (BD), particularly in older adults, is associated with accelerated cognitive decline, including impairments in executive function, working memory, and cognitive flexibility. Our proof-of-concept study by O'Donnell et al. (2022) investigated TILS effects on cognitive performance in euthymic older bipolar patients (ages 60–85). Utilizing a crossover, sham-controlled design, the study administered five weekly sessions of TILS targeting the bilateral prefrontal cortex. Results indicated significant improvements in cognitive flexibility and reductions in impulsivity, suggesting that TILS may mitigate the cognitive deficits characteristic of aging in BD patients. This study represents the first clinical trial examining TILS in this population and underscores its potential as a non-invasive intervention for cognitive enhancement in BD. Furthermore, our subsequent study O'Donnell et al. (2023) demonstrated that TILS produced beneficial mitochondrial (CCO) and oxygenation effects on the PFC of older adults with BD.

Further supporting these findings in BD, Barrett, Davis et al. (2025) conducted another clinical trial assessing the impact of TILS on cognitive function and prefrontal network interactions in individuals with remitted BD. Our study demonstrated that TILS administration led to significant improvements in cognitive performance, as well as task-related connectivity within PFC neural networks, highlighting the neurotherapeutic potential of TILS in BD.

## 4.3 Impact on mild cognitive impairment and dementia

Mild cognitive impairment (MCI) and early-stage dementia are characterized by declines in memory, attention, and executive function. Studies have investigated the potential of tPBM to ameliorate these early cognitive deficits. We first conducted an open-label study to assess the effects of TILS on neurocognitive functions in older adults with subjective memory complaint, a precursor stage to MCI and dementia. The results indicated that TILS produced beneficial effects on cognition and emotional regulation, highlighting its potential as a therapeutic intervention for age-related cognitive decline (Vargas et al., 2017).

PBM studies by other groups also support cognitive-enhancing effects, as documented by our recent systematic review and meta-analysis (Gamboa et al., 2025). For example, a case series led by Naeser and collaborators has reported significant cognitive improvements in individuals with mild to moderate dementia following combined transcranial and intranasal PBM with LEDs. These results highlighted the potential of PBM in enhancing cognitive function and quality of life in dementia patients (Saltmarche et al., 2017). Another study led by Berman et al. (2017) assessed the effects of near-infrared LED therapy on cognitive function in dementia patients. The findings indicated improvements in memory and cognition following PBM treatment, suggesting its viability as a therapeutic approach for neurodegenerative conditions. While the results are encouraging, further dementia research with large-scale randomized controlled trials is needed.

# 5. Mechanistic insights for cognitive enhancement

While the many effects of tPBM have been recently reviewed elsewhere (e.g., Frankowski et al., 2025), the focus of this chapter is on TILS effects discovered by the Texas group, and on our insights about their relevance for cognitive augmentation (Gonzalez-Lima, 2021). Insights from our human studies reveal several key in vivo mechanisms through which TILS and similar forms of tPBM can exert cognitive-enhancing effects. These mechanisms span **cellular bioenergetics, neurovascular coupling, neural network modulation and neuroplasticity**. Below is a brief review of the elucidated mechanisms.

## 5.1 TILS causes cytochrome c oxidase photo-oxidation

The first cornerstone of our mechanistic insights for cognitive enhancement is the targeting of **CCO**, the terminal enzyme in the mitochondrial respiratory chain. Our in vivo studies in humans using broadband near-infrared spectroscopy (bbNIRS) in collaboration with Dr. Hanli Liu and colleagues, demonstrated that TILS photo-oxidizes CCO, the rate-limiting enzyme in the mitochondrial electron transport chain. By absorbing near-infrared light, CCO catalytic activity is enhanced, leading to increased mitochondrial ATP production and improved cellular energy metabolism (Wang et al., 2024).

**CCO Mechanistic Pathway: 1064 nm Light → CCO Activation → ↑ Electron Transport → ↑ ATP → ↑ Neuronal Firing Efficiency → ↑ Cognitive Function**.

- **Target:** Cytochrome c oxidase (CCO) in mitochondria.
- **Effect:** Near-infrared light enhances CCO activity, leading to increased ATP production.
- **Outcome:** Improved neuronal energy metabolism supports enhanced cognitive functions.

This photonic energy promotes a shift toward **a more energetically favorable state in neurons**, particularly in high-demand areas like the PFC during cognitive tasks.

## 5.2 TILS causes hemodynamic changes

The second cornerstone of our mechanistic insights for cognitive enhancement is the augmentation of oxygenated hemoglobin (HbO) in the PFC by TILS. Functional near-infrared spectroscopy (fNIRS) studies have demonstrated that TILS induces significant increases in HbO concentrations in the PFC during cognitive tasks. These hemodynamic changes are associated with enhanced cognitive performance, suggesting improved cerebral oxygenation as a hemodynamic mechanism underlying TILS effects (Holmes et al., 2019).

**Hemodynamic Mechanistic Pathway: TILS → Vasodilation via Nitric Oxide (NO) Release → ↑ Cerebral Blood Flow → ↑ Oxygen Delivery → ↑ Task Performance**.

- **Target:** Prefrontal cortex (PFC).
- **Effect:** TILS increases oxygenated hemoglobin (HbO) levels, indicating enhanced cerebral blood flow.
- **Outcome:** Improved oxygen delivery supports neuronal activity and cognitive performance.

Increased HbO reflects **improved cerebral blood flow and oxygenation**, suggesting that TILS improves **neurovascular coupling**. This is particularly important under cognitive load, where regional PFC demand for oxygen spikes (Fig. 3).

Besides these two cornerstone mechanistic pathways, additional TILS effects relevant to cognitive enhancement in healthy and clinical populations have been observed in our studies, as summarized below.

## 5.3 Neural network modulation

Our team has also explored **brain-wide network effects.** In controlled trials, TILS altered PFC **network connectivity** (Urquhart et al., 2020), enhancing cognitive flexibility and reducing impulsivity in healthy adults and individuals with BD (Barrett, Davis et al., 2025).

- **Target:** Prefrontal-parietal cortical network.
- **Effect:** TILS modulates functional connectivity between brain regions.
- **Outcome:** Modulated neural networks support more efficient information processing and cognitive flexibility.

An insight from these studies is that TILS may **normalize disrupted connectivity** in clinical populations, particularly within **fronto-parietal networks** associated with executive control and emotional regulation and the **default mode network** associated with memory consolidation. In addition, TILS-evoked changes in the power density of alpha and beta brain waves are consistent with the facilitation of cognitive processing

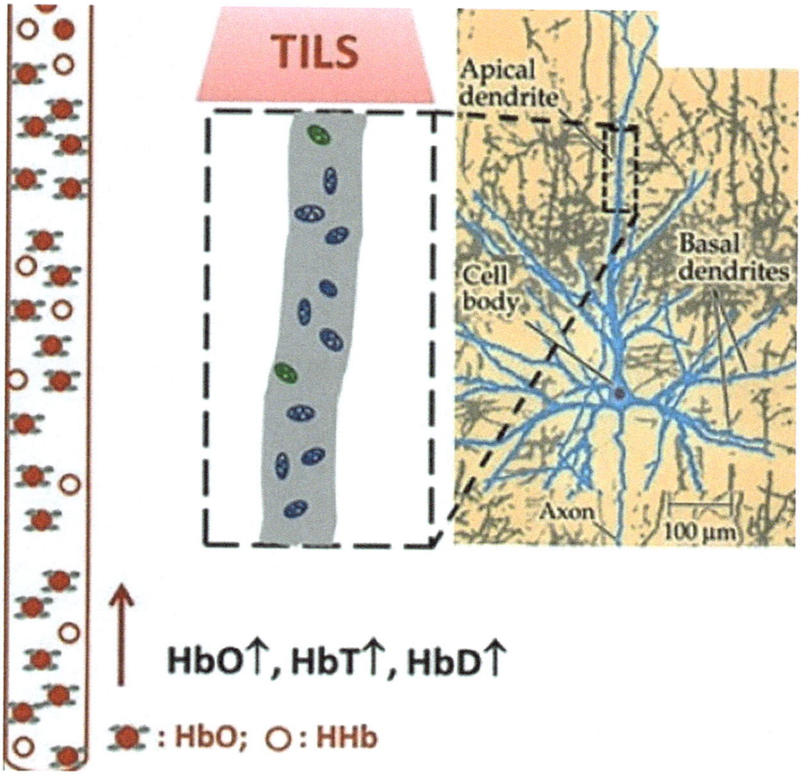


**Fig. 3** Illustration of photons reach mitochondria within the cortical neurons and elicit a hemodynamic response, characterized by increased HbO, total hemoglobin (HbT), and differential hemoglobin (HbD = oxygenated minus deoxygenated hemoglobin). *Modified from Wang et al. (2017).*

(Wang et al., 2019). This insight aligns well with our findings from task-based **whole-head fNIRS, EEG and fMRI**, showing enhanced **functional integration** in attention-related and default mode brain networks post-TILS (Gonzalez-Lima, 2021) (Fig. 4).

## 5.4 Neuroplasticity and anti-inflammatory effects

The PFC is the most commonly targeted area in human TILS studies of cognitive enhancement due to its key role in working memory, sustained attention, and inhibitory control. TILS has been shown to enhance **response accuracy** and **reduce reaction time** in multiple cognitive tasks, as well as improve **learning curves** in PFC-based category learning tasks (Gamboa et al., 2025). These effects suggest increased **synaptic efficacy** and **short-term plasticity** in the PFC, consistent with our mitochondrial and oxygenation-based models of cognitive enhancement (Wang et al., 2024).

In collaboration with Dr. Fabrizio Dos Santos Cardoso and colleagues, we have shown evidence of **structural neuroplasticity** in the dendritic branching of cortical neurons of animals exposed to chronic tPBM daily for two months (Cardoso, Serra et al., 2022). Our animal studies also provide compelling evidence that chronic tPBM has an extensive **anti-inflammatory**

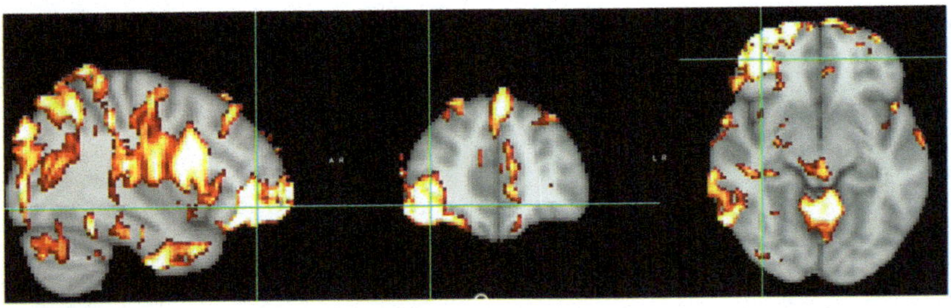

**Fig. 4** TILS of the right anterior prefrontal cortex (green cross) engages a lateral fronto-parieto-occipital network involved in cognitive functions called central executive network and the default mode network. MRI images of bipolar patient after 6 weekly TILS, courtesy of Dr. Jorge Almeida, UT dell medical school.

**Table 2** Summary of TILS mechanistic insights in controlled human studies.

| Mechanism | Evidence from human studies | Outcome |
|---|---|---|
| **Mitochondrial Boost** | CCO photo-activation shown via bbNIRS with behavioral improvements in energy-intensive cognitive tasks | ↑ Cognitive performance |
| **Cerebral Oxygenation** | fNIRS data: ↑ HbO during cognitive tasks | ↑ Prefrontal activation and efficiency |
| **Network Connectivity** | Improved functional integration in fNIRS, EEG and behavioral tests | ↑ Executive function, ↓ Impulsivity |
| **Neuroplasticity and Anti-Inflammatory** | Faster learning and decision-making in younger and older adults | ↑ Learning and memory retention |

action (Cardoso, Mansur et al., 2022) that we linked to specific metabolomic changes in the cerebral cortex (Cardoso et al., 2021). An insight from these and other animal studies is that chronic tPBM effects could be relevant for clinical populations experiencing neuroinflammation (Cardoso, Salehpour et al., 2022). While these preclinical studies underscore another therapeutic potential of tPBM, future chronic studies are needed in clinical populations to verify these potential mechanisms (Table 2).

## 6. Summary and conclusions

The cognitive-enhancing promise of PBM and TILS has emerged at a critical juncture. Aging populations are facing unprecedented increases in dementia and MCI rates. Post-COVID neurocognitive symptoms ("brain fog") have increased interest in safe cognitive therapies. Mental health challenges—from bipolar disorder to depression—are increasingly recognized as involving cognitive dysfunction that is not fully addressed by antidepressants or mood stabilizers. Furthermore, advances in neuroimaging (e.g., bbNIRS, fNIRS, EEG, and fMRI) have allowed us to observe in real time how tPBM alters cortical CCO and hemodynamics, network connectivity, and cognitive task performance in humans.

Our research has been pivotal in advancing understanding of tPBM as a non-invasive method to enhance cognitive functions in humans. Collaborative research by numerous trainees and colleagues cited in the references below has demonstrated that TILS can improve various aspects of cognitive performance, including executive function, attention, and emotional regulation. These non-invasive interventions utilized low-level light to stimulate mitochondrial activity, leading to improved cerebral metabolism and neuronal function. Below is a summary of key conclusions that highlight the cognitive benefits of TILS.

### 6.1 Cognitive and emotional benefits in healthy people

Our clinical trials demonstrated that TILS improves sustained attention, working memory, executive function, and emotional well-being in healthy individuals. Notably, participants exhibited faster reaction times and better performance in delayed match-to-sample tasks following TILS treatment. Additionally, self-reported measures of positive affect increased, suggesting mood enhancement as a secondary benefit.

## 6.2 Applications in neuropsychiatric disorders

In older adults with bipolar disorder, TILS led to significant improvements in cognitive flexibility and reduced impulsivity, highlighting its potential as a therapeutic modality for cognitive deficits associated with psychiatric conditions. These and other similar findings support the incorporation of TILS into treatment regimens for cognitive impairments in aging and neuropsychiatric populations.

## 6.3 Mechanistic insights

Our research has provided compelling evidence that TILS can enhance cognitive functions through mechanisms involving mitochondrial stimulation, improved cerebral oxygenation, and modulation of neuroplasticity. Both preclinical and clinical studies underscore the therapeutic potential of TILS in treating cognitive deficits associated with neuropsychiatric disorders. Future research should focus on optimizing treatment parameters, exploring long-term effects, and conducting large-scale clinical trials to establish standardized protocols for TILS application.

## 6.4 Safety and feasibility

TILS has been consistently reported as safe, with no significant adverse effects observed across many studies. Its non-invasive nature and ease of application make it a promising candidate for widespread clinical use.

Among the researchers who catalyzed the translation from animal to human applications of TILS, the Texas group has played a leading role because our work established:

- That **CCO is a primary photoacceptor in the human brain**.
- That **behavioral enhancements in attention, memory, and affect** are reproducible following TILS in controlled human studies.
- That TILS can be safely and effectively applied across populations—from healthy young adults to older patients with mood or cognitive disorders.

These studies have provided the **first double-blind, placebo-controlled human evidence** that TILS can not only enhance cognition acutely but may also contribute to neurocognitive recovery in clinical populations. Our ongoing research is contributing new insights into the mechanisms and applications of tPBM for cognitive enhancement in neurodevelopmental disorders such as ADHD (Salehpour et al., 2025) and autism (Diaz et al., 2025). While the initial results are encouraging, further research, particularly large-scale randomized controlled trials, is necessary to optimize treatment protocols and fully elucidate the therapeutic potential of these interventions.

## Acknowledgements

I thank the many colleagues who pioneered PBM research and my students and collaborators, without them the TILS studies summarized in this chapter could not have been done. The names of our coauthors are individually recognized in our listed references. I express my special gratitude to my Ph.D. students who contributed to all aspects of our research: Douglas W. Barrett, Julio C. Rojas, Celeste Saucedo, Courtney Alexander O'Donnell, Adam R. Cobb, Eric Zaizar, Li-Da Huang, Xinlong Wang, Jungyun Hwang, Fabrízio dos Santos Cardoso, Patrick O'Connor, Dariella Fernandez, Nicole Jackson, Laura A. Gamboa, Farzad Salehpour, Sarah D. Diaz, Hunter Dutkiewicz, and Jordan A. Schwartz. I thank my Texas faculty collaborators for their important research contributions: Hanli Liu, Fenghua Tian, Michael J. Telch, Andreana P. Haley, and Jorge R. C. Almeida. I also thank John Mitrofanis for his invitation to contribute a chapter to this important book. I hope that our join efforts will contribute to advance the field of PBM by helping in developing research and treatment strategies. I am grateful for the generous funding by federal and private grants from the National Institutes of Health, BRAIN Initiative, Oskar Fischer Project Fund, Elhapa Foundation, and Vanguard Charitable.

## References

Anders, J. J., Lanzafame, R. J., & Arany, P. R. (2015). Low-level light/laser therapy versus photobiomodulation therapy. *Photomedicine and Laser Surgery, 33*(4), 183–184. https://doi.org/10.1089/pho.2015.9848.

Auchter, A., Williams, J., Barksdale, B., Monfils, M. H., & Gonzalez-Lima, F. (2014). Therapeutic benefits of methylene blue on cognitive impairment during chronic cerebral hypoperfusion. *Journal of Alzheimer's Disease, 42*(s4), S525–S535. https://doi.org/10.3233/jad-141527.

Barrett, D. W., Beevers, C. G., & Gonzalez-Lima, F. (2025). Augmenting internet-based cognitive behavioral therapy for major depressive disorder with transcranial infrared laser stimulation. *Biological Psychiatry Global Open Science, 5*(2), 100449. https://doi.org/10.1016/j.bpsgos.2025.100449.

Barrett, D. W., Davis, R. E., Siegel-Ramsay, J. E., Bichlmeier, A., Almeida, J. R. C., & Gonzalez-Lima, F. (2025). Cognitive improvement and prefrontal network interactions in individuals with remitted bipolar disorder after transcranial infrared laser stimulation. *Frontiers in Psychiatry, 16*, 1547230. https://doi.org/10.3389/fpsyt.2025.1547230.

Barrett, D. W., & Gonzalez-Lima, F. (2013). Transcranial infrared laser stimulation produces beneficial cognitive and emotional effects in humans. *Neuroscience, 230*, 13–23. https://doi.org/10.1016/j.neuroscience.2012.11.016.

Berman, M. H., Halper, J. P., Nichols, T. W., H, J., Lundy, A., & Huang, J. H. (2017). Photobiomodulation with near infrared light helmet in a pilot, placebo controlled clinical trial in dementia patients testing memory and cognition. *Journal of Neurology and Neuroscience, 08*(01), https://doi.org/10.21767/2171-6625.1000176.

Bikson, M., Brunoni, A. R., Charvet, L. E., Clark, V. P., Cohen, L. G., Deng, Z.-D., ... Lisanby, S. H. (2018). Rigor and reproducibility in research with transcranial electrical stimulation: An NIMH-sponsored workshop. *Brain Stimulation, 11*(3), 465–480. https://doi.org/10.1016/j.brs.2017.12.008.

Blanco, N. J., Maddox, W. T., & Gonzalez-Lima, F. (2017). Improving executive function using transcranial infrared laser stimulation. *Journal of Neuropsychology, 11*(1), 14–25. https://doi.org/10.1111/jnp.12074.

Blanco, N. J., Saucedo, C. L., & Gonzalez-Lima, F. (2017). Transcranial infrared laser stimulation improves rule-based, but not information-integration, category learning in humans. *Neurobiology of Learning and Memory, 139*, 69–75. https://doi.org/10.1016/j.nlm.2016.12.016.

Cardoso, F. D. S., Salehpour, F., Coimbra, N. C., Gonzalez-Lima, F., & Gomes Da Silva, S. (2022). Photobiomodulation for the treatment of neuroinflammation: A systematic review of controlled laboratory animal studies. *Frontiers in Neuroscience, 16*, 1006031. https://doi.org/10.3389/fnins.2022.1006031.

Cardoso, F. D. S., Serra, F. T., Coimbra, N. C., Gonzalez-Lima, F., & Gomes Da Silva, S. (2022). Transcranial photobiomodulation changes neuronal morphology in the cerebral cortex of rats. *Neuroscience Letters, 781*, 136681. https://doi.org/10.1016/j.neulet.2022.136681.

Cardoso, F., Dos, S., Dos Santos, J. C. C., Gonzalez-Lima, F., Araújo, B. H. S., Lopes-Martins, R.Á. B., & Gomes Da Silva, S. (2021). Effects of chronic photobiomodulation with transcranial near-infrared laser on brain metabolomics of young and aged rats. *Molecular Neurobiology, 58*(5), 2256–2268. https://doi.org/10.1007/s12035-020-02247-z.

Cardoso, F., Dos, S., Mansur, F. C. B., Araújo, B. H. S., Gonzalez-Lima, F., & Gomes Da Silva, S. (2022). Photobiomodulation improves the inflammatory response and intracellular signaling proteins linked to vascular function and cell survival in the brain of aged rats. *Molecular Neurobiology, 59*(1), 420–428. https://doi.org/10.1007/s12035-021-02606-4.

Cassano, P., Petrie, S. R., Mischoulon, D., Cusin, C., Katnani, H., Yeung, A., ... Iosifescu, D. V. (2018). Transcranial photobiomodulation for the treatment of major depressive disorder. The ELATED-2 Pilot Trial. *Photomedicine and Laser Surgery, 36*(12), 634–646. https://doi.org/10.1089/pho.2018.4490.

Cobb, A. R., McTeague, L. M., O'Connor, P., Gonzalez-Lima, F., & Telch, M. J. (2025). Neuromodulation of heart rate and heart rate variability in a randomized controlled trial of tDCS-augmented in vivo exposure for specific fears. *Behaviour Research and Therapy, 186*, 104701. https://doi.org/10.1016/j.brat.2025.104701.

Cobb, A. R., O'Connor, P., Zaizar, E., Caulfield, K., Gonzalez-Lima, F., & Telch, M. J. (2021). Tdcs-Augmented in vivo exposure therapy for specific fears: A randomized clinical trial. *Journal of Anxiety Disorders, 78*, 102344. https://doi.org/10.1016/j.janxdis.2020.102344.

Diaz, S., Guimaraes, G., Moore, N. J., Shah, N., Davis, R., Barrett, D. W., & Gonzalez-Lima, F. (2025). Transcranial photobiomodulation on symptoms of autism spectrum disorder. [Conference presentation]. Texas Academy of Science 128th Annual Meeting, Baylor University, Waco, Texas.

Disner, S. G., Beevers, C. G., & Gonzalez-Lima, F. (2016). Transcranial laser stimulation as neuroenhancement for attention bias modification in adults with elevated depression symptoms. *Brain Stimulation, 9*(5), 780–787. https://doi.org/10.1016/j.brs.2016.05.009.

Frankowski, D. W., Ferrucci, L., Arany, P. R., Bowers, D., Eells, J. T., Gonzalez-Lima, F., ... Lakatta, E. G. (2025). Light buckets and laser beams: Mechanisms and applications of photobiomodulation (PBM) therapy. *Geroscience*. https://doi.org/10.1007/s11357-025-01505-z.

Gamboa, L., Davis, R., Guimarães, G., Thakkar, K., Mirji, A., Dutkiewicz, H., & Gonzalez-Lima, F. (2025). Transcranial photobiomodulation improves cognitive function in healthy and neuropsychiatric populations: A systematic review and meta-analysis. *Manuscript Submitted for Publication to Neuroscience and Biobehavioral Reviews*.

Gonzalez-Lima, F. (Ed.). (1998). *Cytochrome oxidase in neuronal metabolism and Alzheimer's disease* Plenum Press. https://doi.org/10.1007/978-1-4757-9936-1.

Gonzalez-Lima, F. (2021). Neuroprotection and neurocognitive augmentation by photobiomodulation. In I. Opris, M. A. Lebedev, & M. F. Casanova (Eds.). *Modern approaches to augmentation of brain function* (pp. 165–207). Springer International Publishing. https://doi.org/10.1007/978-3-030-54564-2_9.

Gonzalez-Lima, F., & Auchter, A. (2015). Protection against neurodegeneration with low-dose methylene blue and near-infrared light. *Frontiers in Cellular Neuroscience, 9*. https://doi.org/10.3389/fncel.2015.00179.

Gonzalez-Lima, F., Barksdale, B. R., & Rojas, J. C. (2014). Mitochondrial respiration as a target for neuroprotection and cognitive enhancement. *Biochemical Pharmacology, 88*(4), 584–593. https://doi.org/10.1016/j.bcp.2013.11.010.

Hayworth, C. R., Rojas, J. C., Padilla, E., Holmes, G. M., Sheridan, E. C., & Gonzalez-Lima, F. (2010). *In vivo* low-level light therapy increases cytochrome oxidase in skeletal muscle. *Photochemistry and Photobiology, 86*(3), 673–680. https://doi.org/10.1111/j.1751-1097.2010.00732.x.

Holmes, E., Barrett, D. W., Saucedo, C. L., O'Connor, P., Liu, H., & Gonzalez-Lima, F. (2019). Cognitive enhancement by transcranial photobiomodulation is associated with cerebrovascular oxygenation of the prefrontal cortex. *Frontiers in Neuroscience, 13*, 1129. https://doi.org/10.3389/fnins.2019.01129.

Huang, L.-D., Kao, T.-C., Sung, K.-B., & Abraham, J. A. (2021). Simulation study on the optimization of photon energy delivered to the prefrontal cortex in low-level-light therapy using red to near-infrared light. *IEEE Journal of Selected Topics in Quantum Electronics, 27*(4), 1–10. https://doi.org/10.1109/jstqe.2021.3051671.

Hwang, J., Castelli, D. M., & Gonzalez-Lima, F. (2016). Cognitive enhancement by transcranial laser stimulation and acute aerobic exercise. *Lasers in Medical Science, 31*(6), 1151–1160. https://doi.org/10.1007/s10103-016-1962-3.

Mitrofanis, J., & Henderson, L. (2020). How and why does photobiomodulation change brain activity? *Neural Regeneration Research, 15*(12), 2243. https://doi.org/10.4103/1673-5374.284989.

O'Donnell, C. M., Barrett, D. W., Fink, L. H., Garcia-Pittman, E. C., & Gonzalez-Lima, F. (2022). Transcranial infrared laser stimulation improves cognition in older bipolar patients: Proof of concept study. *Journal of Geriatric Psychiatry and Neurology, 35*(3), 321–332. https://doi.org/10.1177/0891988720988906.

O'Donnell, C. M., Barrett, D. W., O'Connor, P., & Gonzalez-Lima, F. (2023). Prefrontal photobiomodulation produces beneficial mitochondrial and oxygenation effects in older adults with bipolar disorder. *Frontiers in Neuroscience, 17*, 1268955. https://doi.org/10.3389/fnins.2023.1268955.

Repantis, D., Schlattmann, P., Laisney, O., & Heuser, I. (2010). Modafinil and methylphenidate for neuroenhancement in healthy individuals: A systematic review. *Pharmacological Research, 62*(3), 187–206. https://doi.org/10.1016/j.phrs.2010.04.002.

Rojas, J. C., Bruchey, A. K., & Gonzalez-Lima, F. (2012). Neurometabolic mechanisms for memory enhancement and neuroprotection of methylene blue. *Progress in Neurobiology, 96*(1), 32–45. https://doi.org/10.1016/j.pneurobio.2011.10.007.

Rojas, J. C., & Gonzalez-Lima, F. (2011). Low-level light therapy of the eye and brain. *Eye and Brain,* 49. https://doi.org/10.2147/eb.s21391.

Rojas, J. C., & Gonzalez-Lima, F. (2013). Neurological and psychological applications of transcranial lasers and leds. *Biochemical Pharmacology, 86*(4), 447–457. https://doi.org/10.1016/j.bcp.2013.06.012.

Rojas, J. C., & Gonzalez-Lima, F. (2017). Transcranial low-level laser light therapy for neurocognitive enhancement. In M. R. Hamblin, M. De Sousa, & T. Agrawal (Eds.). *Handbook of low-level laser therapy* (pp. 1057–1076). Pan Stanford Publishing.

Rojas, J. C., Lee, J., John, J. M., & Gonzalez-Lima, F. (2008). Neuroprotective effects of near-infrared light in an in vivo model of mitochondrial optic neuropathy. *The Journal of Neuroscience, 28*(50), 13511–13521. https://doi.org/10.1523/jneurosci.3457-08.2008.

Salehpour, F., Barrett, D. W., Farzamnia, A., Burugu, V., & Gonzalez-Lima, F. (2025). Transcranial infrared laser stimulation (TILS) improves impulse control in adults with attention-deficit/hyperactivity disorder (ADHD) [Conference Presentation]. Texas Academy of Science 128th Annual Meeting, Baylor University, Waco, Texas.

Saltmarche, A. E., Naeser, M. A., Ho, K. F., Hamblin, M. R., & Lim, L. (2017). Significant improvement in cognition in mild to moderately severe dementia cases treated with transcranial plus intranasal photobiomodulation: Case series report. *Photomedicine and Laser Surgery, 35*(8), 432–441. https://doi.org/10.1089/pho.2016.4227.

Saucedo, C. L., Courtois, E. C., Wade, Z. S., Kelley, M. N., Kheradbin, N., Barrett, D. W., & Gonzalez-Lima, F. (2021). Transcranial laser stimulation: Mitochondrial and cerebrovascular effects in younger and older healthy adults. *Brain Stimulation, 14*(2), 440–449. https://doi.org/10.1016/j.brs.2021.02.011.

Schiffer, F., Johnston, A. L., Ravichandran, C., Polcari, A., Teicher, M. H., Webb, R. H., & Hamblin, M. R. (2009). Psychological benefits 2 and 4 weeks after a single treatment with near infrared light to the forehead: A pilot study of 10 patients with major depression and anxiety. *Behavioral and Brain Functions, 5*(1), 46. https://doi.org/10.1186/1744-9081-5-46.

Shumake, J., Conejo-Jimenez, N., Gonzalez-Pardo, H., & Gonzalez-Lima, F. (2004). Brain differences in newborn rats predisposed to helpless and depressive behavior. *Brain Research, 1030*(2), 267–276. https://doi.org/10.1016/j.brainres.2004.10.015.

Shumake, J., Poremba, A., Edwards, E., & Gonzalez-Lima, F. (2000). Congenital helpless rats as a genetic model for cortex metabolism in depression. *Neuroreport, 11*(17), 3793–3798. https://doi.org/10.1097/00001756-200011270-00040.

Tian, F., Varghese, J., Tran, A. P., Fang, Q., & Gonzalez-Lima, F. (2020). Effects of wavelength on transcranial laser stimulation: A Monte Carlo simulation study based on standard brain model (conference presentation). In M. R. Hamblin, J. D. Carroll, & P. Arany (Eds.). *Mechanisms of photobiomodulation therapy XV* (pp. 10) SPIE. https://doi.org/10.1117/12.2545286.

Urquhart, E. L., Wanniarachchi, H., Wang, X., Gonzalez-Lima, F., Alexandrakis, G., & Liu, H. (2020). Transcranial photobiomodulation-induced changes in human brain functional connectivity and network metrics mapped by whole-head functional near-infrared spectroscopy in vivo. *Biomedical Optics Express, 11*(10), 5783. https://doi.org/10.1364/boe.402047.

Valla, J., Berndt, J. D., & Gonzalez-Lima, F. (2001). Energy hypometabolism in posterior cingulate cortex of alzheimer's patients: Superficial laminar cytochrome oxidase associated with disease duration. *The Journal of Neuroscience, 21*(13), 4923–4930. https://doi.org/10.1523/jneurosci.21-13-04923.2001.

Vargas, E., Barrett, D. W., Saucedo, C. L., Huang, L.-D., Abraham, J. A., Tanaka, H., ... Gonzalez-Lima, F. (2017). Beneficial neurocognitive effects of transcranial laser in older adults. *Lasers in Medical Science, 32*(5), 1153–1162. https://doi.org/10.1007/s10103-017-2221-y.

Vieira, W. F., Iosifescu, D. V., McEachern, K. M., Gersten, M., & Cassano, P. (2023). Photobiomodulation. *Psychiatric clinics of North America, 46*(2), 331–348. https://doi.org/10.1016/j.psc.2023.02.013.

Wang, X., Dmochowski, J. P., Zeng, L., Kallioniemi, E., Husain, M., Gonzalez-Lima, F., & Liu, H. (2019). Transcranial photobiomodulation with 1064-nm laser modulates brain electroencephalogram rhythms. *Neurophotonics, 6*(02), 1. https://doi.org/10.1117/1.nph.6.2.025013.

Wang, X., Gonzalez-Lima, F., & Liu, H. (2024). Transcranial infrared laser stimulation. In E. M. Wassermann, A. V. Peterchev, U. Ziemann, S. H. Lisanby, H. R. Siebner, & V. Walsh (Eds.). *The oxford handbook of transcranial stimulation* (pp. 276–298)(2nd ed.). Oxford University Press. https://doi.org/10.1093/oxfordhb/9780198832256.013.10.

Wang, X., Tian, F., Reddy, D. D., Nalawade, S. S., Barrett, D. W., Gonzalez-Lima, F., & Liu, H. (2017). Up-regulation of cerebral cytochrome-c-oxidase and hemodynamics by transcranial infrared laser stimulation: A broadband near-infrared spectroscopy study. *Journal of Cerebral Blood Flow & Metabolism, 37*(12), 3789–3802. https://doi.org/10.1177/0271678x17691783.

Zaizar, E. D., Papini, S., Gonzalez-Lima, F., & Telch, M. J. (2021). Singular and combined effects of transcranial infrared laser stimulation and exposure therapy on pathological fear: A randomized clinical trial. *Psychological Medicine, 53*(3), 908–917. https://doi.org/10.1017/s0033291721002270.

CHAPTER SIX

# Light on nonmotor symptoms in Parkinson's disease patients: The impact on quality of life

**Catherine Hamilton[a,*] and Frank Nicklason[b]**
[a]Well Red Pty Ltd, Launceston, Tasmania, Australia
[b]Geriatric Medicine, Royal Hobart Hospital, Hobart, Tasmania, Australia
*Corresponding author. e-mail address: catherine@wellred.com.au

## Contents



## Abstract

Non-motor symptoms of Parkinson's disease contribute substantially to a reduction in health-related quality of life for patients and carers. Whilst some non-motor symptoms are dopamine-responsive, current treatment options for many are limited and often unsatisfactory. We report on early observations of non-motor symptoms in seven Parkinson's disease patients treated with daily transcranial red and near infrared photobiomodulation, over periods ranging from six to thirty-four months. Six of seven patients used home-made devices, including bucket light hats with unpulsed 630 nm-850 nm LED strips, and/or a commercial helmet delivering 40Hz-pulsed sequential 670 nm and 810 nm light. One patient used a home-made 660nm intranasal device. Progress was assessed by the subjects themselves, their spouse and

*International Review of Neurobiology*, Volume 189
ISSN 0074-7742, https://doi.org/10.1016/bs.irn.2026.01.017

attending clinicians. A range of non-motor symptoms appeared to improve during photobiomodulation treatment, especially fatigue, sleep, apathy, mood, anxiety, concentration/attention, anhedonia and olfaction. There were also improvements in motor signs and in the "sense of self". Photobiomodulation was well-tolerated, with no adverse effects and high patient compliance. In summary, daily photobiomodulation helped elicited and maintain improvements in otherwise difficult to treat symptoms.

## 1. Introduction

"If a man has lost a leg or an eye, he knows he has lost a leg or an eye; but if he has lost a self—himself—he cannot know it, because he is no longer there to know it." (Sacks, 2014).

When James Parkinson first published a description of an unusual constellation of signs and symptoms in 1817, his definition was focussed on its 'involuntary tremulous motion with lessened muscular power, in parts not in action and even when supported; with a propensity to bend the trunk forwards, and to pass from a walking to a running pace' (Parkinson, 1817). He noted two symptoms unrelated to movement: constipation, requiring 'stimulating medicines of very considerable power' and, at the late stages of disease progression, sleep disturbance due to 'tremulous motion of the limbs' (Parkinson, 1817). He commented that "the senses and intellect" were "uninjured".

Some 200 years after this first description of what has become known as Parkinson's disease there has been a profound change in the understanding of the disease and particularly of the treatment, reflective mainly of the dramatic improvement in motor signs with the use of either dopamine replacement drug therapy and/or deep brain stimulation at high frequency (Tolosa et al.,1998). There is increasing recognition of the number, variety, onset and impact of nonmotor symptoms (Parkinson, 1817; Lees, 2017; Poewe, 2008).

In the section that follows, we will consider; (1) the nonmotor symptoms of Parkinson's disease; (2) animal models of nonmotor deficits in Parkinson's disease; (3) the treatment of nonmotor symptoms in the clinical setting, and; (4) observations of seven patients using transcranial photobiomodulation.

## 2. Nonmotor symptoms of Parkinson's disease

Applying the concept of 'quality of life' to Parkinson's disease began in 1998, with acceptance that, to date, it was likely that many of the symptoms and signs that caused the most distress in the life of a person with Parkinson's (PwP) were not always clinically recognised (Martínez-Martín, 1988).

The disease definition was refined to acknowledge and include prodromal symptoms, that precede the onset of the classical features of Parkinson's disease (Berg et al., 2014). These include sleep disturbance, excessive daytime sleepiness, mood changes, hyposmia, fatigue, and pain (Mahlknecht & Poewe, 2013; Pont-Sunyer et al., 2015; Postuma et al., 2015; Postuma & Berg, 2019). Tools have been developed to more systematically evaluate nonmotor symptoms (Berg et al., 2014; Chaudhuri et al., 2006; Titova & Chaudhuri, 2018). There is increasing recognition of the impact of nonmotor symptoms on the PwP's health-related quality of life and carer burden (Kadastik-Eerme et al., 2015; Tessitore et al., 2018; Kudlicka et al., 2014; Grün et al., 2016). The impact of living with Parkinson's disease is now acknowledged as a complex adjustment of motor and nonmotor factors with fatigue, mood swings, cognitive impairment, loss of identity, and dignity all having a major influence (Sjödahl Hammarlund et al., 2018). It is a lot more than just the shake and the shuffle.

The classification of nonmotor symptoms has varied over time. Following is a summary of the major nonmotor symptoms associated with Parkinson's disease: (Pfeiffer, 2016, Titova et al., 2017, FN):

**a)** Sleep disturbances: including insomnia, restless legs syndrome, REM sleep behaviour disorder, excessive daytime sleepiness, and obstructive sleep apnoea, fatigue.
**b)** Behavioural change: including depression, apathy, anxiety, irritability, panic attacks, visual hallucinations, phantosmia, anhedonia.
**c)** Cognitive: including inattention, slowed auditory and visual processing and thinking, impaired multi-tasking, dementia, impulse control.
**d)** Sensory abnormalities: including olfaction, ageusia, visual disturbances, pain, paraesthesia.
**e)** Autonomic dysfunction: including drooling, dysphagia, constipation, sexual and urinary dysfunction, orthostatic hypotension, seborrhoea, hyperhidrosis, thermoregulatory dysfunction.
**f)** Fatigue

While some of the nonmotor symptoms above make their presence known (eg constipation), many are subtle and patients can have no awareness of their presence (such as apathy and anhedonia).

## 3. Animal models of nonmotor symptoms in Parkinson's disease

Neurotoxins 1-methyl-4-phenyl-1,2,3,6-tetrahydropyridine (MPTP) and 6-hydroxydopamine (6-OHDA) are the most used toxins for modelling Parkinson's disease in animals and their effects are primarily motor, however some nonmotor effects have been noted. For instance, nonhuman primates with MPTP-induced bilateral parkinsonism have demonstrated frontostriatal cognitive deficits (Schneider & Kovelowski, 1990). Nocturnal sleep was lighter with reduced rapid eye movement and reduced slow wave sleep in nonhuman primates treated with MPTP (Almirall et al., 1999). In mice treated with MPTP, changes in sleep architecture have been demonstrated, with longer periods of wakefulness and paradoxical sleep episodes, together with an increase in paradoxical sleep (Monaca et al., 2004).

A recent review of 6OHDA-treated rodents has focussed on their nonmotor deficits (Prates-Rodrigues et al., 2025). Variability in particular experimental methods made it difficult to compare, but many of the studies quoted in the review reported nonmotor deficits including memory impairment, autonomic deficits, olfactory deficit, constipation as well as reduced faecal propulsion (Bonito-Oliva et al., 2014) . The conclusion was, though, that there needed to be standardisation of (6-OHDA) use before its value in studying nonmotor symptoms could be shown. Rotenone is a less-used neurotoxin used to establish chronic Parkinson's disease models in animals. In this model, some clear nonmotor deficits including lethargy, depressive behaviour and gastrointestinal dysfunction have been observed (He et al., 2024).

Transgenic animal models are able to reproduce nonmotor deficits that occur at an early stage, including hyposmia and gastrointestinal dysfunction (Blandini & Armentero, 2012). Such models, that generate an abnormal aggregation of α-synuclein protein within the neuronal cell bodies, have been shown to elicit abnormalities in cognition, anxiety-like behaviour and reduction in olfaction (Aniszewska et al., 2022; Gomez-Benito et al., 2020).

Other animal models that have shown some nonmotor effects are GBA-deficiency mouse (memory and cognition dysfunction), over-expression of LRRK2 gene mutation in transgenic mice (gastrointestinal and olfactory dysfunction), DJ-1 knockout (cognitive dysfunction), Parkin knockout (anxiety and cognitive dysfunction), VMAT-2 deficient mice (anxiety and depression-like behaviour, gastrointestinal, sleep and olfactory dysfunction) (Titova et al., 2017).

While animal models are showing potential for research into some nonmotor symptoms, they are unable to replicate the wide range of nonmotor symptoms thus far identified in PwP (Emborg, 2007). They can have their uses, though, and animal models are selected depending on the researcher's focus, be it on specific aspects of pre-clinical, prodromal, early or advanced stage disease (Taguchi et al., 2020).

## 4. Treatment of nonmotor symptoms in the clinical setting and the use of photobiomodulation

Treatment with dopaminergic drug therapy (L Dopa preparations, the dopamine agonists and monoamine oxidase B inhibitors are covered by this broader term), although very effective in treating the motor signs of the disease, at least initially, has a less favourable impact on many of the non-motor symptoms. Indeed, as it stands, conventional treatment options for nonmotor deficits are symptomatic and limited (Todorova et al., 2014; Sprenger & Poewe, 2013). In this context, therapeutic use of red and infrared lights ($\lambda$ = 600–1070 nm), known as photobiomodulation delivered transcranially has been shown to have clinically beneficial effects on patients with cognitive impairment (Saltmarche et al., 2017), neurological disorders (Hamblin, 2016), traumatic brain injury (Hamblin, 2018; Henderson et al., 2015; Naeser et al., 2011) and depression (Caldieraro & Cassano, 2019).

Photobiomodulation has been shown to improve cell function and cell survival in a range of animal models of Parkinson's disease from rodents to non-human primates (Hamblin, 2016; Mitrofanis, 2019, El Khoury et al., 2019; Moro et al., 2013, Darlot et al, 2016). While the mechanisms are covered in detail in other chapters of this book, it is useful to note that the primary mechanism for these positive effects is the absorption of light by a chromophore within a photoreceptor molecule, the main, but not the only photo-acceptor being cytochrome c oxidase residing in the mitochondrial membrane (Karu, 2010; Hamblin, 2018; Sommer, 2019). Light absorption stimulates mitochondrial activity and increases energy available to the cell through an increase in ATP production, fueling an up-regulation of cellular activities leading to neurogenesis and synaptogenesis as well as nitric-oxide stimulation of angiogenesis in surrounding tissues. There is also evidence on the key issue of neuroprotection with photobiomodulation in animal models; an effective neuroprotective treatment for the disease, one that stope the progression of the disease process, is not currently available for patients. While photobiomodulation is most effective when directly applied to the cell, it is also effective through the indirect or abscopal mechanism, presumably through the light activation of factors or cells which then gain access to the brain through the circulation (Johnstone et al. 2016; Mitrofanis, 2019).

There are several early investigations of the effectiveness of photobiomodulation on Parkinson's disease patients. Though these investigations have focused more on motor signs (Santos et al., 2019), improvements in a range of nonmotor symptoms were noted in some of the reports (Hamilton et al., 2018, 2019). Parkinson's disease patients using daily transcranial lights were assessed five years after starting transcranial lights. Compared to their initial assessment, there were motor improvements (walking speed, stride length, timed up-and-go test), and most had improvements in nonmotor symptoms cognition, sleep quality, sense of smell. Most participants had either an unchanged or an improved quality of life score, which is remarkable, given the progressive nature of the disease (Hamilton et al., 2018, 2019; Liebert et al., 2021, 2024). No adverse effects were reported in any of these clinical reports.

## 5. Materials and Methods

In this section, we will describe the impact of daily transcranial photobiomodulation on the nonmotor symptoms in seven patients previously diagnosed with Parkinson's disease. Our observations were quite striking, with all of the nonmotor symptoms experienced by these patients improving with daily photobiomodulation application. These observations, as with our previous papers (Hamilton et al, 2018, Hamilton et al., 2019a), came not from any systematic research study nor randomized clinical trial with placebo controls, but from a series of observations made by the patients, carers and, in particular, their attending medical practitioners.

### 5.1 Materials and methods

The use of photobiomodulation was voluntary and consent was obtained from each patient for inclusion of his/her case in this report. Disease progress and symptom change was assessed by the patient, the spouse or carer and/or the treating medical practitioner. CH regularly visited the homes of six of the seven patients, and spent time in conversation with all patients and family members face-to-face or by telephone. In four cases

(SN, MH, TU, FS), FN was the treating specialist. Two of the patients have been using photobiomodulation for 34 months, providing useful information about its longer-term potential.

Most of the home-made bucket light helmets were based on the "Eliza bucket" design shown in Fig. 1, published in redlightsonthebrain.blog (Hamilton, 2017). This has been described in an earlier publication (Hamilton et al., 2019a).

Patient SN initially used a modified Eliza bucket with 630 nm and 850 nm LED strips, while patient ML used a home-made intranasal light device shown in Fig. 2. The device was 7 cm long with a continuous 660 nm LED light.

The commercial device, shown in Fig. 3, was a Well Red Coronet Duo (Well Red Pty Ltd, 2020). This device had eight arms, each with two rows of individual LED lights, 670 nm and 810 nm, programmed to deliver twelve minutes of 40 Hz pulsed 670 nm followed immediately by twelve minutes of 40 Hz pulsed 810 nm. The details for the Bucket and Coronet devices are given in Table 1.

Patients were asked to ensure that the lower part of the back of the head received good light coverage. Some patients used two different devices over the course of photobiomodulation, starting with a bucket Eliza device then changing to a Coronet device. This is described in each case report.

**Fig. 1** Home-made "Eliza bucket" light helmet.

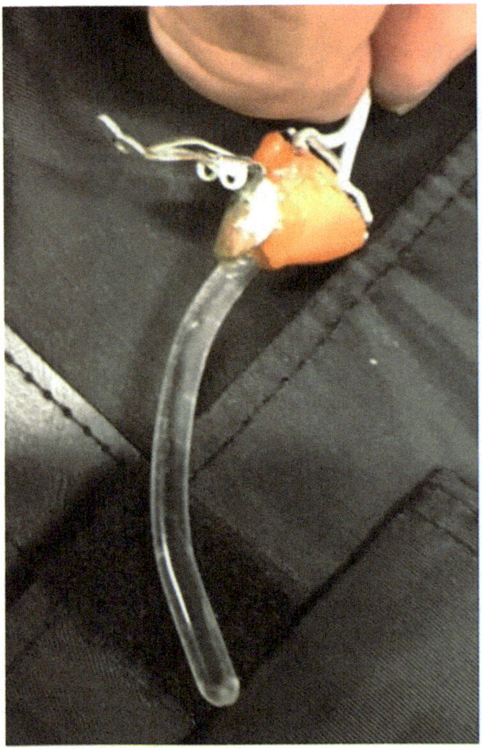

**Fig. 2** ML's home-made intranasal device.

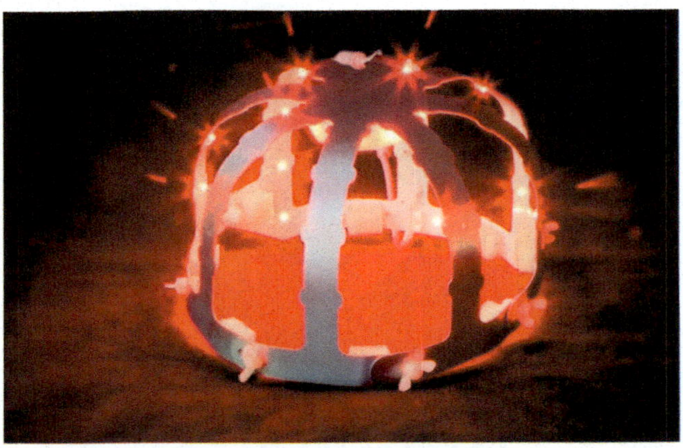

**Fig. 3** Two-wavelength Well Red Duo Coronet in 670nm part of the cycle.

**Table 1** Transcranial light device parameters.

| | **Bucket** | **Coronet** |
|---|---|---|
| Manufacturer | C&D Hamilton | Well Red Pty Ltd |
| Model identifier | Eliza | Coronet Duo |
| Year produced | 2016 | 2020 |
| No. & type of emitters | 670 nm and 810 nm (n = 150); 850 nm (n = 120). All LED[a] strips | LED[a] - 670 nm 40<br>LED[a] - 810 nm 40 |
| Wavelength & bandwidth | 670 nm, 810 nm, 850 nm. Bandwidth unknown. In sequence. | 670 nm +/− 10 nm<br>810 nm +/− 10 nm<br>In sequence. |
| Pulse mode | Continuous wave | 670 nm–40 Hz; 12 % duty cycle<br>810 m–40 Hz–20 % duty cycle |
| Irradiance (mW/cm$^2$) | Unknown | 670 nm–4.73<br>810 nm–6.1 |
| Exposure duration (sec) | 670 nm, 810 nm, 850 nm: 600–900 | 670 nm–720 per session<br>810 nm–720 per session |
| Radiant exposure (J/cm$^2$) | 670 nm = 6.96 W used,<br>810 nm = 26.4 W used<br>850 nm = 6 W used<br>Note: efficiency unknown | 670 nm–3.4 per session<br>810 nm = 4.4 per session |
| Radiant energy (J) | Unknown | 670 nm–1192 per session<br>810 nm–1538 per session |
| Area irradiated (cm$^2$) | 350 | 350 |
| Application technique | Transcranial | Transcranial |
| No. and frequency of treatment sessions | 1-2 Daily, ongoing | 1-3 Daily, ongoing |

[a]LED, light-emitting diode.

## 5.2 Results: Case descriptions

In each of the seven cases presented below, the nonmotor symptoms and motor signs that were present prior to photobiomodulation will be considered. This will be followed by a consideration of the effect of photobiomodulation on the nonmotor symptoms in particular. Patient comments are included in the descriptions.

a. Patient SN

SN was a 78-year old female,a highly articulate retired speech pathologist. A life-long textile artist, SN was expert in spinning, felting and free embroidery. She was well-known for her new designs and complex knitting. SN relished the creative process, using trial and error to conquer technical challenges. She was diagnosed with Parkinson's disease twelve months earlier at a specialist clinic.

On diagnosis she started using an Eliza bucket hat, then eight months later changed to a Coronet device.

**nonmotor:** SN dated her earliest symptoms some eight years prior to diagnosis. She experienced progressively less interest in creative design and increased difficulty following complex knitting patterns. SN's generalized anxiety became more overt over time and she was "always tired". Previously, SN would spend any spare moments during the day in the garden, but prior to starting photobiomodulation she found excuses to avoid it. SN described herself changing from a person who did things straight away to someone with "mañana syndrome"; things could always be done tomorrow.

**motor:** Four years prior to diagnosis, SN's left hand became noticeably affected, she misjudged distances, frequently knocking and dropping things. Craft skills diminished. Two years later, these changes had developed further and SN's left arm movement and control diminished. Fine finger movements were increasingly difficult, especially holding and manipulating a fork. She described difficulty coordinating movements to towel dry her back and put her left arm into a jacket. When typing, her fingers slipped off and/or 'refused' to release from the key. SN did not experience tremor. Her balance became problematic and she had two bad falls.

**after twelve months of photobiomodulation:** Once diagnosed, given SN's history of medication sensitivities, she opted to try photobiomodulation. SN commenced twice-daily 20–30 minute sessions of transcranial light first using a 630 nm/850 nm bucket device, then eight months later, the Coronet device. After starting photobiomodulation, SN noticed improvements in nonmotor symptoms and motor signs (Tables 2 and 3). An example of SN's overall improvement in nonmotor symptoms occurred about eight months after starting photobiomodulation, while she was on holiday on a remote island. As she walked along a coastal track, being buffeted by a strong wind, she experienced "exhilaration", describing it as, "the best fun I'd had in years". SN recognized that she had, during the insidious progression of Parkinson's disease, completely lost the "capacity for joy". SN had not previously noticed the disappearance of this capacity. She only discovered its long absence because "it was coming back". No adverse effects occurred from daily use of photobiomodulation.

**Table 2** Patient SN - nonmotor symptoms. (√ improvement; - no change and X deterioration).

| Nonmotor symptoms | 12 months after photobiomodulation |
|---|---|
| Fatigue | √ |
| Apathy | √ |
| Mood | √ |
| Anxiety | √ |
| Anhedonia | √ |
| Concentration/attention | √ |
| Creativity | √ |
| Sense of self | √ |
| Development of new nonmotor symptoms | None |

**Table 3** Patient SN - motor signs. (√ improvement, - no change and X deterioration).

| Motor signs | 12 months after photobiomodulation |
|---|---|
| Gait | √ |
| Arm swing | √ |
| Fine finger control | √ |
| Balance | √ |
| Development of new motor signs | None |

**summary:** This case is not unusual in the way that the nonmotor symptoms of Parkinson's disease were the first to appear, predating motor signs by some four years. This case is perhaps most remarkable in the clarity of SN's observations and the quality of her descriptions of her abilities and feelings both prior to diagnosis and following photobiomodulation use. The evidence in the reduced creative output during the prodromal years is striking. SN's experience of the return of the "capacity for joy" was a revelation to her, as was her realization that she had been unaware of its loss. SN developed no new motor signs nor nonmotor symptoms over the twelve months of photobiomodulation use. SN has never used dopaminergic medication and her improvements have been maintained.

b. Patient MH

MH, a 65-year old male diagnosed seven years earlier, has been previously described (Hamilton et al., 2019a, 2019b). Five years after diagnosis, MH started using an Eliza bucket device, initially ten minutes for each wavelength twice daily and then eight months later, increasing to fifteen minutes for each wavelength twice daily. Two years later, he changed to the Coronet device, using it twice daily. His daily medications included levodopa/carbidopa.

**nonmotor:** MH reported that apathy or what he called the "couldn't be bothereds" was the worst thing for him. MH was previously a man who was always busy and never still. Prior to using photobiomodulation, MH mainly sat, rarely initiating any activities. He and his wife had, for many years, been caravanning for 5 months of every year, driving some 20,000 kilometres each trip. After the Parkinson's disease diagnosis, MH's wife had taken on more of the driving, as he was finding it difficult to concentrate and follow directions. MH became less confident making phone calls, saying that he couldn't "think laterally" in the conversation. He routinely asked his wife to make phone calls on his behalf. As well as apathy, MH reported fatigue, difficulty maintaining thoughts, hyposmia, difficulty sleeping, and ongoing low mood and difficulty coping with life (Table 4). MH described difficulty putting his hand into a plastic bag, for example, to unpack bagged fruit, saying that his hand would "get stuck".

**motor:** Prior to starting photobiomodulation, MH experienced resting tremor, impaired fine manual skills, reduced facial animation, gait change with reduced stride-length, and hesitant and laboured speech (Table 5).

**after 34 months of photobiomodulation:** One month after starting photobiomodulation MH reported that he had resumed his usual activities

**Table 4** Patient MH - nonmotor symptoms. (√ improvement; - no change and X deterioration).

| **Nonmotor symptoms** | **34 months after photobiomodulation** |
|---|---|
| Fatigue | √ |
| Apathy | √ |
| Sleep | √[a] |
| Mood | √ |
| Anxiety | √ |
| Concentration/attention | √ |
| Olfaction | √ |
| Social confidence | √ |
| Development of new nonmotor symptoms | None |

[a]Improvement was sustained for 12 months, after which intermittent disturbed sleep recurred.

**Table 5** Patient MH - motor signs. (√ improvement, - no change and X deterioration).

| **Motor signs** | **34 months after photobiomodulation** |
|---|---|
| Tremor | √ |
| Gait | √ |
| Fine finger control | √ |
| Speech | √ |
| Facial animation | √ |
| Development of new motor signs | None |

and could think more clearly. The "couldn't be bothereds" had gone, and MH was back to being busy and active and he felt much more socially confident.

MH began to smell things again, especially what was cooking for dinner. Improvements in his sleep, speech and gait became evident, together with increased facial animation. MH also reported that he had much more energy and reduced tremor. His fine manual skills improved. MH was now able to tie a fly onto a fishing line, do up shirt buttons without assistance and manipulate plastic bags without difficulty (Tables 4 and 5). After having done little of the driving the previous year, MH resumed his former full driving duties, towing a new, heavier, longer and less manoeuverable caravan. MH told his specialist, FN, “It's like the disease has been stopped”. (Fig. 4).

One year after starting photobiomodulation, MH experienced intermittent disturbed sleep which has not deteriorated further. As well, he developed drug-related dystonia. Ten months earlier MH had orthopaedic surgery on his left foot. That and subsequent surgery were unsuccessful; he is now on daily opioids, unable to walk unaided, his activities severely curtailed, and the annual caravan trip had to be cancelled. This has made it difficult to assess aspects of his Parkinson's disease. MH's Parkinson's disease medication remained unchanged until thirty-three months after starting photobiomodulation, when a trial of amantadine started then ceased after a few weeks. No adverse effects occurred from daily use of photobiomodulation.

**summary:** Motor signs and nonmotor symptoms improved, though sleep disturbance re-appeared intermittently, but has not progressed. The orthopaedic problems have obstructed MH's progress, but the improvements in apathy, mood and anxiety have been maintained, despite the orthopaedic difficulties. Medication has remained unchanged.

**Fig. 4** MH wearing Eliza bucket hat when caravanning.

c. Patient SB

SB, a 78-year-old female, was diagnosed with Parkinson's disease. She had a past history of polio which left her with gait abnormalities. Six months after diagnosis, SB started photobiomodulation, using the Coronet twice-daily.

**nonmotor:** SB complained of fatigue and long-standing sleeping difficulties. She experienced no sense of sleepiness at night and could not sleep without medication (Table 6). SB had been taking 15mg zopiclone nightly for over two decades.

**motor:** Prior to starting photobiomodulation SB had right arm and chin tremor, gait disturbance, muscle rigidity and reduced fine motor function of the right hand, difficulties with balance, and Pisa syndrome in which she was unable to maintain an erect sitting and standing posture. For the previous 4 years SB always wore shoes when showering because clawing of her toes made it difficult for her to balance and walk barefooted. She had intermittent difficulty with articulation and little facial animation. Levodopa/carbidopa had improved stiffness and mobility, but her hand shook when she reached her right arm out, for example to shake hands. She needed a walking stick outside her home, even to walk a short distance to the garage (Table 7).

**after six months of photobiomodulation:** Two weeks after starting photobiomodulation SB walked more easily. She began twice-daily walks without using her walking stick, prompting unsolicited comments from others about improvement in her mobility.

There was a reduction in the clawing of her toes enabling her to walk in bare feet and take a shower without wearing shoes. The tremor also improved, for example she was able to carry a cup of tea in her right hand from her unit to a neighbour's unit without spilling the contents. The Pisa syndrome disappeared. SB's gait also improved (Table 7). Although she was not able to walk faster (because of persisting polio-related problems), SB was able to walk with a much freer rhythm. Soon after starting

**Table 6** Patient SB - nonmotor symptoms. (√ improvement; - no change and X deterioration).

| Nonmotor symptoms | 6 months after photobiomodulation |
|---|---|
| Fatigue | √ |
| Sleep | √ |
| Development of new nonmotor signs | Yes – urinary frequency |

**Table 7** Patient SB - motor signs. (√ improvement, - no change and X deterioration).

| Motor signs | 6 months after photobiomodulation |
|---|---|
| Tremor | √ |
| Gait | √ |
| Fine finger control | √ |
| Pisa syndrome | √ |
| Balance | √ |
| Speech | √ |
| Facial animation | √ |
| Rigidity | √ |
| Development of new motor signs | None |

photobiomodulation, she began feeling sleepy at night, a novel experience for her. She slowly reduced her sleeping medication, and settled on one quarter to one half of her pre-photobiomodulation dose. SB's weight dropped from 76kg to 65kg, which pleased her. SB felt that it reflected her greater physical activity. There has been no change in Parkinson's disease medication and she has maintained the reduction in zopliclone.

Three months after starting photobiomodulation, SB began experiencing urinary frequency which she described as inconvenient. It has not progressed further, nor has it improved (Table 6). No adverse effects occurred from daily use of photobiomodulation.

**summary:** Over the six months of photobiomodulation, all the improvements in nonmotor symptoms and motor signs have been maintained. One new nonmotor symptom, urinary frequency developed over this period but medication is unchanged. Increased energy levels, sleep quality, balance and being able to walk in bare feet are the improvements most valued by SB.

d. Patient PN

PN, a 65-year-old male diagnosed four years earlier, has been previously described (Hamilton et al., 2019b, 2019a) Nine months after diagnosis, PN started using an Eliza bucket device, ten minutes for each wavelength twice daily. Eighteen months later, PN changed to the Coronet, initially once daily and, after three months, twice daily.

**nonmotor:** Previously a chatty, social "ideas man", with a drive to "get things done", PN became increasingly quiet, slow and withdrawn, and less engaged in family and social activities (Table 8). He described his new state as feeling like he was inside "a shell". PN was always tired, he had difficulty sleeping, had lost facial animation, and was increasingly anxious. When upset, which occurred more often than it used to, his Parkinson's disease symptoms were exacerbated and slow to settle. A life-long home-renovator, he had stopped using tools partly because of fine motor control problems but more because he had no energy or interest. He avoided his workshop, which had previously been a major feature of his daily routine. PN lost the sense of "excitement" that he used to get from individual activity and engaging with others about things of mutual interest, "I didn't notice that I'd lost it, and no-one told me that it had gone." Because of these many changes, PN's wife had taken over domestic tasks that had always been his responsibility (Table 8).

**motor:** Prior to starting photobiomodulation, PN had a resting tremor, akinesia, impaired fine motor skills and facial animation, impaired gait with loss of arm swing, hesitant speech, intermittent drooping of the left side of his mouth, difficulty swallowing and persistent cough (Table 9).

**after 34 months of photobiomodulation:** PN noticed improvements in energy levels and motivation within a month of starting photobiomodulation. These improvements continued, along with reduction in anxiety, improved sleep, and a better gait pattern. Speech and facial animation improved and there was loss of both the facial droop and cough. An improvement in swallowing was reported. PN's fine motor skills improved and he resumed his home renovation activities. PN once again enjoyed social situations and was back to his "chatterbox" self. All previous domestic responsibilities have been resumed. PN has regained the sense of "excitement" (Tables 8 and 9), planning to purchase and renovate a house. PN's tremor initially improved but it increased thirty-two months after starting photobiomodulation, resulting in an increase in dopaminergic medication, the first increase in nearly four years. PN experienced no adverse effects from daily use of photobiomodulation.

**summary:** Over the 34 months of photobiomodulation, all improvements in nonmotor symptoms have been maintained, as have those in the motor signs, other than his tremor. PN is now aware of the degree of apathy he had been experiencing prior to his using photobiomodulation. He no longer feels that he is in "a shell" and is sanguine about the return of

**Table 8** Patient PN - nonmotor symptoms. (√ improvement; - no change and X deterioration).

| Nonmotor symptoms | 34 months after photobiomodulation |
|---|---|
| Fatigue | √ |
| Apathy | √ |
| Sleep | √ |
| Mood | √ |
| Anxiety | √ |
| Anhedonia | √ |
| Concentration/attention | √ |
| Decision-making | √ |
| Sense of self | √ |
| Social confidence | √ |
| Development of new nonmotor symptoms | None |

**Table 9** Patient PN - motor signs. (√ improvement; - no change and X deterioration).

| Motor signs | 34 months after photobiomodulation |
|---|---|
| Tremor | √[a] |
| Gait | √ |
| Fine finger control | √ |
| Speech | √ |
| Facial animation | √ |
| Mouth droop | √ |
| Swallowing | √ |
| Cough | √ |
| Development of new motor signs | None |

[a]Tremor initially improved, but by month 30 the improvement was lost.

the tremor. PN most values the improvements in energy and motivation, his renewed enjoyment in social activities and resuming serious renovating.

e. Patient TU

TU, a 74-year old male has been previously described (Hamilton et al., 2019a). An artist, he was diagnosed with Parkinson's disease at a specialist clinic two and a half years previously. Levodopa was not tolerated and was discontinued. Soon after diagnosis, TU started using an Eliza bucket device, ten minutes each wavelength once daily. One year after starting photobiomodulation, TU changed to the Coronet device, also used once daily.

**nonmotor:** Prior to starting photobiomodulation TU experienced Rapid Eye Movement (REM) sleep disruption with occasional dream enactment (Table 10).

**motor:** Resting tremor in the right hand and mild to moderate loss of facial animation (Table 11).

**after 24 months of photobiomodulation:** Improved sleep with cessation of dream enactment was reported by TU and his partner. Improvements were also noted in facial animation and tremor. There was no change in medication. No adverse effects occurred from daily use of photobiomodulation.

**Table 10** Patient TU - nonmotor symptoms. (√ improvement; - no change and X deterioration).

| Nonmotor symptoms | 24 months after photobiomodulation |
|---|---|
| Sleep | √ |
| Mood | √ |
| Development of new nonmotor symptoms | None |

**Table 11** Patient TU - motor signs. (√ improvement, - no change and X deterioration).

| Motor signs | 24 months after photobiomodulation |
|---|---|
| Tremor | √ |
| Facial animation | √ |
| Development of new motor signs | None |

**summary:** After two years of photobiomodulation, TU has maintained all improvements in nonmotor symptoms and motor signs. He has not developed new nonmotor symptoms nor motor signs and takes no medication for his Parkinson's disease. He feels that his Parkinson's disease is not progressing. He has not only maintained his pre-diagnosis level of activity but has taken on new tasks that are physically, intellectually and artistically challenging.

f. Patient ML

ML, a 76-year-old male and retired general surgeon, was diagnosed with Parkinson's disease fifteen years previously. ML's sister has advanced Parkinson's, resides in a nursing home and uses intra-jejunal levodopa gel system for better symptom control. ML's clinical history has been previously described (Hamilton et al., 2019a).

Medications taken prior to commencing photobiomodulation were levodopa/carbidopa 100/25 mg 1.5 tables 3rd hourly from 8am to 9pm, rasagiline 1 mg/day, pramipexole 0.75 mg, the antidepressant venlafaxine 37.5 mg/day and isradipine. An enthusiastic researcher of Parkinson's disease treatments, ML imported isradipine following suggestions of neuroprotective effects in animal studies. ML had taken isradipine for over ten years. Flecainide was used for paroxysmal atrial fibrillation.

Eighteen months earlier and despite healthy skepticism about photobiomodulation, ML, an amateur electronics buff, made himself an unpulsed 660 nm intranasal device, 7 cm long (Fig. 5). He could insert the device deep into one nostril. ML used his invention for about 20 min on most, but not all, days. ML was motivated to use this rather daunting looking device based on anatomical considerations as it provided a light source closer to the basal ganglia.

**nonmotor:** Prior to starting photobiomodulation ML experienced fatigue, anxiety, depression, impaired thinking, impaired memory, REM sleep disruption with dream enactment, restless legs, hyposmia as well as mild urinary frequency and constipation. His clinical consultations consisted of a somewhat disheartening catalogue of many clinical features of Parkinson's disease which were not responding well to his medication. Typically, ML described "feeling flat", functioning below par intellectually, having difficulty concentrating and lacking physical energy (Table 12).

**motor:** Tremor, cogwheel rigidity, impaired facial movements and impaired gait (Table 13).

**after 18 months of photobiomodulation:** ML felt remarkably better soon after starting photobiomodulation. On the first consultation, several

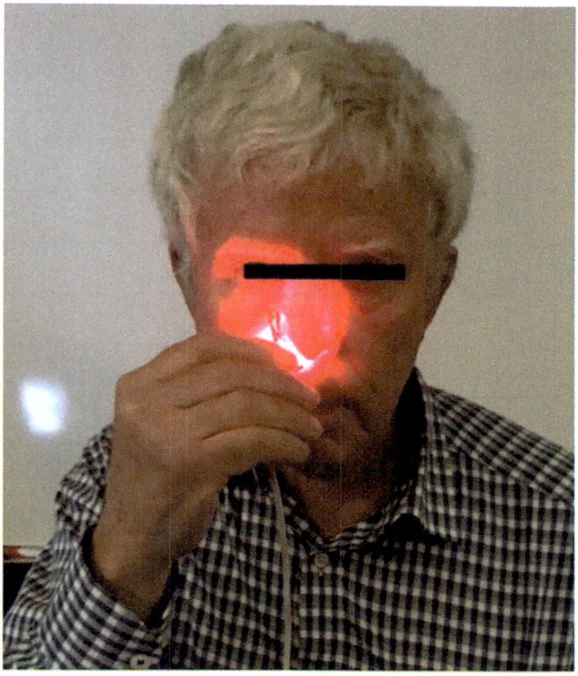

**Fig. 5** ML using his home-made intranasal device.

months after commencing photobiomodulation, ML stated: "I feel so much better, it's ridiculous how good I feel. My Parkinson's disease symptoms have virtually evaporated" (Tables 12 and 13). Attending doctors and ML's wife, who has a strong scientific background, noticed strikingly improved mood, facial animation and energy levels. Sleep quality improved, leg restlessness stopped and dream enactment reduced in frequency. Urinary frequency and constipation both improved. ML's motor signs also improved. Levodopa/carbidopa 100/25 mg has been reduced from 750 mg/day to 400 mg/day.

Symptom and sign improvements have been maintained for over 18 months since commencement of photobiomodulation. No adverse effects have been reported.

**summary:** ML's improvements in motor signs and non–motor symptoms resulted in extraordinary gains in the feeling of wellbeing and quality of life. ML's clinical consultations became cheerful occasions, ML describing himself feeling "on top of the world".

**Table 12** Patient ML - nonmotor symptoms. (√ improvement; - no change and X deterioration).

| **Nonmotor symptoms** | **18 months after photobiomodulation** |
|---|---|
| Fatigue | √ |
| Apathy | √ |
| Sleep | √ |
| Mood | √ |
| Anxiety | √ |
| Concentration/attention | √ |
| Hyposmia | √ |
| Memory | √ |
| Urinary frequency | √ |
| Constipation | √ |
| Restless legs | √ |
| Development of new nonmotor symptoms | None |

**Table 13** Patient ML - motor signs. (√ improvement, - no change and X deterioration).

| **Motor signs** | **18 months after photobiomodulation** |
|---|---|
| Tremor | √ |
| Gait | √ |
| Facial animation | √ |
| Cogwheel rigidity | √ |
| Development of new motor signs | None |

g. Patient FS

FS, a 67-year-old female, was diagnosed with Parkinson's disease at a specialist clinic eleven years previously. Ten years after diagnosis, FS commenced twice daily photobiomodulation using a Coronet device. After a few weeks, she decided to increase her "headlights" to three times daily. Her medication was Levodopa/carbidopa 100/25 5 times daily and Sinemet Slow Release.

**nonmotor:** As the disease progressed, FS had become constantly tired with REM sleep disruption and restless legs. Previously untroubled by anxiety, FS became more easily upset and had difficulty making decisions. Her husband had health problems which rendered him unable to walk, and having a substantial carer role added to FS's burden. FS was a keen singer and a stalwart member of a community choir. Both voice quality and breath control had deteriorated and FS feared being unable to continue singing. The choir director had noticed the reduction in voice quality as well as FS's loss of facial animation and reduced attention span. FS described a constant internal tremor as a continued and distressing buzzing sensation, "my whole body felt like I was a bee in a bottle trying to escape constantly" (Table 14).

**motor:** Prior to starting photobiomodulation FS experienced tremor, hypokinesia, loss of arm swing and rigidity. She had to take her four-times daily dopaminergic medication exactly on time, otherwise her symptoms

**Table 14** Patient FS - nonmotor symptoms. (√ improvement; - no change and X deterioration).

| Nonmotor symptoms | 10 months after photobiomodulation |
|---|---|
| Fatigue | √ |
| Apathy | √ |
| Sleep | √ |
| Mood | √ |
| Anxiety | √ |
| Concentration/attention | √ |
| Decision-making | √ |
| Sense of self | √ |
| Social confidence | √ |
| "Buzzing" | √ |
| Restless legs | √ |
| Development of new nonmotor symptoms | None |

would rapidly deteriorate. Mobility had declined to the extent that FS had difficulty crossing the road, as she could not be sure that she could get to the other side without freezing (Table 15).

**after ten months of photobiomodulation:** FS noticed improvements in sleep quality and fatigue levels within the first few weeks (Tables 14 and 15). She wore a sleep-tracking device which showed that the periods of deep sleep had increased in duration. Anxiety and indecisiveness became much less pronounced. Seven months after starting photobiomodulation FS made the decision to sell the family home and purchase a nearby home more suited to her and her husband's physical needs. She had been unable to face this and other major decisions prior to using photobiomodulation. FS noticed that the internal tremor, the buzzing sensation, had reduced and now rarely bothered her. FS's singing voice regained strength and control; this change was observed by the choir director and other choir members. Gait improved and FS was able to walk across the road without fear of freezing. FS had a small reduction in daily dopaminergic medication with levodopa/carbidopa 100/25 reduced from 5 to 4

**Table 15** Patient FS - motor signs. (√ improvement, - no change and X deterioration).

| Motor signs | 10 months after photobiomodulation |
|---|---|
| Tremor | √ |
| Gait | √ |
| Facial animation | √ |
| Speech | √ |
| Freezing | √ |
| Development of new motor signs | No |

**Table 16** Patient details.

| Patients | SN | MH | SB | PN | TU | ML | FS |
|---|---|---|---|---|---|---|---|
| Age (years) | 78 | 65 | 78 | 65 | 74 | 76 | 67 |
| Gender (M/F) | F | M | F | M | M | M | F |
| Years since diagnosis | 1 | 7 | 1 | 5 | 2 | 15 | 11 |

**Table 17** Number of months each photobiomodulation device was used by each patient.

| Months | SN | MH | SB | PN | TU | ML | FS |
|---|---|---|---|---|---|---|---|
| Eliza bucket | 8 | 24 | – | 24 | 14 | – | – |
| Coronet | 4 | 10 | 6 | 10 | 10 | – | 10 |
| Intranasal device | – | – | – | – | – | 18 | – |
| **TOTAL MONTHS** | **12** | **34** | **6** | **34** | **24** | **18** | **10** |

**Table 18** Nonmotor symptoms before and after photobiomodulation. (√ denotes improvement; - denotes no change; and X denotes deterioration).

| Nonmotor | SN | MH | SB | PN | TU | ML | FS |
|---|---|---|---|---|---|---|---|
| Fatigue | √ | √ | √ | √ | | √ | √ |
| Apathy | √ | √ | | √ | | √ | √ |
| Sleep | | √[a] | √ | √ | √ | √ | √ |
| Mood | √ | √ | | √ | | √ | √ |
| Anxiety | √ | √ | | √ | | √ | √ |
| Anhedonia | √ | | | √ | | | √ |
| Concentration/attention | √ | √ | | √ | | √ | √ |
| Decision-making | | | | √ | | | √ |
| Creativity | √ | | | | | | |
| Olfaction | | √ | | | | √ | |
| Sense of self | √ | | | √ | | | √ |
| Social confidence | | √ | | √ | | | √ |
| Memory | | | | | | √ | |
| Urinary frequency | | | | | | √ | |
| Constipation | | | | | | √ | |
| Restless legs | | | | | | √ | √ |
| "Buzzing" | | | | | | | √ |
| New symptoms | No | No | Yes[b] | No | No | No | No |

[a]Improvement was sustained for 12 months, after which intermittent disturbed sleep recurred.
[b]Urinary frequency.

**Table 19** Motor signs before and after photobiomodulation. (√ denotes improvement; - denotes no change; and X denotes deterioration).

| Motor signs | SN | MH | SB | PN | TU | ML | FS |
|---|---|---|---|---|---|---|---|
| Tremor | | √ | √ | √[a] | √ | √ | √ |
| Gait | √ | √ | √ | √ | | √ | √ |
| Arm swing | √ | | | | | | |
| Fine finger movements | √ | √ | | √ | | √ | |
| Pisa syndrome | | | | | | | |
| Speech | | √ | | √ | | | √ |
| Facial animation | | √ | | √ | √ | √ | √ |
| Cog-wheel rigidity | | | | | | √ | |
| Rigidity | | | √ | | | | |
| Balance | √ | | √ | | | | |
| Freezing | | | | | | | √ |
| Facial droop | | | | √ | | | |
| Swallowing | | | | √ | | | |
| Cough | | | | √ | | | |
| New Motor signs | None | None | None | None | None | None | None |

[a]Initial improvement maintained for 32 months.

**Table 20** Change of daily dopaminergic medication after photobiomodulation.

| Daily dopaminergic medication | SN | MH | SB | PN | TU | ML | FS |
|---|---|---|---|---|---|---|---|
| No change | √ | √ | √ | | √ | | |
| Increase | | | | √[a] | | | |
| Decrease | | | | | | √ | √ |

[a]32 months after starting photobiomodulation.

times daily. Significantly, FS can now be late with medication and not have adverse repercussions. No adverse effects occurred from daily use of photobiomodulation.

**summary:** FS described feeling "like my old self again." Over the ten months of photobiomodulation, all improvements in nonmotor symptoms and motor signs have been maintained, and no new symptoms and signs have developed. In FS's opinion, the nonmotor improvements have been the "best thing" about photobiomodulation, especially in her sense of self, energy, sleep, anxiety, decision-making and loss of the buzzing sensation. The other key gain was to be able to cross the road without worrying about freezing.

In summary, the seven patients comprised four males and three females with ages ranging from 65 to 78 years living with Parkinson's for 1 to 15 years (Table 16).

The use of transcranial photobiomodulation devices ranged from six to thirty-four months (Table 17).

Tables 18 and 19 summarize the pre- and post-photobiomodulation nonmotor symptoms and motor signs respectively for all seven patients. Dopaminergic medication was unchanged in three patients, reduced in two and increased in two at the end of the observation period (Table 20).

## 7. Discussion

The key set of clinical observations in this report was that daily use of photobiomodulation improved the PwP's reports of all the nonmotor symptoms - for example, fatigue, apathy, sleep disturbance, mood, anxiety, anosmia and cognition - in the seven PwP. These observations assume increasing significance when considering that current therapy available to treat nonmotor symptoms is often ineffective or associated with side-effects.

We found that photobiomodulation also helped improve some motor signs, but not as consistently as for nonmotor symptoms.

In the discussion that follows, focus will be on the photobiomodulation-induced improvements in each of the main nonmotor symptoms, together with the possible mechanisms that generate them.

## 7.1 photobiomodulation improved nonmotor symptoms in all patients

The main nonmotor symptoms of the disease that showed improvement after photobiomodulation in the seven patients were fatigue, apathy, sleep, mood and anxiety, olfaction and cognitive function. Each of these will be considered in turn below.

**fatigue:** Six of the seven patients described fatigue as a significant symptom prior to photobiomodulation. All six reported that fatigue improved, some quite dramatically after photobiomodulation use (Table 18). The onset of fatigue often predates motor symptoms in Parkinson's disease and is strongly predictive of a poor health-related quality of life (Pont-Sunyer et al., 2015; Postuma and Berg, 2019; Martinez-Martin et al., 2011). Although highly prevalent, the symptom of fatigue tends not to generate enough attention in clinical settings, and reliably effective treatments are lacking (Pfeiffer, 2016). Fatigue is essentially the "elephant in the Parkinson's disease clinic consulting room". The two subjects using photobiomodulation the longest (MH and PN) have not reported a return to previous levels of fatigue. The impact of increased energy levels was important to the patients and their carers, not the least being resumption of activities that had previously seemed impossible to undertake ever again. That photobiomodulation appears to improve an inevitable and untreatable symptom of Parkinson's disease, is of major clinical importance.

**apathy:** Apathy in Parkinson's disease was first described in 1895 by Edouard Brissaud as "a state of mixed indifference and lack of reaction to external as well as internal events" (Prange et al., 2018). In recent decades, apathy appears likely to be a distinct symptom - a lack of motivation resulting from reduced goal-directed behaviour, reduced goal-directed cognition and decreased emotional engagement (Marin, 1991). An electrophysiological study of emotional processing in non-demented patients with Parkinson's disease, indicated that apathy is separate to anhedonia or depression (Dietz et al., 2013). Like other nonmotor symptoms, the onset of apathy is subtle and can predate diagnosis by many years. In the mix of other more overt Parkinson's disease motor signs, apathy is often not recognized by the PwP, carer or clinician until later in the disease process, and once apathy has appeared, it stays. In a community prevalence study, 38% of participants with Parkinson's disease exhibited apathy, some in combination with either depression and dementia or both, and some with apathy alone (Pedersen et al., 2009). Another study found the incidence to be 40% (den Brok et al., 2015). The presence of apathy was found to be significantly associated with more severe motor signs, higher depression scores and reduced cognitive function. In studies of factors affecting quality of life in people with Parkinson's disease, apathy was identified as a major contributor (Kadastik-Eerme et al., 2015; Prakash et al., 2016; Santangelo et al., 2014). Apathy magnifies disability, handicap, and carer stress.

Anosognosia, the condition in which the person is unaware of having a disability or health problem, as a result of brain changes, is commonly seen in people with neurological disorders; apathy in Parkinson's disease is a prime example of anosognosia in action (Maggi et al., 2023). For those people who have little or no awareness of their apathetic state, there is also a diminution in executive function making it difficult for them to identify mismatches between feelings and behaviour, and between what is expected and what is actually achieved. (Maggi et al.,2023). Reduction in executive function is also relevant to other nonmotor symptoms, especially cognition, mood, anxiety and the sense of self.

In our report, five of the seven patients had developed apathy prior to commencing photobiomodulation (Table 18). In all five, apathy disappeared after daily use of photobiomodulation. While one patient had been acutely aware of his increased apathy levels prior to starting photobiomodulation, in fact regarding it as the worst symptom of all, the other patients had little appreciation of their apathy, and its impact on their lives. Indeed, it was only after using photobiomodulation that these patients recognized the importance of what they had lost. The two subjects using photobiomodulation the longest (MH and PN) have not reported a return to previous levels of apathy.

Given the prevalence of and lack of specific treatments for apathy in Parkinson's disease, plus the evidence suggesting that apathy may be a risk factor for the development of more severe motor signs, depression and dementia (Pedersen et al., 2009), the observed effect of photobiomodulation on apathy is also of major clinical interest.

**sleep:** Sleep disturbance is highly prevalent in Parkinson's disease. It is multi-factorial and can be difficult to treat. Sleep problems are predictive factors for poor health-related quality of life for Parkinson's disease patients and contribute substantially to carer-burden (Sprenger & Poewe, 2013). In this report, six of the seven patients reported disturbed sleep (Table 18). After photobiomodulation, all six had improvements in sleep, including reduction or cessation of episodes of rapid eye movement (REM) sleep dream enactment. One patient experienced an intermittent resumption of dream enactment but this has remained stable.

**mood and anxiety:** Low mood is very common in Parkinson's disease, with diagnoses ranging from dysthymia to major depressive disorder (Poewe, 2007, 2008). Antidepressants can be helpful in treating major depression in people with Parkinson's disease, but can have adverse effects (Sprenger & Poewe, 2013). Five of the patients had low mood and anxiety prior to using photobiomodulation. During photobiomodulation all five experienced improvements in mood and anxiety levels, two patients for the extended period of 34 months (Tables 4 and 8). All five patients reported an improvement in coping with the kinds of situations which previously had been anxiety triggers. Transcranial photobiomodulation can improve mood and anxiety in Parkinson's disease patients without adverse effects (Caldiero and Cassano, 2019).

**olfaction:** Partial or total loss of the sense of smell occurs in some 90% of people with Parkinson's disease (Pfeiffer, 2016). Not infrequently, patients are unaware of any changes in olfaction (another example of anosognosia). Hyposmia can predate motor signs by many years, and once established, hyposmia/anosmia does not spontaneously improve. There is no effective treatment available to restore smell sense. Of the seven patients, two reported hyposmia prior to commencing photobiomodulation (Table 18). Following photobiomodulation, both patients reported some improvement in olfaction, even though one was using the intranasal device, while the other used a transcranial device. On the one hand, self-reporting of olfaction is not regarded as reliable, as patients can both over-estimate and under-estimate their ability to smell (Postuma et al., 2012). On the other hand, there is increasing clinical feedback from Parkinson's disease patients spontaneously reporting improvements in olfaction after starting daily photobiomodulation (FN and CH; personal; clinical observations).

**cognitive function:** Concentration, attention, memory and decision-making were impaired in five of the seven Parkinson's disease patients. All five experienced improvements following photobiomodulation (Table 18), with a major positive impact on their lives.

Photobiomodulation has been shown to improve cognitive function in older adults as well as those with other neurological conditions (Saltmarche et al., 2017; Chan et al., 2018; Hamblin, 2016). Cognitive impairment is feared by Parkinson's disease patients and carers alike and increases the likelihood of transfer to nursing home care. From our observations, photobiomodulation appears to improve an otherwise untreatable and high-cost symptom of Parkinson's disease.

**Anhedonia and 'sense of self':** Four of the patients had improvements in anhedonia, the inability to find pleasure in normally pleasurable things (Table 18). None had been aware of the loss of this capacity for pleasure or joy until it had returned. To lose joy in life is a tragedy and to regain it is profound. Patient SN's description of regaining the "capacity for joy" after having not been aware that it had been lost is poignant and telling, as is patient FS's delight in feeling that she had returned to her old self again. Three of the seven patients reported a return of a "sense of self" described by one as "feeling like me again".

The sense of self is likely related to autobiographical memory, the ability to recall episodes experienced in the past, as well as metacognition, the ability to think about one's own thinking. (Simpson et al., 2023). Memory change occurs with ageing, but autobiographical memory for the things that happened to oneself, is more stable. With neurodegenerative changes, the ability to find and retain the memories that have been key to the definition of oneself is increasingly impaired. These changes occur without the knowledge of the individual, as the capacity to recognise the loss is impaired.

For these patients, reversing anhedonia and regaining a sense of self are valued and transformational changes. They are game-changers.

**Other nonmotor symptoms:** Three patients regained enjoyment in social activities after photobiomodulation. One subject reported improvement in the perception of constant "buzzing" or inner restlessness (Todorova et al., 2014). Also described as an "internal tremor" this symptom does not always make it onto the list of possible nonmotor symptoms in Parkinson's disease, despite being a distressing and distracting symptom. Other nonmotor symptoms observed to improve after photobiomodulation were urinary frequency, restless leg syndrome, balance and constipation.

**longevity of impact:** In this case report, for the two patients with the longest use of photobiomodulation, almost all the observed improvements in nonmotor symptoms were sustained over the period, up to 34 months (Tables 4 and 8). It was striking that the photobiomodulation-induced improvements in nonmotor symptoms can be maintained for such extended periods, provided continued daily use was continued. In a small clinical trial of Parkinson's disease patients using photobiomodulation, it was noted that cessation of photobiomodulation resulted in a loss of any improvements; on resumption of photobiomodulation, the improvements returned (Liebert, 2019). This has been reported by others using transcranial lights for Parkinson's disease (Hamilton, 2025).

**improvement in motor signs:** We also observed some improvements in motor signs in all seven patients, especially in gait, tremor, fine finger control, speech, facial animation, swallowing and cough. In one patient, the improvement in tremor reduced after 32 months of photobiomodulation (Table 9). These observations were very similar to those described in a recent placebo-controlled clinical trial that studied motor function responses to photobiomodulation (Santos et al., 2019).

**adverse effects:** There were no adverse effects observed in any of the seven patients. While there would have been some LED light exposure to the eyes, especially patient ML, there were no reports of adverse effects in vision. Retinal research suggests that photobiomodulation may be a potential therapy to regulate inflammation associated with retinal degeneration (Lu et al., 2018).

We should note that it is likely that our observations in the seven patients were due to their use of daily photobiomodulation and not to anything else, such as changes in medication (Table 20). Indeed, during their period of photobiomodulation therapy, five patients had no changes in medication, while two had a reduction. One patient (PN) had an increase in dopaminergic medication at the end of the observation period, but many of his improvements were evident before this increase, suggestive that the improvements were photobiomodulation- rather than drug-induced.

## 7.2 possible mechanisms for improvements in nonmotor symptoms

The mechanisms that generate the improvements in nonmotor symptoms described in this report are not clear, although it is tempting to speculate. Given the wide-ranging set of symptoms - for example, fatigue, apathy, sleep disturbance, mood, anxiety, anosmia, confidence and cognition, all associated with distinct functional areas of the brain - together with the fact that the transcranially-applied photobiomodulation can only penetrate 20–30 mm through body tissues, it is likely that there was an activation of the different areas of the cerebral cortex associated with these functions.

The cortex lies within 20–30 mm of the cranial surface, well within reaching distance of transcranially-applied photobiomodulation. Indeed, transcranial photobiomodulation has been shown to influence cortical activity substantially (Naeser et al., 2014; Naeser & Hamblin, 2015; Vargas et al., 2017; Chao, 2019). Many of the nonmotor symptoms, from apathy to sleep disturbance and from mood to cognition, may have been improved after activation of different regions of the prefrontal cortex. Olfaction may have been improved by stimulation of the posterior orbitofrontal cortex, which may have been reached by photobiomodulation by both intranasal and transcranial devices.

The activation of different cortical areas is not necessarily disease-modifying nor neuroprotective; it does not slow or stop the degeneration of the deep lying dopaminergic neurones in the midbrain substantia nigra pars compacta. These diseased neurones are not within the reach of transcranial photobiomodulation (Johnstone et al., 2016; Mitrofanis, 2019). It is possible, however that there may be a neuroprotective aspect to this treatment through the circulation; that photobiomodulation may activate free-floating mitochondria within blood plasma, immune and/or stem cells that then help the survival of the diseased midbrain dopaminergic neurones (Johnstone et al., 2016; Mitrofanis, 2019; Al Amir Dache et al., 2020). While this "indirect" or abscopal effect of photobiomodulation is known to act systemically in animal models of the disease, its significance and consistency in clinical cases is not known (Johnstone et al., 2016; Mitrofanis, 2019). However, the focus on Parkinson's disease as a dopaminergic motor syndrome, while understandable given the long and successful history of dopamine replacement treatment, has the adverse effect of limiting wider consideration of Parkinson's disease symptomatology. The observations of the seven patients in this report support the notion that Parkinson's signs and symptoms arise from wider neurological dysfunction: it is not all about dopamine and the basal ganglia.

## 7.3 The observers' comments

Although not part of the standard tabulation of results, the experience of the present authors is useful to describe. As mentioned earlier, our initial two reports focused on the motor signs (Hamilton et al., 2019b), which is unsurprising, reflective of the classical clinical and therapeutic focus of Parkinson's disease. In fact, there seemed little point in considering nonmotor symptoms, given their usual grim trajectory. As experienced clinicians, FN and CH had not previously observed consistent improvements in nonmotor symptoms such as fatigue, apathy, anhedonia, cognitive decline and olfaction in Parkinson's disease patients. Similarly, sleep disruption, mood and anxiety are not reliably nor easily remediable in Parkinson's disease patients. Whilst the severity of symptoms can be tempered with medication, this can be at the cost of troublesome and unwanted effects. Observations regarding improvement in nonmotor features of Parkinson's disease are based not solely on the reports of the PwP;frequently it is the spouse or other close consort that makes that positive statement.

It has, in fact, been a privilege to observe the way daily transcranial photobiomodulation can affect lives affected by Parkinson's disease so positively. In some ways it appeared as an awakening, the unmasking of symptoms – some previously identified and some completely unknown to patient and clinician – with the consequential recognition of the oppression these symptoms had been causing. As nonmotor symptoms lifted, so did patient confidence in their physical and mental abilities. Fear and uncertainty about the future plague people with Parkinson's disease, and most keep this to themselves and tend not to mention it in clinical settings (Trahair et al., 2025). Anticipatory fear – with no evidence of any reason to be anything but fearful about the future - results in a loss of confidence. As the disease progresses, confidence in all aspects of life continues to decline, and hope is lost. To see patients with Parkinson's disease regain confidence and hope for the future was remarkable. One of the authors (FN) remarks that the 'fieldwork' done by CH over several years getting to know PwP and their carers in extraordinary detail contrasts with the busy outpatient clinic settings where people are more often, more hurriedly, evaluated. "CH is more like a neurological anthropologist than the harried clinician- a careful, astute, and empathic observer with less time-constraint." This interaction has allowed conversations about how it feels to move from life with nonmotor symptoms to life without them. Such conversations have the potential to reveal much more about patient experience with nonmotor symptoms, and what the recognition and reduction of these symptoms can mean for PwP and carers. This is a potential new area of Parkinson's research..

In summary, these case reports indicate the potential of transcranial photobiomodulation as a treatment option for nonmotor symptoms of Parkinson's disease. Currently, clinical practice in Parkinson's disease is to recognize these nonmotor features and try to reduce the impact through largely supportive means in the absence of a reliable and effective therapy.

Patients and health practitioners frequently assume that the natural history of all these symptoms is an inexorable, inevitable and costly decline. That photobiomodulation appeared to improve these irreversible nonmotor symptoms without adverse effects and with a high degree of patient compliance is of great clinical importance. Finally, it should be noted that unlike the poorly maintained improvements seen with a classical placebo effect, most of the photobiomodulation-induced improvements we observed were durable and persisted with continued treatment during the observation period. Indeed, the presence of some nonmotor symptoms had been completely unknown to the patient.

## 8. Conclusions

Our observations strongly suggest transcranial photobiomodulation as a potential treatment option for the nonmotor symptoms in Parkinson's disease. Further, we suggest that such improvements, once apparent, continue with daily photobiomodulation for up to thirty-four months. Photobiomodulation is well tolerated with high patient compliance; no adverse effects were reported. That many PwP persist using photobiomodulation long-term is further evidence that it does useful things that are noticed and valued. These observations take on a greater significance when considering that nonmotor symptoms, such as fatigue, apathy, sleep disturbance, mood, anxiety, anosmia, confidence and cognition, are currently difficult to reliably treat, and that these symptoms are most responsible for poor quality of life for both patients and carers. These observations also open up a new research pathway into nonmotor symptoms.

Further investigation of this novel treatment option is required, given the increasing global incidence and prevalence of Parkinson's disease. A large-scale, sham-controlled, clinical trial of at least twelve months duration is required, to explore the validity of our observations.

## Acknowledgments

We thank David Hamilton, Ron Brown and Light Ahead Inc for assistance with the provision of the photobiomodulation devices.

## Conflicts of Interest

Catherine Hamilton is a co-founder of Well Red Pty Ltd and is the author of redlightsonthebrain.blog. The authors declare no other conflict of interest.

## References

Al Amir Dache, Z., Otandault, A., Tanos, R., et al. (2020). Blood contains circulating cell-free respiratory competent mitochondria. *The FASEB Journal, 34*(3), 3616–3630. https://doi.org/10.1096/fj.201901917RR.

Almirall, H., Pigarev, I., de la Calzada, M. D., Pigareva, M., Herrero, M. T., & Sagales, T. (1999). Nocturnal sleep structure and temperature slope in MPTP treated monkeys. *Journal of Neural Transmission, 106*(11), 1125–1134. https://doi.org/10.1007/s007020050228.

Aniszewska, A., Bergström, J., Ingelsson, M., & Ekmark-Lewén, S. (2022). Modeling Parkinson's disease-related symptoms in alpha-synuclein overexpressing mice. *Brain and Behavior, 12*(7), e2628. https://doi.org/10.1002/brb3.2628.

Berg, D., Postuma, R. B., Bloem, B., Chan, P., Dubois, B., Gasser, T., Goetz, C. G., et al. (2014). Time to redefine Parkinson's disease introductory statement of the MDS task force on the definition of Parkinson's disease. *Movement Disorders, 29*(4), 454–462. https://doi.org/10.1002/mds.25844.

Blandini, F., & Armentero, M.-T. (2012). Animal models of Parkinson's disease. *The FEBS Journal, 279*(7), 1156–1166. https://doi.org/10.1111/j.1742-4658.2012.08491.x.

Bonito-Oliva, A., Débora, M., & Gilberto, F. (2014). A mouse model of nonmotor symptoms in Parkinson's disease: Focus on pharmacological interventions targeting affective dysfunctions. *Frontiers in Behavioral Neuroscience, 8*.

Caldieraro, M. A., & Cassano, P. (2019). Transcranial and systemic photobiomodulation for major depressive disorder: A systematic review of efficacy, tolerability and biological mechanisms. *Journal of Affective Disorders, 243*, 262–273. https://doi.org/10.1016/j.jad.2018.09.048.

Chan, A. S., Lee, T. L., Yeung, M. K., & Hamblin, M. R. (2018). Photobiomodulation improves the frontal cognitive function of older adults. *International Journal of Geriatric Psychiatry*. https://doi.org/10.1002/gps.5039.

Chao, L. L. (2019). Effects of home photobiomodulation treatments on cognitive and behavioral function, cerebral perfusion, and resting-state functional connectivity in patients with dementia: A pilot trial. *Photobiomodulation, Photomedicine, and Laser Surgery, 37*(3), 133–141.

Chaudhuri, K. R., Healy, D. G., & Schapira, A. H. V. (2006). Nonmotor symptoms of Parkinson's disease: Diagnosis and management. *The Lancet Neurology, 5*(3), 235–245. https://doi.org/10.1016/S1474-4422(06)70373-8.

Darlot, F., Moro, C., El Massri, N., Chabrol, C., Johnstone, D. M., Reinhart, F., Agay, D., et al. (2016). Near-infrared light is neuroprotective in a monkey model of Parkinson disease. *Annals of Neurology, 79*(1), 59–75. https://doi.org/10.1002/ana.24542.

den Brok, M. G. H. E, van Dalen, J. W., van Gool, W. A., Moll van Charante, E. P., de Bie, R. M. A., & Richard, E. (2015). Apathy in Parkinson's disease: A systematic review and meta-analysis. *Movement Disorders, 30*(6), 759–769. https://doi.org/10.1002/mds.26208.

Dietz, J., Bradley, M. M., Jones, J., Okun, M. S., Perlstein, W. M., & Bowers, D. (2013). The late positive potential, emotion and apathy in Parkinson's disease. *Neuropsychologia, 51*(5), 960–966. https://doi.org/10.1016/j.neuropsychologia.2013.01.001.

El Khoury, H., Mitrofanis, J., & Henderson, L. A. (2019). Exploring the effects of near infrared light on resting and evoked brain activity in humans using magnetic resonance imaging. *Neuroscience* NSC 19340.

Emborg, M. E. (2007). Nonhuman primate models of Parkinson's Disease. *ILAR Journal, 48*(4), 339–355. https://doi.org/10.1093/ilar.48.4.339.

Gómez-Benito, M., Granado, N., García-Sanz, P., Michel, A., Dumoulin, M., & Moratalla, R. (2020). Modeling Parkinson's disease with the alpha-synuclein protein. *Frontiers in Pharmacology, 11*. https://doi.org/10.3389/fphar.2020.00356.

Grün, D., Pieri, V., Vaillant, M., & Diederich, N. J. (2016). Contributory factors to caregiver burden in Parkinson disease. *Journal of the American Medical Directors Association, 17*(7), 626–632. https://doi.org/10.1016/j.jamda.2016.03.004.

Hamblin, M. R. (2016). Shining light on the head: Photobiomodulation for brain disorders. *BBA Clinical, 6*, 113–124. https://doi.org/10.1016/j.bbacli.2016.09.002.

Hamblin, M. R. (2018). Photobiomodulation for traumatic brain injury and stroke. *Journal of Neuroscience Research, 96*(4), 731–743. https://doi.org/10.1002/jnr.24190.

Hamilton, C. (2017). "DIY Red Light Hat." Https://redlightsonthebrain.blog/. DIY Red Light Hat, Https://Redlightsonthebrain.Blog/ (blog), accessed 1 December 2019. http://www.wellred.com.au, accessed 20/6/2025.

Hamilton, C. L., El Khoury, H., Hamilton, D., Nicklason, F., & Mitrofanis, J. (2019a). The 'Buckets': Early observations on the use of red and infrared light helmets in Parkinson's disease patients. *Photobiomodulation, Photomedicine, and Laser Surgery*. https://doi.org/10.1089/photob.2019.4663.

Hamilton, C., Hamilton, D., Nicklason, F., & Mitrofanis, J. (2019b). Chapter 33 - Transcranial photobiomodulation therapy: Observations from four movement disorder patients. In R. Michael, Hamblin, & H. Ying-Ying (Eds.). *Photobiomodulation in the Brain* (pp. 463–472). Academic Press. https://doi.org/10.1016/B978-0-12-815305-5.00033-.

He, S., Ru, Q., Chen, L., Xu, G., & Wu, Y. (2024). Advances in animal models of Parkinson's disease. *Brain Research Bulletin, 215*, 111024. https://doi.org/10.1016/j.brainresbull.2024.111024.

Henderson, T. A., Morries, L., & Cassano, P. (2015). Treatments for traumatic brain injury with emphasis on transcranial near-infrared laser phototherapy. *Neuropsychiatric Disease and Treatment,* 2159. https://doi.org/10.2147/NDT.S65809.

Johnstone, D. M., Moro, C., Stone, J., Benabid, A.-L., & Mitrofanis, J. (2016). Turning on lights to stop neurodegeneration: The potential of near infrared light therapy in Alzheimer's and Parkinson's disease. *Front Neurosci, 9*, 500.

Kadastik-Eerme, L., Rosenthal, M., Paju, T., Muldmaa, M., & Taba, P. (2015). Health-related quality of life in Parkinson's disease: A cross-sectional study focusing on non-motor symptoms. *Health and Quality of Life Outcomes, 13*(1), https://doi.org/10.1186/s12955-015-0281-x.

Karu, T. I. (2010). Multiple roles of cytochrome c oxidase in mammalian cells under action of red and IR-A radiation. *IUBMB Life, 62*(8), 607–610. https://doi.org/10.1002/iub.359.

Kudlicka, A., Clare, L., & Hindle, J. V. (2014). Quality of life, health status and caregiver burden in Parkinson's disease: Relationship to executive functioning. *International Journal of Geriatric Psychiatry, 29*(1), 68–76. https://doi.org/10.1002/gps.3970.

Lees, A. (2017). An essay on the shaking palsy. *Brain, 140*(3), 843–848. https://doi.org/10.1093/brain/awx035.

Liebert, A., "South Australian Clinical Trial," 2019, personal communication, November 2019.

Liebert, A., Bicknell, B., Laakso, E.-L., Tilley, S., Heller, G., Kiat, H., & Herkes, G. (2024). Improvements in clinical signs and symptoms of Parkinson's disease using photobiomodulation: A five-year follow-up. *BMC Neurology, 24*(1), 381. https://doi.org/10.1186/s12883-024-03857-z.

Lu, Y.-Z., Fernando, N., Natoli, R., Madigan, M., & Valter, K. (2018). 670nm light treatment following retinal injury modulates Müller cell gliosis: Evidence from in vivo and in vitro stress models. *Experimental Eye Research, 169*, 1–12. https://doi.org/10.1016/j.exer.2018.01.011.

Maggi, G., Vitale, C., Delle Curti, A., Amboni, M., & Santangelo, G. (2023). Unawareness of apathy in Parkinson's disease: The role of executive dysfunction on symptom recognition. *Brain Sciences, 13*(6), 964. https://doi.org/10.3390/brainsci13060964.

Mahlknecht, P., & Poewe, W. (2013). Defining premotor Parkinson's disease: A window of opportunity for neuroprotection? *Neurodegenerative Disease Management; London, 3*(1), 61–70. http://dx.doi.org.ezproxy1.library.usyd.edu.au/10.2217/nmt.12.74.

Marin, R. S. (1991). Apathy: A neuropsychiatric syndrome. *The Journal of Neuropsychiatry and Clinical Neuroscience, 3*, 243–254.

Martinez-Martin, P., Rodriguez-Blazquez, C., Kurtis, M. M., Chaudhuri, K. R., & on Behalf of the NONMOTOR SYMPTOMSS Validation Group (2011). The impact of nonmotor symptoms on health-related quality of life of patients with Parkinson's disease: Nonmotor symptoms and HRQ O L in Parkinson's disease. *Movement Disorders, 26*(3), 399–406. https://doi.org/10.1002/mds.23462.

Martínez-Martín, P. (1998). An introduction to the concept of 'quality of life in Parkinson's disease. *Journal of Neurology, 245*(1), S2–6. https://doi.org/10.1007/PL00007733.

Mitrofanis, J. (2019). *Run in the light: Exploring exercise and photobiomodulation in Parkinson's disease [Internet]*. Morgan & Claypool Publishers, https://doi.org/10.1088/2053-2571/ab2f70.

Monaca, C., Laloux, C., Jacquesson, J.-M., Gelé, P., Maréchal, X., Bordet, R., ... Derambure, P. (2004). Vigilance states in a Parkinsonian model, the MPTP mouse. *European Journal of Neuroscience, 20*(9), 2474–2478. https://doi.org/10.1111/j.1460-9568.2004.03694.x.

Moro, C., Torres, N., El Massri, N., Ratel, D., Johnstone, D. M., Stone, J., ... Mitrofanis, J. (2013). Photobiomodulation preserves behaviour and midbrain dopaminergic cells from MPTP toxicity: Evidence from two mouse strains. *BMC Neuroscience, 14*, 40. https://doi.org/10.1186/1471-2202-14-40.

Naeser, M. A., & Hamblin, M. R. (2015). Traumatic brain injury: A major medical problem that could be treated using transcranial, red/near-infrared LED photobiomodulation. *Photomedicine and Laser Surgery, 33*(9), 443–446.

Naeser, M. A., Saltmarche, A., Krengel, M. H., Hamblin, M. R., & Knight, J. A. (2011). Improved cognitive function after transcranial, light-emitting diode treatments in chronic, traumatic brain injury: Two case reports. *Photomedicine and Laser Surgery, 29*(5), 351–358. https://doi.org/10.1089/pho.2010.2814.

Naeser, M. A., Zafonte, R., Krengel, M. H., Martin, P. I., Frazier, J., Hamblin, M. R., et al. (2014). Significant improvements in cognitive performance post-transcranial, red/near-infrared light-emitting diode treatments in chronic, mild traumatic brain injury: Open-protocol study. *Journal of Neurotrauma, 31*(11), 1008–1017.

Parkinson, J. (1817). An Essay on the Shaking Palsy, Project Gutenberg, 2007, https://www.gutenberg.org/files/23777/23777-h/23777-h.htm, accessed 20/02/2025.

Pedersen, K. F., Larsen, J. P., Alves, G., & Aarsland, D. (2009). Prevalence and clinical correlates of apathy in Parkinson's disease: A community-based study. *Parkinsonism & Related Disorders, 15*(4), 295–299. https://doi.org/10.1016/j.parkreldis.2008.07.006.

Pfeiffer, R. F. (2016). Nonmotor symptoms in Parkinson's disease. *Parkinsonism & Related Disorders, 22*, S119–S122. https://doi.org/10.1016/j.parkreldis.2015.09.004.

Poewe, W. (2007). Depression in Parkinson's disease. *Journal of Neurology, 254*(S5), 49–55. https://doi.org/10.1007/s00415-007-5008-4.

Poewe, W. (2008). Nonmotor symptoms in Parkinson's disease. *European Journal of Neurology, 15*(1), 14–20. https://doi.org/10.1111/j.1468-1331.2008.02056.x.

Pont-Sunyer, C., Hotter, A., Gaig, C., Seppi, K., Compta, Y., Katzenschlager, R., Mas, N., et al. (2015). The onset of nonmotor symptoms in Parkinson's disease (The ONSET PARKINSON'S DISEASE Study): The onset Parkinson's disease study. *Movement Disorders, 30*(2), 229–237. https://doi.org/10.1002/mds.26077.

Postuma, R. B., Aarsland, D., Barone, P., Burn, D. J., Hawkes, C. H., Oertel, W., & Ziemssen, T. (2012). Identifying prodromal Parkinson's disease: Pre-motor disorders in Parkinson's disease. *Movement Disorders, 27*(5), 617–626. https://doi.org/10.1002/mds.24996.

Postuma, R. B., & Berg, D. (2019). Prodromal Parkinson's disease: The decade past, the decade to come. *Movement Disorders, 34*(5), 665–675. https://doi.org/10.1002/mds.27670.

Postuma, R. B., Berg, D., Stern, M., Poewe, W., Olanow, C. W., Oertel, W., Obeso, J., et al. (2015). MDS clinical diagnostic criteria for Parkinson's disease. *Movement Disorders: Official Journal of the Movement Disorder Society, 30*(12), 1591–1601. https://doi.org/10.1002/mds.26424.

Prakash, K. M., Nadkarni, N. V., Lye, W.-K., Yong, M.-H., & Tan, E.-K. (2016). The impact of nonmotor symptoms on the quality of life of Parkinson's disease patients: A longitudinal study. *European Journal of Neurology, 23*(5), 854–860. https://doi.org/10.1111/ene.12950.

Prange, S., Pagonabarraga, J., Krack, P., Kulisevsky, J., Sgambato, V., Tremblay, L., ... Broussolle, E. (2018). Historical crossroads in the conceptual delineation of apathy in Parkinson's disease. *Brain, 141*(2), 613–619. https://doi.org/10.1093/brain/awx362.

Prates-Rodrigues, M., Schweizer, B. L. A., de Paula Gomes, C., Ribeiro, Â. M., Padovan-Neto, F. E., Masini, D., & Lopes-Aguiar, C. (2025). Challenges and opportunities in exploring nonmotor symptoms in 6-hydroxydopamine models of Parkinson's disease: A systematic review. *Journal of Neurochemistry, 169*(2), e70008. https://doi.org/10.1111/jnc.70008.

Sacks O. (2014). "The Man Who Mistook His Wife for a Hat: Picador Classic", p.42, Pan Macmillan, US.

Saltmarche, A. E., Naeser, M. A., Ho, K. F., Hamblin, M. R., & Lim, L. (2017). Significant improvement in cognition in mild to moderately severe dementia cases treated with transcranial plus intranasal photobiomodulation: Case series report. *Photomedicine and Laser Surgery, 35*(8), 432–441. https://doi.org/10.1089/pho.2016.4227.

Santangelo, G., Barone, P., Cuoco, S., Raimo, S., Pezzella, D., Picillo, M., Erro, R., et al. (2014). Apathy in untreated, de novo patients with Parkinson's disease: Validation study of apathy evaluation scale. *Journal of Neurology, 261*(12), 2319–2328. https://doi.org/10.1007/s00415-014-7498-1.

Santos, L., Olmo-Aguado, S. del, Valenzuela, P. L., Winge, K., Iglesias-Soler, E., Argüelles-Luis, J., ... Lucia, A. (2019). Photobiomodulation in Parkinson's disease: A randomized controlled trial. *Brain Stimulation, 12*(3), 810–812. https://doi.org/10.1016/j.brs.2019.02.009.

Schneider, J. S., & Kovelowski, C. J. (1990). Chronic exposure to low doses of MPTP. I. Cognitive deficits in motor asymptomatic monkeys. *Brain Research, 519*(1), 122–128. https://doi.org/10.1016/0006-8993(90)90069-N.

Simpson, S., Eskandaripour, M., & Levine, B. (2023). Effects of healthy and neuropathological aging on autobiographical memory: A meta-analysis of studies using the autobiographical interview. *The Journals of Gerontology: Series B, 78*(10), 1617–1624. https://doi.org/10.1093/geronb/gbad077.

Sjödahl Hammarlund, C., Westergren, A., Åström, I., Edberg, A.-K., & Hagell, P. (2018). The impact of living with Parkinson's disease: Balancing within a web of needs and demands. *Parkinson's Disease, 2018*, 1–8. https://doi.org/10.1155/2018/4598651.

Sommer, A. P. (2019). Mitochondrial cytochrome c oxidase is not the primary acceptor for near infrared light—it is mitochondrial bound water: The principles of low-level light therapy. *Annals of Translational Medicine,* 7(1), https://doi.org/10.21037/atm.2019.01.43.

Sprenger, F., & Poewe, W. (2013). Management of motor and nonmotor symptoms in Parkinson's disease. *CNS Drugs,* 27(4), 259–272. https://doi.org/10.1007/s40263-013-0053-2.

Taguchi, T., Ikuno, M., Yamakado, H., & Takahashi, R. (2020). Animal model for prodromal Parkinson's disease. *International Journal of Molecular Sciences, 21*(6), 1961. https://doi.org/10.3390/ijms21061961.

Tessitore, A., Marano, P., Modugno, N., Pontieri, F. E., Tambasco, N., Canesi, M., Latorre, A., et al. (2018). Caregiver burden and its related factors in advanced Parkinson's disease: Data from the PREDICT study. *Journal of Neurology, 265*(5), 1124–1137. https://doi.org/10.1007/s00415-018-8816-9.

Titova, N., & Chaudhuri, K. R. (2018). Non-motor Parkinson disease: New concepts and personalised management. *Medical Journal of Australia, 208*(9), 404–409. https://doi.org/10.5694/mja17.00993.

Titova, N., Qamar, M. A., & Chaudhuri, K. R. (2017). The nonmotor features of Parkinson's disease. *International Review of Neurobiology, 132*, 33–54. https://doi.org/10.1016/bs.irn.2017.02.016.

Todorova, A., Jenner, P., & Ray Chaudhuri, K. (2014). Nonmotor Parkinson's: Integral to motor Parkinson's, yet often neglected. *Practical Neurology, 14*(5), 310–322. https://doi.org/10.1136/practneurol-2013-000741.

Tolosa, E., Martí, M. J., Valldeoriola, F., & Molinuevo, J. L. (1998). History of levodopa and dopamine agonists in Parkinson's disease treatment. *Neurology, 50*(6 6), S2–10. https://doi.org/10.1212/wnl.50.6_suppl_6.s2. PMID: 9633679.

Trahair, E. D., Steiger, D., Lowe, N., Rapoport, R., Kelliher, C., Benvengo, S., ... Mantri, S. (2025). Fears and uncertainties of people with Parkinson's disease. *Journal of Parkinson's Disease, 15*(3), 595–602. https://doi.org/10.1177/1877718X251328565.

Vargas, E., Barrett, D. W., Saucedo, C. L., Huang, L.-D., Abraham, J. A., Tanaka, H., et al. (2017). Beneficial neurocognitive effects of transcranial laser in older adults. *Lasers in Medical Science, 32*(5), 1153–1162.

Well Red Pty Ltd, (2020). https://wellred.com.au/, accessed 12 March 2026.

## Further reading

Hamblin, M. R. (2018). Mechanisms and Mitochondrial redox signaling in photobiomodulation. *Photochemistry and Photobiology, 94*(2), 199–212. https://doi.org/10.1111/php.12864.

Johnstone, D. M., El Massri, N., Moro, C., Spana, S., Wang, X. S., Torres, N., Chabrol, C., et al. (2014). Indirect Application of near infrared light induces neuroprotection in a mouse model of Parkinsonism - an abscopal neuroprotective effect. *Neuroscience, 274*, 93–101. https://doi.org/10.1016/j.neuroscience.2014.05.023.

Santangelo, G., Trojano, L., Barone, P., Errico, D., Grossi, D., & Vitale, C. (2013). Apathy in Parkinson's disease: Diagnosis, neuropsychological correlates, pathophysiology and treatment. *Behavioural Neurology, 27*(4), 501–513.

CHAPTER SEVEN

# Exploring the effect of red and near-infrared light on neurodegenerative disease: A focus on amyotrophic lateral sclerosis, Charcot's devastating disease of the motor system

**Audrey Valverde[a], Hélène Blasco[b], Philippe Corcia[b], Olivier Herault[c], Pierre Magistretti[d], Laurent Herault[a], Corrado Cali[a,e,f], Luca Bartesaghi[e,f], Jonathan Stone[g], and John Mitrofanis[a,e,f,h,*]**

[a]Université Grenoble Alpes, Fonds Clinatec, Grenoble, France
[b]Inserm U930, Universite Francois-Rabelais, Laboratoire de Biochimie et Biologie Moleculaire, CHRU de Tours, Tours, France
[c]UMR Inserm U1069, Niche, Nutrition, Cancer et métabolisme Oxydatif (N2COx) UFR de Médecine, Tours, France
[d]King Abdullah University of Science and Technology, Thuwal, Saudi Arabia
[e]Department of Nueuroscience "Rita Levi Montalcini", University of Turin, Turin, Italy
[f]Neuroscience Institute Cavalieri Ottolenghi, Orbassano, Turin, Italy
[g]School of Medical Science, University of Sydney, Sydney, Australia
[h]Institute of Ophthalmology, University College London, London, United Kingdom
*Corresponding author. e-mail address: john.mitrofanis@me.com

## Contents



*International Review of Neurobiology,* Volume 189
ISSN 0074-7742, https://doi.org/10.1016/bs.irn.2026.01.013

## Abstract

All neurodegenerative diseases, from Alzheimer's disease to amyotrophic lateral sclerosis (ALS), are characterised by a relentless and progressive degeneration of neurones. The degenerating neurones suffer from mitochondrial dysfunction, glutamate excitotoxicity, metabolic disorder and atypical protein aggregations; there is also widespread neuroinflammation and damage to the neurovascular unit across the nervous system. Unfortunately, there is no current treatment option that addresses all, if not many, of these striking abnormalities, one that stops or even slows the progression of the disease (ie neuroprotective).

In this chapter, we explore the potential effectiveness of red and near infrared light (R-NIr) on ALS, one of the most devastating of all the neurodegenerative diseases. This condition impacts the motor system, from the cerebral cortex and brainstem to the spinal cord, as well as many skeletal muscles. Individuals suffer greatly and the survival period after onset of the first signs is often very short, averaging just over 2 years, as against 4–8 years in dementia. We outline two main reasons why R-NIr may have positive outcomes in ALS; (1) R-NIr has been shown to be neuroprotective in many other neurodegenerative diseases, improving cell function and survival, and; (2) unlike many other treatments attempted previously, R-NIr addresses many, if not all features of pathology associated with ALS.

In summary, we suggest that R-NIr, with its multi-modal effect, could be a valuable treatment option for patients with ALS, particularly if the treatment is started early, before the development of excessive cellular damage.

## Abbreviations

**ATP** adenosine triphosphate.
**ALS** amyotrophic lateral sclerosis.
**BBB** blood brain barrier.
**BDNF** brian derived neurotrophic factor.
**R-Nir** red and near infrared light.
**ROS** reactive oxygen species.

## 1. Introduction

The neurodegenerative diseases encompass a range of neurological conditions, from Alzheimer's to Parkinson's disease and from Huntington's disease to amyotrophic lateral sclerosis (ALS). Their common ground is a relentless and progressive degeneration of neurones across specific areas of the central nervous system. The degeneration is often slow in development, manifesting in distinct clinical signs and/or symptoms in older individuals. Most cases are idiopathic, with no readily identifiable cause, with only a small number being familial, with a clear family history and number of genetic mutations being identified. In terms of molecular and cellular changes, there are a number of similarities, including; dysfunction to small organelles called mitochondria (with traffic disruption, mtDNA dysregulation and mitophagy or lysosome with autophagy initiation complex); development of atypical misfolding protein assemblies (misfoldings of different types of protein within the neurones that appear to propagate across the brain); widespread neuroinflammation (with activation of the resident glial cells, astrocytes and microglia), and; much vascular dysfunction (an often overlooked cause with damage to endothelial cell integrity and the blood brain barrier (BBB), leading to the development of extensive micro-bleeds within the neural tissue) (Corcia et al., 2023, 2025; de Calbiac et al., 2024; Hong et al., 2025; Johnstone et al., 2016; Nogueira-Machado et al., 2025; Stone et al., 2024; Wilson et al., 2023).

Although these different conditions have similarities, and hence fall under the umbrella term of neurodegenerative diseases, they do have some key differences. There are distinct differences in particular, in their topographical origin in the central nervous system and in their initial signs and symptoms; for example, while Parkinson's disease pathology begins among neurones in the midbrain and impacts the motor system, Alzheimer's disease pathology begins in the cortex (entorhinal, frontal and temporal cortex) then hippocampus and impacts cognitive systems, initially memory. Further, the particular type of cellular pathology evident in the neurones can be very different with, for example, different proteins developing abnormal assemblies in each case, $\alpha$-synuclein in Parkinson's disease, $\beta$-amyloid in Alzheimer's disease and TDP-43 in ALS (Busche & Hyman, 2020; Calabresi et al., 2023; Corcia et al., 2023, 2025; Jo et al., 2020; Johnstone et al., 2016; Wilson et al., 2023).

Unfortunately, for any of these neurodegenerative conditions, there is no current treatment option that stops or even slows the disease process;

at best, there are some new strategies with pharmaceutical agents, nanoparticle or gene therapy and surgical treatments available, that all generate symptomatic improvements, but nothing that is considered disease-modifying or neuroprotective (Buccellato et al., 2023; Corcia et al., 2023, 2025; Hariz & Blomstedt, 2022; Johnstone et al., 2016; Lamptey et al., 2022; Sudhakar & Richardson, 2019; Wilson et al., 2023). One of the key problems with previous attempts to develop an effective treatment for each of the neurodegenerative disease is that most have been largely uni-modal, targeting only a single feature of the pathology in many of the cases. What could be more successful is the development of a treatment that targets more than one feature of the pathology, a multi-modal approach.

In recent times, the use of red and near-infrared light (R-NIr), known also as photobiomodulation, has been shown to generate a range of beneficial cellular outcomes, including many if not all those associated with the neurodegenerative diseases, from improving mitochondrial activity to breaking down abnormal protein aggregations and from reducing neuroinflammation and vascular dysfunction (Shen et al., 2024). Indeed, many preclinical studies have reported that R-NIr treatment improves the behaviour and patterns of cell function and survival in two of the main neurodegenerative diseases, namely Alzheimer's and Parkinson's disease. In essence, R-NIr treatment, with its multi-modal effect, addressing a number of features of pathology, could be viewed as a potentially valuable treatment option for the neurodegenerative diseases (Hamblin, 2016; Mitrofanis, 2019).

In this chapter, we explore the potential of using R-NIr treatment for one of the most devastating – at least in terms of disease severity and progression – of all the neurodegenerative diseases, ALS. What makes this condition, known also as Charcot's disease, Charcot Marie Tooth disease Lou Gehrig's disease or motoneurone disease, so devastating is that after diagnosis, that in itself is often difficult and delayed, prognosis is poor and survival times are short (Corcia et al., 2023, 2025; Grad et al., 2017; Hardiman et al., 2017; Kiernan et al., 2011; Saberi et al., 2015; Younger & Brown, 2023).

In the section that follows, we explore the main features of ALS, both clinical and cellular mechanisms, followed by the current treatment options available to patients. We then discuss the potential of R-NIr treatment for this condition.

# 3. The condition

## 3.1 General

One of the fathers of modern neurology, French neurologist Jean-Martin Charcot, is credited with one of the first series of reports of ALS in the mid-1800s and, in fact, he coined the name "amyotrophic lateral sclerosis". The "a-myo-trophic" part is from the Greek and means "no muscle nourishment", the "lateral" part represents a main region where many cells and fibres in the spinal cord degenerate and the "sclerosis" part indicates the scarring or hardening of this spinal cord region (Grad et al., 2017; Hulisz, 2018).

Charcot's initial reports described the disease as a rather straight-forward, simple condition, stating that "the diagnosis as well as the anatomy and physiology of the condition ALS is one of the most completely understood conditions in the realm of clinical neurology." Since that time however, ALS has been recognised as anything but that, being a most complex and heterogenous disorder. There is much variability, for example, in clinical signs, gene expression, disease progression and severity. This challenging situation has prompted a number of authors to consider ALS more of a syndrome, rather than a single disorder; it has also, no doubt, contributed to the slowness in developing a viable neuroprotective treatment option (Corcia et al., 2023, 2025; Grad et al., 2017; Hardiman et al., 2017; Hulisz, 2018; Kiernan et al., 2011; Rizea et al., 2024; Saberi et al., 2015; Younger & Brown, 2023).

## 3.2 Clinical features

ALS is characterised by the degeneration of both upper and lower motoneurones (Fig. 1). Upper motoneurones stay entirely within the central nervous system, projecting from the cerebral cortex to the brainstem or spinal cord. The major signs after lesion include clumsiness and slowing of movement, exaggerated tendon reflexes and muscle spasticity. Lower motoneurones on the other hand, project from the central brainstem or spinal cord, out to the periphery, namely to the skeletal muscles. Major signs after lesion to these neurones include paralysis, diminished tendon reflexes, muscle fasciculations, flaccidity and muscular atrophy.

ALS is relentlessly progressive. Approximately 50 % of patients die within 30 months of occurrence of first motor weakness and only about 20 % survive anywhere between 5 years and 10 years thereafter. Ultimately, in later stages of the disease, most patients need some assistance with day-to-day activities. Dysphagia is a common problem, resulting in

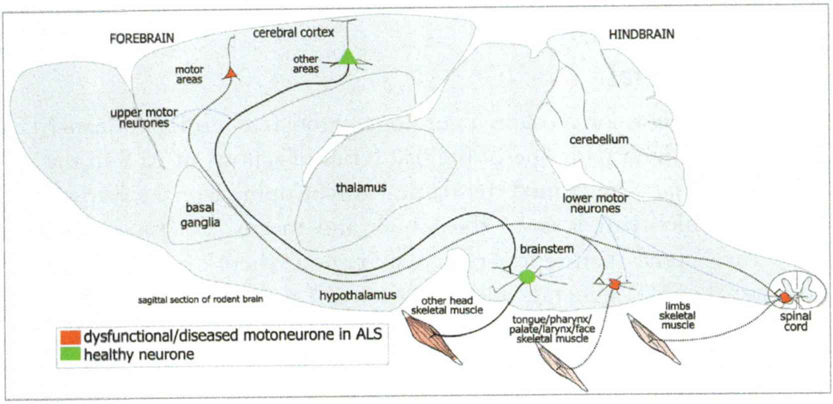


**Fig. 1** Schematic diagram of sagittal section through the brain indicating the major regions of diseased/dysfunctional neurones in ALS (red cells with dashed axons). The main zones of degeneration are regions associated with motor function; most other regions of the brain remain unaffected (green cells with solid axons).

malnutrition and weight loss. Once the respiratory muscles become affected, dyspnoea (difficulty breathing) develops and becomes progressively worse over time, resulting in respiratory failure and consequent death (Kiernan et al., 2011).

The initial presentation of ALS does vary between patients, with some presenting either with;.

1. classical ALS, split into bulbar and spinal ALS according to the site of onset of the disease. Bulbar ALS represents about 25–30 % of cases, the remaining correspond to spinal ALS. At clinical examination, there are combinations of upper and lower motoneurone signs in two to four of the following anatomical regions (bulbar, cervical, thoracic, lumbar), prompting the terms probable ALS or definite ALS according to the revised El Escorial criteria.
2. primary lateral sclerosis, with upper motoneurone signs only after 4 years of evolution, and;
3. progressive muscle atrophy, with lower motoneurone signs only.

The clinical hallmark of ALS is the presence of both upper and lower motoneurone signs involving brainstem and several spinal cord sites. The spread of pathology between upper and lower motoneurones is generally rather even, but there is often a skew towards lower motoneurones (Corcia et al., 2023, 2025; Grad et al., 2017; Hardiman et al., 2017; Hulisz, 2018; Kiernan et al., 2011; Rizea et al., 2024; Saberi et al., 2015; Younger & Brown, 2023).

It should be noted that in more recent times, ALS has been linked to non-motor regions of the central nervous system, in particular the frontotemporal areas of the cerebral cortex. The recent identification of TDP-43-positive ubiquitinated cytoplasmic inclusions (see below) in almost all cases of ALS, and more than half of the patients with frontotemporal dementia, has prompted the idea that the two neurodegenerative disorders have overlap and are linked closely. Indeed, some patients with frontotemporal dementia have been reported to develop ALS, and it has been argued that the more subtle cognitive deficits may have been overlooked in many ALS patients, particularly when they have such striking motor signs (Hardiman et al., 2017; Kiernan et al., 2011).

## 3.3 The types and risks

There are two main forms of ALS: (1) sporadic, that make up most of the cases (90–95 %). These have some links to several genetic mutations (see below), but have no family history. A number of studies have associated a series of potential environmental risk factors for the development of ALS including, for example; jobs that deal with manual labor, military service, football, pesticides, smoking, heavy metals and electric shock, and; (2) familial, that make up a much smaller number of cases (5–10 %; Corcia et al., 2023, 2025; Grad et al., 2017; Hardiman et al., 2017; Hulisz, 2018; Kiernan et al., 2011; Rizea et al., 2024; Saberi et al., 2015; Smeyers et al., 2021; Younger & Brown, 2023).

To date, nearly 17 genes have been identified as being involved in familial ALS (Goutman et al., 2022), and key cellular functions such as protein homeostasis, RNA/DNA metabolism and repair, endosomal and vesicle transport, axon structure and function and energy production mitochondria function and transport (through damage to mitochondria; see Table 1). Of these genetic mutations, perhaps the best known are SOD1 (Monteiro Neto et al., 2025), that encodes for copper/zinc ion-binding and removal of toxic elements, C9orf72 plays an important role in autophagy, lysosomal homeostasis and mitochondrial complex I assembly (Tang et al., 2020) immune regulation and TARDBP (Balendra et al., 2025), that encodes for regulating to key intracellular signalling and transport and maintaining homeostasis (eg TDP-43 protein). All of these genetic mutations have developed into transgenic mouse models of the disease to identify the implications of these mutations in signalling pathways in the context of ALS (Corcia et al., 2023, 2025; Grad et al., 2017; Hardiman et al., 2017; Hulisz, 2018; Kiernan et al., 2011; Rizea et al., 2024; Saberi et al., 2015; Smeyers et al., 2021; Younger & Brown, 2023).

**Table 1** The major genetic mutations associated with ALS.

| Gene | Role | Pathology | Mutations % | References |
|---|---|---|---|---|
| ***SOD1*** | Encodes Cu/Zn superoxide dismutase; to maintain neuronal redox homeostasis; regulates aerobic glycolysis and PDH activity | Encodes Cu metabolic dysregulation (lipid, glycolisys), modification senescence phenotype and affected the glutamate clearance | 20 % of familial forms & 1–2 % of sporadic cases | Das and Svendsen (2015), Dutta et al. (2018), Monteiro Neto et al. (2025) |
| ***C9ORF72*** | Mitochondrial function, autophagy and vesicular traffic, mTORC1 regulation | Expansion of GGGGCC hexanucleotide repeat; causes RNA toxicity, autophagy and vesicular traffic disruption, mitochondria dysfunction, inflammation | ~40 % of familial & ~8 % sporadic cases | Kotni et al. (2016), Marogianni et al. (2019), Tang et al. (2020) |
| ***TARDBP*** | Encodes TDP-43 protein; RNA metabolism | Nuclear-to-cytoplasmic mislocalization, aberrant protein expression and aggregation, posttranslational modifications, and impaired clearance of aggregates. | 3–5 % of familial & <1 % sporadic cases | Shang and Huang (2016), Balendra et al. (2025) |
| ***FUS*** | Encodes FUS/TLS protein; Roles in gene expression regulation, RNA metabolism, and DNA repair | RNA & DNA dysfunction (splicing, miRNA genesis, silencing), increased cortical neurone activity, motoneurone hypoexcitability, reduced axonal or neuritic organelle transport, changes in synaptic morphology, glial release of inflammatory cytokines, endoplasmic reticulum stress, and mitochondrial dysfunction | 4 % of familial & <1 % sporadic cases | Sproviero et al. (2012), Moens et al. (2025) |

| | | | | |
|---|---|---|---|---|
| ***UBQLN2*** | Roles in cellular proteostasis (proteasome, autophagy systems), stress responses, ERAD… | Sequestration and proteasome misassembly, impairing clearance, acidification block, fusion impairment, mitochondrial deficits, endoplasmic reticulum stress. | Not defined | Renaud et al. (2019), Benson et al. (2021), Lin et al. (2022) |
| ***ATXN2*** | Roles in cellular processes, including mRNA maturation and transcription, cytoskeleton reorganization, calcium-mediated signaling… | Not described but length polyQ repeats of 27–33 in ATAXIN-2 (ATXN2) increase risk for ALS and associated at Parkinson's disease | Not defined | Magaña et al. (2013), van Damme et al. (2011), Rajabi et al. (2024) |
| ***NEK1*** | Kinase activity, cell cycle, neuronal function and genome maintenance | Delayed DNA repair, CHK1/2 phosphorylation failure, genomic instability, cell cycle alteration | 1.2 % familial, 1 % of sporadic cases | Kenna et al. (2016), Jiang et al. (2023) |
| ***KIF5A*** | Microtubule-based motor protein; axonal transport: granules, proteins and mitochondria transports | Neurofilament, VAPB, FUS abnormal transport, impaired transport of mitochondria | Rare variant | Hirokawa et al. (2010), Karle et al. (2012), Nicolas et al. (2018) |
| ***PFN1*** | Encodes profilin 1 protein; | Actin dynamics alteration and axon outgrowth inhibition | Rare variant | Witke (2004), Wu et al. (2012), Lindamood et al. (2025) |

The overall population-based lifetime risk of ALS has been estimated at about 1:350 for men and 1:400 for women, with peak age at onset being the 7th decade for sporadic disease and around 50 for familial disease (Corcia et al., 2023, 2025; Grad et al., 2017; Hardiman et al., 2017; Hulisz, 2018; Kiernan et al., 2011; Rizea et al., 2024; Saberi et al., 2015; Smeyers et al., 2021; Younger & Brown, 2023).

## 3.4 Disease origin and cellular features in the nervous system

Although there have been a few theories, it is still not clear where or why precisely the disease starts. There is the "dying-forward" hypothesis that proposes that the cerebral cortex is the starting point. The descending cortical upper motoneurones, that are mainly excitatory and use the neurotransmitter glutamate, become abnormally active in ALS. This then leads to an excessive amount of glutamate being released at the synapse with the spinal cord lower motoneurones, causing much toxicity and dysfunction, leading subsequently to their death (see below; Baker, 2014; Eisen et al., 2023). Conversely, the "dying-back" hypothesis proposes that ALS begins within the skeletal muscle or at the neuromuscular junction. There are then diminished levels of neurotrophic factors that leads to a degeneration of the synaptically-linked neurones. Finally, there is the idea that each of these hypotheses, the dying-forward and dying-back, can occur independently, resulting in both upper and lower motoneurone degeneration (Dadon-Nachum et al., 2011; Loeffler et al., 2016). It should be noted that the pattern of disease spreading across each region of the nervous system can be very complex. The disease progresses from one area to another in no predictable pattern. For example, lower motoneurone degeneration across the spinal cord may be contiguous between levels or non-contiguous, skipping a level or two (Corcia et al., 2025; Kiernan et al., 2011).

In general, most ALS patients show little sign of gross abnormalities in the brain. Histologically, there are clear signs of degeneration however, from the motor cortex, across the descending axonal pathways to spinal cord and brainstem and those crossing to the other side of the cortex, among the hypoglossal nerves innervating many muscles of the tongue, as well as the ventral roots of the spinal cord innervating muscles of the body. There is a considerable loss of the lower motoneurones of the spinal cord (~50 %; α cells and others) and lower cranial motor nuclei of the brainstem, together with a variable amount of upper motoneurones in the cortex (the large pyramidal cells of Betz). For the past decade or so, ubiquitinated inclusions have also been useful in confirming an ALS diagnosis (Leigh et al., 1991). They appeared as either dense, rounded or irregular ubiquitin-immunoreactive cytoplasmic inclusions (known as "dense bodies"), or as loosely arranged bundles of filamentous material ("skein-like" inclusions). There is also evidence of a reduction in size of all these neurones and contain small round- to oval-shaped eosinophilic intracellular inclusions called Bunina bodies found mainly in the cytoplasm, that are composed of cystatin C and transferrin (Okamoto et al., 2008). Many studies have also reported a widespread neuroinflammation (Parisi et al., 2013), particularly across the grey and white matter of the spinal cord (see below). There are many microscopic holes, particularly in the grey matter of the spinal cord, resulting in a sponge-like (spongiosis) appearance of the tissue. Finally, and more recently, neurones have been shown to contain dense, round, ubiquitinated protein inclusions in the cytoplasm, such as SOD1 and TDP-43 (see below; Grad et al., 2017; Hardiman et al., 2017; Kiernan et al., 2011; Saberi et al., 2015).

## 4. The cellular mechanisms of disease

The precise mechanisms that underpin the pathology of ALS are not clear. Over the years, evidence has been accumulated for a number of key factors, however. Given the heterogeneity of the disease (see above), these mechanisms may not be mutually exclusive and some may contribute more than others in the different cases. Some of the major factors are considered below (Fig. 2).

### 4.1 Protein inclusions

A pathological hallmark of ALS is the aggregation and accumulation of ubiquitinated protein inclusions in the motoneurones (Fig. 2). Such aggregations and accumulations are considered to result from a breakdown of pathways, for example the ubiquitin-proteasome system and/or autophagy, that clear degraded proteins from cells. Unless these are cleared, cell homeostasis is compromised severely. SOD1, and in particular TDP-43 (together with their fragments and phosphoTDP-43), are examples of proteins that make up protein inclusions in the cytoplasm (see above) and their build-up in cells can lead to much cell dysfunction and damage (Hardiman et al., 2017; Kiernan et al., 2011; Rizea et al., 2024; Tsekrekou et al., 2024).

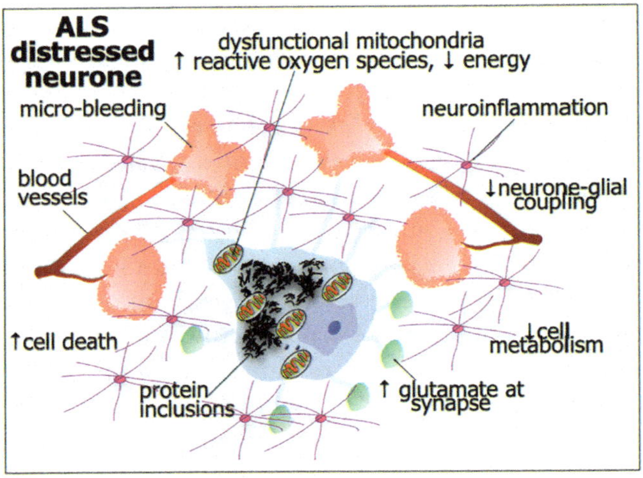


**Fig. 2** Schematic diagram of the major abnormal cellular functions associated with diseased/dysfunctional motoneurones in ALS. There is also much surrounding inflammation and many sites of micro-bleeding.

It is worth noting that abnormal protein inclusions occur also in other neurodegenerative disorders, for example β-amyloid plaques and tau protein in Alzheimer's Disease and α-synuclein in Parkinson's Disease. It is not clear however, what precisely triggers the breakdown of protein clearance and the formation of the inclusions (Debnath et al., 2016; Hardiman et al., 2017; Kiernan et al., 2011; Rizea et al., 2024; Yerbury & Cashman, 2020).

## 4.2 Glutamate excitotoxicity

Over the years, this mechanism has been one of the most discussed. Glutamate is the main excitatory neurotransmitter used by motoneurones in the cortex, and their over-activation leads to excessive levels of glutamate at the synapse with motoneurones in the brainstem and spinal cord, leading to their dysfunction and subsequent death (Fig. 2). This is known as glutamate-induced excitotoxicity and can lead to several disruptive elements that damage many intracellular structures and pathways. These may include: an abnormal flow of calcium ions into the neurones that can activate multiple enzyme cascades; the generation of free radicals, and; the up-regulation of pro-inflammatory mediators (Hardiman et al., 2017; Kiernan et al., 2011; Madji Hounoum et al., 2017; Tolochko et al., 2025).

The precise mechanisms remain unclear, however. It should be noted that motoneurones are particularly sensitive to glutamate toxicity induced by an increase in intracellular calcium, as they have a lower calcium buffering capacity than other types of neurones and some of their receptors (α-amino-3-hydroxy-5-methyl-4-isoxazolepropionic acid) are more calcium permeable. Further, a glutamate re-uptake transporter molecule (excitatory amino acid transporter 2) found in astrocytes is impaired in ALS and this is likely to an increase in glutamate levels and associated toxicity at the synapse (Hardiman et al., 2017; Kiernan et al., 2011; Madji Hounoum et al., 2017).

## 4.3 Neuroinflammation

In ALS, as with the other neurodegenerative diseases, neuroinflammation is a key factor for promoting and intensifying the progression of pathology. The astrocytes become reactive and the microglia are activated releasing a variety of pro-inflammatory cytokines and reactive oxygen species (ROS) and there is T lymphocyte infiltration (Fig. 2). These inflammation patterns have been shown to occur at the pre-symptomatic stage, even before the loss of the motoneurones (Hooten et al., 2015; Hardiman et al., 2017; Liu & Wang, 2017; Provenzano et al., 2023).

The dying motoneurones and astrocytes release mis-folded proteins (eg SOD1) and these are thought to activate microglia by purinergic receptors. Once activated, microglia have a phenotype (M2) that is neuroprotective, aiding in the repair and regeneration of the damaged motoneurones. As the disease progresses, the motoneurones release so-called "danger signals" that trigger a change in microglia phenotype (M1); this change prompts the release of ROS and pro-inflammatory cytokines that are detrimental to the survival of the neurones, exacerbating the process of degeneration (Beers et al., 2011; Liao et al., 2012). In addition, certain mutations such as C9orf72, trigger an alteration on macrophagic and microglial functions, in both peripheral organs (eg spleen and lymph nodes) and the central nervous system (O'Rourke et al., 2016). For the astrocytes, they contribute to the degeneration of the motoneurones in ALS also, by releasing inflammatory mediators, such as prostaglandin and nitric oxide. Further, as with the motoneurones themselves, astrocytes express gene mutations associated with ALS, such SOD1 (Provenzano et al., 2022). These SOD1-expressing astrocytes prompt the death of motoneurones but not, evidently, of other types of neurones such as GABAergic interneurones or those in the dorsal root ganglion. But early in the disease process, as with the microglia, the astrocytes are thought to be protective and help the survival of the damaged motoneurones (Yamanaka et al., 2008). A similar pattern seems to be evident in the infiltrating T cells. They enter the neural tissue very early in the disease and seem, at least initially, to be helpful. They have been shown to be neuroprotective or even enhance neurogenesis. Later, the T cells appear to damage motoneurones by cell-cell contact or cytokine secretion, or contribute to their further damage by exacerbating the activation of the microglia (Nicholas et al., 2024; Zaccai et al., 2024).

In terms of topography, microgliosis and T cell infiltration tend to occur in neural regions associated with the motoneurone degeneration, namely motor cortex, along the corticospinal tract, and in the ventral horn of the spinal cord; astrogliosis occurs in the same general regions, albeit a little more diffusely (Hooten et al., 2015; Hardiman et al., 2017; Liu & Wang, 2017; Provenzano et al., 2023).

Hence, in summary, neuroinflammation appears to have dual functional state that may be time dependent, evolving with disease progression; from a neuroprotective role at disease onset to a toxic role, with an altered cytokine release profile, at the end-stage disease (Hooten et al., 2015; Hardiman et al., 2017; Liu & Wang, 2017; Provenzano et al., 2023).

## 4.4 Mitochondrial dysfunction

A dysfunction to small cell organelles called mitochondria forms a central and early feature of the pathology in ALS (Fig. 2). Mitochondria are involved in the production of energy (adenosine triphosphate; ATP), through the process of oxidative phosphorylation, that drives many intracellular functions. They are involved also in cellular metabolism and signalling, calcium homeostasis and immune responses. Their proper function and stability plays a key role in the cell homeostasis.

Many previous studies have reported impaired mitochondrial function in the motoneurones of both ALS patients and animal models. The mitochondria have an abnormal vacuolated and swollen appearance and altered cristae organisation (Wiedemann et al., 2002). Such features have been associated with the disruption to the transport of mitochondria along the axons (Granatiero & Manfredi, 2019), a function that very much worsens with disease progression. What also worsens is intracellular signalling with other organelles (eg endoplasmic reticulum; Larrea et al., 2025; Miller & Gomez-Suaga, 2023), the calcium-buffer function, the process of oxidative phosphorylation, a decreasing in membrane potential, an increase in reactive oxygen species (ROS) levels, impaired mitophagy by optineurin sequestering (Tak et al., 2020), an impact on ATP consumption (Onesto et al., 2016), dysfunction in mitochondrial dynamics in diminishing mitochondrial length and by interaction between TDP-43 and prohibitin 2/MNF2 (Davis et al., 2018). This leads ultimately to oxidative stress and further mitochondrial damage and dysfunction, a generation of misfolded proteins and protein aggregates, and subsequent motoneurone degeneration (Genin et al., 2023; Hardiman et al., 2017; Obrador et al., 2021; Smith et al., 2019; Zhao et al., 2022).

Mitochondria are of particular importance to all types of neurones. They have high energy requirements, so a constant supply of ATP is critical. Further, a healthy balance of calcium levels within the neurones is central to their normal function and patterns of activity. Most neurones in the central nervous system are long-lived, often throughout the life of the individual, and with age, become far more susceptible to stress and dysfunction arising from mitochondrial damage. It is hence not surprising that mitochondrial dysfunction has been linked not only to ALS, but to all of the neurodegenerative diseases. Motoneurones are particularly sensitive to oxidative stress and prone to mitochondrial damage; many of the defective genes associated with ALS (see above), and their associated proteins that develop abnormal aggregations, are associated with mitochondrial dysfunction (Genin et al., 2023; Hardiman et al., 2017; Smith et al., 2019; Zhao et al., 2022).

In summary, mitochondrial damage appears as an early sign of the disease process in ALS and with disease progression, they become more and more impaired leading ultimately to cell death. Their dysfunction appears central to the core mechanism of motoneurone degeneration in ALS.

## 4.5 Metabolism and neurone-glial coupling

Brain metabolism in ALS is disrupted profoundly across both neuronal and glial populations, undermining the finely-tuned metabolic interplay that normally maintains central nervous system homeostasis. Motoneurones, that are highly energy-dependent, show widespread perturbations in several amino acid metabolic pathways, including those involving phenylalanine, tryptophan, purine, arginine, proline, aspartate, and glutamate (Gautam et al., 2025). These pathways are not only essential for the synthesis of key neurotransmitters but also play pivotal roles in energy metabolism, redox homeostasis, and intracellular signalling. Their disruption threatens motoneurone viability, particularly under conditions of sustained bioenergetic stress (Blasco et al., 2020; Madji Hounoum et al., 2017; Madji Hounoum et al., 2018).

Astrocytes, the primary glial cell type responsible for neuronal metabolic support, also experience substantial metabolic dysregulation in ALS.

Alterations have been documented in pathways such as purine and pyrimidine metabolism, lysine degradation and glycerophospholipid turnover – all critical for membrane stability and optimal mitochondrial function. Dysfunction in these processes weakens astrocytic capacity to maintain cellular homeostasis, particularly in the face of the oxidative and inflammatory stress that characterises the ALS microenvironment (Cassina et al., 2021). A hallmark of neurone-glia metabolic interaction is the astrocyte-neurone lactate shuttle (Pellerin & Magistretti, 1994). Under physiological conditions, astrocytes produce ATP for their own energetic needs via aerobic glycolysis of glucose (also called the Warburg effect). As a consequence, glycolysis produces L-lactate as endproduct, that astrocytes export to motoneurones via monocarboxylate transporters. Motoneurones utilise L-lactate then for oxidative phosphorylation and ATP generation. However, ALS-associated mutations – such as those in SOD1 and TARDBP – impair astrocytic mitochondrial function and glycolytic efficiency, thereby reducing lactate output (Madji Hounoum et al., 2017). This decline in extracellular lactate diminishes the metabolic support available to motoneurones, making them more vulnerable to energy failure and oxidative damage.

Oligodendrocytes also contribute to the metabolic ecosystem of the central nervous system by providing lactate along axonal projections. In ALS, their capacity to support axonal metabolism is similarly impaired, compounding the metabolic insufficiency. This oligodendroglial dysfunction can result in disrupted myelination and axonal degeneration, further exacerbating motoneurone loss and neuronal network instability (Provenzano et al., 2023).

Beyond lactate shuttling, other metabolic networks are also affected, including those regulating glucose uptake, glycolytic flux and mitochondrial oxidative phosphorylation. In both motoneurones and astrocytes, a pathological metabolic shift occurs from mitochondrial respiration toward increased reliance on glycolysis – a likely compensatory mechanism in response to mitochondrial dysfunction (see above). This metabolic reprogramming is associated with excessive ROS production, oxidative stress and ultimately, cell death (Tefera et al., 2021).

In summary, ALS pathogenesis involves complex metabolic dysfunctions across neuronal and glial compartments. Motoneurones suffer from impaired metabolic pathway function, reducing their resilience under high-energy demand. In parallel, astrocytic and oligodendrocytic metabolic derangements undermine neurone-glial coupling and lactate-based energy support. Together, these abnormalities create a deleterious cycle of metabolic stress, oxidative damage, and neurodegeneration (Barone et al., 2023). Therapeutically, this convergence of metabolic disturbances highlights promising intervention points. Strategies aimed at restoring metabolic cooperation – such as enhancing lactate production, improving mitochondrial function (see above) or targeting redox balance – hold potential for mitigating energy deficits and slowing ALS progression (Blasco et al., 2016, 2020).

## 4.6 Neurovascular disorder

An interesting, alternative analysis to the pathogenesis associated with ALS is that it stems from vascular dysfunction (Fig. 2; Garbuzova-Davis et al., 2011). The neurovascular unit is a complex structure made up of neurones, glia, endothelial cells, basement membrane, mural cells (smooth muscle and pericytes). These cells work together to maintain cell homeostasis by providing a constant and consistent supply of oxygen and glucose. In addition, the neurovascular unit offers a series of barriers for the brain from the periphery that limit the entry of toxins, macromolecules and pathogens that may disrupt the delicate balance of environment in the brain. These barriers are the BBB, blood-spinal cord barrier and blood-cerebrospinal fluid barrier. The endothelial cells interact with pericytes that regulate cerebral blood flow, immune trafficking, trans-endothelial fluid transporter, permeability and vasculature integrity. The pericytes in particular are triggered to influence the cerebral blood flow by nearby neuronal and glial cells. Hence, the neurovascular unit allows for a control of exchanges of various substances between the blood and the central nervous system that help proper neurone function and survival (Eve et al., 2018; Garbuzova-Davis et al., 2011; Sorrentino et al., 2022).

ALS, both in animal models and in patients, has been associated with an impaired BBB, blood–spinal cord barrier and/or blood–cerebrospinal fluid barrier. These impairments develop at early stages, in fact, seemingly before motoneurone degeneration; there is a 30–45 % blood flow reduction measured before sign onset in animal models. And the impairments worsen with disease progression. There is much ultrastructural change in the vessels surrounding degenerating motoneurones in both the brainstem and spinal cord, in particular, degenerating endothelial cells (including damage to their mitochondria) and astrocytes (and swelling of their end-feet), together with extracellular oedema. In addition, there is a 10–15 % reduction in total capillary length, as well as an increase in various matrix metalloproteinases, that are thought to lead to a breakdown of the collagen supporting the basement membrane (Eve et al., 2018; Garbuzova-Davis et al., 2011; Sorrentino et al., 2022).

Some of the earliest studies examining vascular integrity in ALS showed that the blood vessels become very leaky. Subsequent studies, in both ALS animal models and patients, have reported that the central nervous system is characterised by numerous sites of micro-bleeding, indicative of the capillary rupture. The micro-bleeds result in many blood borne compounds – for example various proteins (eg C3 and C4), IgGs (that are very toxic to motoneurones) and albumins – entering the extracellular matrix. Further, the micro-bleeds lead to free-floating red blood vessels within the extracellular matrix. These extravasated red blood cells release free iron, a product of haemoglobin degradation, that is extremely toxic to surrounding neurones, prompting degeneration through a generation of free radicals. Indeed, the widespread micro-bleeds are often identified by detection of ferric iron deposits. It is of some importance that, following transplantation of human endothelial progenitor cells and a subsequent repair of the blood-spinal cord barrier and reduction in micro-bleeds in transgenic ALS mice, the survival rate of the motoneurones increases substantially (Eve et al., 2018; Garbuzova-Davis et al., 2011; Sorrentino et al., 2022).

The idea that vascular impairment contributes to neurone degeneration in ALS has also been proposed for the other neurodegenerative diseases, including Alzheimer's disease. In particular, it has been suggested that micro-bleeds kill neurones by four main mechanisms; (1) the hypoxia generated by the loss of blood supply; (2) glutamate-mediated excitotoxicity; (3) the neurotoxicity of haemoglobin, and; (4) immune-mediated destruction of neurones infected by opportunistic pathogens. Each of the last three mechanisms are prompted by the haemorrhage entering and spreading into the neuropil (Cullen et al., 2005, 2006; Stone, 2008; Stone et al., 2024).

In summary, the vascular impairments associated with ALS, evident in both animal models and patients, develop very early in the disease and worsen with disease progression. These impairments cause much damage to the motoneurones and appear a key feature of the overall disease process (Eve et al., 2018; Garbuzova-Davis et al., 2011; Sorrentino et al., 2022).

## 5. Diagnosis and current treatment options

The diagnosis of ALS is mainly clinical; unfortunately, there is no specific diagnostic test for ALS. In general, ALS is diagnosed clinically if there are both upper and lower motoneurone deficits in at least one body region or lower motoneuron deficit in two regions. One of the most encouraging avenues of research has been the identification of biomarkers in ALS, as to reduce delay in diagnosis. To this end, metabolomics and proteomics analysis have been performed on blood and cerebrospinal fluid, as well as targeted explorations, and there are some early indications of positive outcomes. In particular, neurofilament light chain, that expresses severity of neurone degeneration, is elevated in ALS (Benatar et al., 2024; Corcia et al., 2023).

In terms of current treatment options, there have been a large number of pharmaceutical agents that have been tested over the years, with very mixed results. For example, the drug riluzole, speculated to reduce glutamatergic excitotoxicity, increased the survival period of patients by only three months, while edaravone, thought to act as an anti-oxidant agent, has produced variable outcomes at various phases of clinical trial (Corcia et al., 2023; Hardiman et al., 2017; Kiernan et al., 2011). Given the heterogeneity of ALS, a unimodal treatment paradigm is somewhat unrealistic; no single drug or method is going to address the constellation of pathology (see section above) and signs of the disease (Bensimon et al., 2025; Corcia et al., 2025).

## 6. A treatment paradigm: The light

Red to near infrared light (R-NIr; $\lambda = 600$–$1300$ nm) applied to body tissues – referred to often as photobiomodulation – has been shown to have a major influence on the function and homeostasis of body cells. Many studies have shown it to be neuroprotective in several animal models of neurological disease, including the neurodegenerative diseases, stroke and traumatic brain injury and there are some early indications that it can be effective in humans. It has an impeccable safety record, with little or no evidence of side effects or toxicity on body cells and the devices are easy to use and inexpensive. In the sections that follow, the putative mechanisms behind R-NIr are considered, together with what is known of its impact on neurodegenerative disease and ALS (Hamblin, 2016).

### 6.1 The mechanisms of red and near infrared light (R-NIr)

Over the last seventy years or so, there have been countless reports indicating that the bulk of the beneficial effects of R-NIr centres on its ability to drive the mitochondria (Fig. 3). The mitochondria have many receptors – for example, including flavinic and pyridinic rings, lipids, aromatic amino acids,

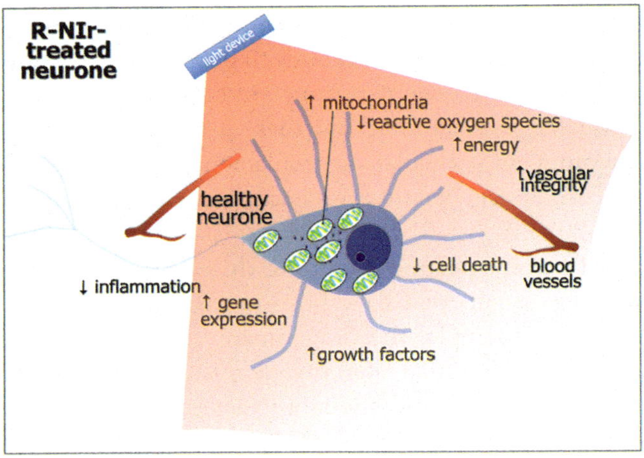


**Fig. 3** Schematic diagram of the major cellular functional improvements associated with R-NIr treatment of diseased/dysfunctional neurones. There are also considerable reductions in surrounding inflammation and improvements in vascular integrity.

cytochrome c oxidase, interfacial water, water-gated ion channels – across the R-NIr wavelengths and these receptors, after being stimulated by R-NIr, trigger the production of more energy (ATP) for the cell to function, together with a reduction in ROS, that can become toxic. R-NIr treatment also induces more long-term cellular changes, by activating the expression of various functional and protective genes. There is an increase in growth factors (eg brain derived growth factor, BDNF), that promotes a regrowth and repair of cells, together with a reduction in neuroinflammation (both astrogliosis and microgliosis). There is also a repair of vascular endothelial integrity and improvement of blood flow, as well as an increase in permeability of the BBB. In essence, light increases the resilience of cells, in particular the neurones, improving their overall intrinsic function and making them more resistant to distress (Fig. 3; Hamblin, 2016; Mitrofanis, 2019; San Miguel et al., 2019; Stone et al., 2024; Zhou et al., 2021).

It should be noted that all the above-outlined mechanisms depend on R-NIr falling directly on the cells. This is regarded as the most effective way in which R-NIr treatment generates beneficial outcomes. There is however, another mechanism of action. R-NIr has been shown – quite remarkably – to be beneficial to cell function and survival (eg in the brain) even when it is applied to a more distant location (eg thigh); that is, when it is not applied directly to the dysfunctional or diseased cells. This indirect stimulation is thought to activate circulating cells or molecules, or even free-floating mitochondria, within the cardiovascular or lymphatic systems that then leads to beneficial effects in the distressed neurones in the brain. Indeed, many studies in preclinical animal models of disease have shown the there are R-NIr-induced changes in cytokine expression, with increases in anti-inflammatory markers (eg IL-10, TGF-β) and decreases in pro-inflammatory ones (eg IL-6, IL-17, IL-1, TNF-α, NF-kβ, PGE2, iNOS, IL-12, IL-23). The indirect stimulation, although offering benefits, is considered less effective than direct application (Johnstone et al. 2016; Mitrofanis, 2019).

## 6.2 The impact of R-NIr treatment in neurological disease: Preclinical and clinical

There have been a number of preclinical and clinical studies examining the impact of R-NIr treatment in a range of neurological diseases. These are outlined briefly below, with a focus on the other types of neurodegenerative disease.

preclinical: Many preclinical studies have reported R-NIr-induced beneficial outcomes on animal models of neurodegenerative diseases (eg Alzheimer's and Parkinson's disease), multiple sclerosis, traumatic brain injury and stroke (Hamblin, 2016). All of these studies have indicated improvements in behaviour – for example, cognition and locomotion – as well as increased cell function and survival (ie neuroprotection) and reductions in neuroinflammation (Hamblin, 2016; Mitrofanis, 2019). For example, in transgenic mouse models of Alzheimer's disease, many studies have shown that there are R-NIr-induced reductions in β-amyloid plaques and aggregations, neurofibrillary tangles of hyperphosphorylated tau protein, synaptic vulnerability, gliosis and oxidative stress, and increases in neurogenesis, mitochondrial function and ATP levels (Hamblin, 2016, 2019). In addition, after internalisation of β-amyloid into human neuroblastoma cells, R-NIr treatment has been shown to increase ATP levels and cell number, while reducing β-amyloid aggregates. Further, when applied to normal mice, R-NIr prompts a much faster clearance of β-amyloid out of the brain and into the lymphatic system. Finally, there are improvements in a range of cognitive behavioural tests in nearly all of the transgenic mouse models of Alzheimer's disease tested after R-NIr treatment (Hamblin, 2016, 2019). For Parkinson's disease, in various toxin-induced rodent and non-human primate models, as well as transgenic mouse models, R-NIr treatment improves the number of dopaminergic cells in the midbrain, together with their terminations in the striatum; there are also reports of reductions in gliosis and improvements in patterns of functional activity across the basal ganglia. Finally, there are consistent R-NIr-induced improvements in locomotive behaviour (rodents and non-human primates) and clear reductions in clinical signs (in non-human primates) (Mitrofanis, 2019).

clinical: In general, there have been only a few larger scale placebo-controlled clinical trials exploring the effect of R-NIr on various types of neurological conditions in humans, from depression to traumatic brain injury and from stroke to Alzheimer's disease; the majority of studies have been either small scale trials (usually $< 20$ patients) or case reports. Further, the bulk of studies have been rather short-term, extending for, on average about three months only. Notwithstanding these limitations, nearly all reports indicate positive outcomes, with improvements being reported in overall brain function (eg with electroencephalogram or functional magnetic resonance) and in a range of cognitive tests and motor scores (Caldieraro & Cassano, 2019; Cassano et al., 2018; Figueiro Longo et al., 2020; Hamilton et al., 2019; Liebert et al., 2021; Naeser et al., 2023; Salehpour et al., 2023; Saltmarche et al., 2017).

## 6.3 What is the effect of R-NIr on ALS?

To date, previous studies examining the effect of R-NIr on ALS are minimal, to say the least (Bedlack et al., 2022; Moskvin, 2024); there are two preclinical studies and a single clinical case report on one patient.

For the preclinical studies, the first explored the impact of R-NIr on behaviour and histology in SOD1 mice (Moges et al., 2009). The mice were treated three times a week ($\lambda = 810$ nm; dose $= 12\,J/cm^2$), starting well before any signs of deficits, over three different regions, primary motor cortex, the cervical and lumbar regions of the spinal cord. These authors found that R-NIr had no impact on overall survival rate of the animals nor on the number of motoneurones in the spinal cord and a limited effect on motor behaviour. However, there was a marked R-NIr-induced reduction in inflammation (astrogliosis, expression of glial fibrillary acidic protein) in the spinal cord. The lack of overwhelmingly positive results from this study may reflect that the dose and frequency of application was too low, that the brainstem was not considered, and that the region of analysis was limited to the spinal cord only.

The second preclinical study explored the effect of R-NIr in canine degenerative myelopathy, a naturally occurring disease caused by mutations in the canine form of SOD1 and that shares many similarities to ALS

(Miller et al., 2020). These animals received R-NIr treatment once or twice a week ($\lambda = 904/980$ nm; dose = 8–21 J/cm$^2$), starting at variable times after the onset of motor signs, across the spinal cord. These authors found that some dogs with the higher doses, had better survival rates compared to the lower doses. Again, as with the previous preclinical study, a higher dose regime, applied more frequently, may well have yielded more positive outcomes.

For the single clinical case report (an abstract from a conference proceeding), a 69 year old man with 4 years of progressive limb onset ALS, started NIr treatment ($\lambda = 810/890$ nm; dose = 4–15 J/cm$^2$), over the forehead, vertebral column, armpit, sternum, umbilicus and pelvis. The R-NIr treatment was applied in three cycles, each consisting of two sessions per day for 10 days, with 40 days between cycles. The patient reportedly showed improved breathing and limb strength. These benefits did not last long however, regressing 20–30 days after each cycle. As with the preclinical studies, one can not help but think that a higher dose of R-NIr, applied more frequently and consistently, would have results in more favourably outcomes (Longo et al., 2009).

Hence, although these early studies provide some indication of positive outcomes in ALS after R-NIr treatment, there is a clear need for more extensive explorations, particularly concerning the issue of dosage, at both preclinical and clinical levels.

## 6.4 When would be the optimal time to apply R-NIr in ALS?

There is a long standing view across the scientific and medical communities that, regardless of the neurological disorder, the earlier the treatment starts, then the better the potential outcome. The strategy here is to try and save the neurones as early in disease process as possible, before they develop too much damage. Once the damage progresses beyond a critical point, there is little or no chance of saving the neurones (Porciatti & Chou, 2022). The mitochondria appear to be central factor here; if they still have some functional viability, they can still be saved by a therapeutic intervention (Mitrofanis et al., 2024).

One of the key factors that has frustrated the development of an effective neuroprotective treatment, not only for ALS but also for the other neurodegenerative diseases as well, is that diagnosis is often very late in the disease process, seemingly well after much mitochondrial and neuronal damage has occurred already. Hence, a clear strategy would be to apply treatment as early as possible, before the onset of clinical signs and before too much cellular damage. In animal models, this could be achieved relatively easily, because the onset of disease process is defined and the treatment could be applied well before any cell damage starts. In patients, this is not as easy, because most cases are sporadic and time of onset is not known. Nevertheless, treatment could be applied at the earliest indications of disease as to give the remaining motoneurones the best chance of survival. This limitation in patients highlights the need for an early biomarker, one that could flag the onset of disease at a very early stage. One is hopeful for such a marker to be found soon.

## 7. Conclusions

ALS is one of the most devastating of all the neurodegenerative diseases. What makes it so devastating is that after diagnosis, prognosis is poor and survival times are short. It impacts, rather selectively, the motor system of the central nervous system. There is much cellular dysfunction including; mitochondrial abnormalities, development of abnormal protein aggregations, glutamate excitotoxicity, neurone-glia coupling breakdown, widespread neuroinflammation and compromised vascular integrity. Unfortunately, there is no current treatment that address many, not even two or three, of these dysfunctions, one that stops or even slows the progression of the disease, that is, is neuroprotective.

Unlike many of the treatment options that have been tested previously, mainly pharmaceutical ones, R-NIr appears to address more than one, in fact, many of the features of pathology of ALS. These include:

1. R-NIr has been shown to improve behaviour and be neuroprotective in animal models of other, closely-related neurodegenerative diseases, namely Alzheimer's and Parkinson's disease;
2. R-NIr targets the mitochondria, a central niche of dysfunction in ALS, increasing ATP energy production and reducing ROS toxicity, thereby improving overall neurone function and survival;
3. R-NIr reduces the abnormal protein aggregates in cells, a major factor in ALS development;
4. R-NIr reduces neuroinflammation, a key feature of pathology of ALS, and;
5. R-NIr can improve vascular endothelial cell integrity and improve blood circulation, features that are highly disruptive in ALS.

R-NIr treatment, with its multi-modal effect, addressing, many if not all, features of ALS pathology, could be viewed as a valuable treatment option for this condition.

## Acknowledgements

We thank Fonds Clinatec and Fondation COVEA France for their generous funding of our work.

## Funding

Fonds Clinatec and Fondation COVEA France.

## Author contributions

All authors contributed to the writing of the manuscript.

## Competing interest

none.

## References

Baker, M. R. (2014). ALS—dying forward, backward or outward? *Nature Reviews Neurology, 10*, 660. https://doi.org/10.1038/nrneurol.2013.221-c1.

Balendra, R., Sreedharan, J., Hallegger, M., Luisier, R., Lashuel, H. A., Gregory, J. M., & Patani, R. (2025). Amyotrophic lateral sclerosis caused by TARDBP mutations: From genetics to TDP-43 proteinopathy. *Lancet Neurology, 24*, 456–470. https://doi.org/10.1016/S1474-4422(25)00109-7.

Barone, M., Leo, A. D., & de van der Schueren, M. A. E. (2023). Malnutrition assessment by global leadership initiative on malnutrition criteria in patients with amyotrophic lateral sclerosis. *Nutrition (Burbank, Los Angeles County, Calif.), 109*, 111997. https://doi.org/10.1016/j.nut.2023.111997.

Bedlack, R., Barkhaus, P., Barnes, B., Bereman, M., Bertorini, T., Carter, G., ... Wicks, P. (2022). ALSUntangled #60: Light therapy. *Amyotrophic Lateral Sclerosis and Frontotemporal Degeneration, 23*, 315–319. https://doi.org/10.1080/21678421.2021.1883668.

Beers, D. R., Zhao, W., Liao, B., Kano, O., Wang, J., Huang, A., ... Henkel, J. S. (2011). Neuroinflammation modulates distinct regional and temporal clinical responses in ALS mice. *Brain, Behavior, and Immunity, 25*, 1025–1035. https://doi.org/10.1016/j.bbi.2010.12.008.

Benatar, M., Ostrow, L. W., Lewcock, J. W., Bennett, F., Shefner, J., Bowser, R., ... Wuu, J. (2024). Biomarker qualification for neurofilament light chain in ALS: Theory and practice. *Annals of Neurology, 95*, 211–216. https://doi.org/10.1002/ana.26860.

Bensimon, G., Leigh, P. N., Tree, T., Malaspina, A., Payan, C. A., Pham, H.-P., ... MIROCALS Study Group (2025). Efficacy and safety of low-dose IL-2 as an add-on therapy to riluzole (MIROCALS): A phase 2b, double-blind, randomised, placebo-controlled trial. *Lancet, 405*, 1837–1850. https://doi.org/10.1016/S0140-6736(25)00262-4.

Benson, B. C., Shaw, P. J., Azzouz, M., Highley, J. R., & Hautbergue, G. M. (2021). Proteinopathies as hallmarks of impaired gene expression, proteostasis and mitochondrial function in amyotrophic lateral sclerosis. *Frontiers in Neuroscience, 15*, 783624. https://doi.org/10.3389/fnins.2021.783624.

Blasco, H., Lanznaster, D., Veyrat-Durebex, C., Hergesheimer, R., Vourch, P., Maillot, F., ... Corcia, P. (2020). Understanding and managing metabolic dysfunction in amyotrophic lateral sclerosis. *Expert Review of Neurotherapeutics, 20*, 907–919. https://doi.org/10.1080/14737175.2020.1788389.

Blasco, H., Patin, F., Madji Hounoum, B., Gordon, P. H., Vourc'h, P., Andres, C. R., & Corcia, P. (2016). Metabolomics in amyotrophic lateral sclerosis: How far can it take us? *European Journal of Neurology, 23*, 447–454. https://doi.org/10.1111/ene.12956.

Buccellato, F. R., D'Anca, M., Tartaglia, G. M., Del Fabbro, M., Scarpini, E., & Galimberti, D. (2023). Treatment of Alzheimer's disease: Beyond symptomatic therapies. *International Journal of Molecular Sciences, 24*, 13900. https://doi.org/10.3390/ijms241813900.

Busche, M. A., & Hyman, B. T. (2020). Synergy between amyloid-β and tau in Alzheimer's disease. *Nature Neuroscience, 23*, 1183–1193. https://doi.org/10.1038/s41593-020-0687-6.

Calabresi, P., Mechelli, A., Natale, G., Volpicelli-Daley, L., Di Lazzaro, G., & Ghiglieri, V. (2023). Alpha-synuclein in Parkinson's disease and other synucleinopathies: From overt neurodegeneration back to early synaptic dysfunction. *Cell Death & Disease, 14*, 176. https://doi.org/10.1038/s41419-023-05672-9.

Caldieraro, M. A., & Cassano, P. (2019). Transcranial photobiomodulation for major depressive and anxiety disorders and for posttraumatic stress disorder. *Photobiomodulation in the brain*. Elsevier, 479–487. https://doi.org/10.1016/B978-0-12-815305-5.00035-X.

Cassano, P., Petrie, S. R., Mischoulon, D., Cusin, C., Katnani, H., Yeung, A., ... Iosifescu, D. V. (2018). Transcranial photobiomodulation for the treatment of major depressive disorder. The ELATED-2 pilot trial. *Photomedicine and Laser Surgery, 36*, 634–646. https://doi.org/10.1089/pho.2018.4490.

Cassina, P., Miquel, E., Martínez-Palma, L., & Cassina, A. (2021). Glial metabolic reprogramming in amyotrophic lateral sclerosis. *Neuroimmunomodulation, 28*, 204–212. https://doi.org/10.1159/000516926.

Corcia, P., Lunetta, C., Vourc'h, P., Pradat, P., & Blasco, H. (2023). Time for optimism in amyotrophic lateral sclerosis. *European Journal of Neurology, 30*, 1459–1464. https://doi.org/10.1111/ene.15738.

Corcia, P., Piras, R., & Lunetta, C. (2025). Why is the treatment and management of amyotrophic lateral sclerosis so difficult? *Expert Review of Neurotherapeutics, 25*, 1–3. https://doi.org/10.1080/14737175.2024.2415002.

Cullen, K. M., Kócsi, Z., & Stone, J. (2006). Microvascular pathology in the aging human brain: Evidence that senile plaques are sites of microhaemorrhages. *Neurobiology of Aging, 27*, 1786–1796.

Cullen, K. M., Kócsi, Z., & Stone, J. (2005). Pericapillary haem-rich deposits: Evidence for microhaemorrhages in aging human cerebral cortex. *Journal of Cerebral Blood Flow and Metabolism: Official Journal of the International Society of Cerebral Blood Flow and Metabolism, 25*, 1656–1667. https://doi.org/10.1038/sj.jcbfm.9600155.

Dadon-Nachum, M., Melamed, E., & Offen, D. (2011). The "dying-back" phenomenon of motor neurons in ALS. *Journal of Molecular Neuroscience: MN, 43*, 470–477. https://doi.org/10.1007/s12031-010-9467-1.

Das, M. M., & Svendsen, C. N. (2015). Astrocytes show reduced support of motor neurons with aging that is accelerated in a rodent model of ALS. *Neurobiology of Aging, 36*, 1130–1139. https://doi.org/10.1016/j.neurobiolaging.2014.09.020.

Davis, S. A., Itaman, S., Khalid-Janney, C. M., Sherard, J. A., Dowell, J. A., Cairns, N. J., & Gitcho, M. A. (2018). TDP-43 interacts with mitochondrial proteins critical for mitophagy and mitochondrial dynamics. *Neuroscience Letters, 678*, 8–15. https://doi.org/10.1016/j.neulet.2018.04.053.

de Calbiac, H., Renault, S., Haouy, G., Jung, V., Roger, K., Zhou, Q., ... Kabashi, E. (2024). Poly-GP accumulation due to C9orf72 loss of function induces motor neuron apoptosis through autophagy and mitophagy defects. *Autophagy, 20*, 2164–2185. https://doi.org/10.1080/15548627.2024.2358736.

Debnath, K., Shekhar, S., Kumar, V., Jana, N. R., & Jana, N. R. (2016). Efficient inhibition of protein aggregation, disintegration of aggregates, and lowering of cytotoxicity by green tea polyphenol-based self-assembled polymer nanoparticles. *ACS Applied Materials & Interfaces, 8*, 20309–20318. https://doi.org/10.1021/acsami.6b06853.

Dutta, K., Patel, P., & Julien, J.-P. (2018). Protective effects of Withania somnifera extract in SOD1G93A mouse model of amyotrophic lateral sclerosis. *Experimental Neurology, 309*, 193–204. https://doi.org/10.1016/j.expneurol.2018.08.008.

Eisen, A., Vucic, S., & Mitsumoto, H. (2023). History of ALS and the competing theories on pathogenesis: IFCN handbook chapter. *Clinical Neurophysiology Practice, 9*, 1–12. https://doi.org/10.1016/j.cnp.2023.11.004.

Eve, D. J., Steiner, G., Mahendrasah, A., Sanberg, P. R., Kurien, C., Thomson, A., ... Garbuzova-Davis, S. (2018). Reduction of microhemorrhages in the spinal cord of symptomatic ALS mice after intravenous human bone marrow stem cell transplantation accompanies repair of the blood-spinal cord barrier. *Oncotarget, 9*, 10621–10634. https://doi.org/10.18632/oncotarget.24360.

Figueiro Longo, M. G., Tan, C. O., Chan, S., Welt, J., Avesta, A., Ratai, E., ... Gupta, R. (2020). Effect of transcranial low-level light therapy vs sham therapy among patients with moderate traumatic brain injury: A randomized clinical trial. *JAMA Network Open, 3*, e2017337. https://doi.org/10.1001/jamanetworkopen.2020.17337.

Garbuzova-Davis, S., Rodrigues, M. C. O., Hernandez-Ontiveros, D. G., Louis, M. K., Willing, A. E., Borlongan, C. V., & Sanberg, P. R. (2011). Amyotrophic lateral sclerosis: A neurovascular disease. *Brain Research, 1398*, 113–125. https://doi.org/10.1016/j.brainres.2011.04.049.

Gautam, P., Yadav, R., Vishwakarma, R. K., Pathak, A., & Singh, C. (2025). Metabolic dysregulation in amyotrophic lateral sclerosis: Insights from $^{1}$H NMR-based metabolomics in a tertiary care center in India. *Metabolic Brain Disease, 40*, 196. https://doi.org/10.1007/s11011-025-01616-8.

Genin, E. C., Abou-Ali, M., & Paquis-Flucklinger, V. (2023). Mitochondria, a key target in amyotrophic lateral sclerosis pathogenesis. *Genes (Basel), 14*, 1981. https://doi.org/10.3390/genes14111981.

Goutman, S. A., Hardiman, O., Al-Chalabi, A., Chió, A., Savelieff, M. G., Kiernan, M. C., & Feldman, E. L. (2022). Emerging insights into the complex genetics and pathophysiology of amyotrophic lateral sclerosis. *Lancet Neurology, 21*, 465–479. https://doi.org/10.1016/S1474-4422(21)00414-2.

Grad, L. I., Rouleau, G. A., Ravits, J., & Cashman, N. R. (2017). Clinical spectrum of amyotrophic lateral sclerosis (ALS). *Cold Spring Harbor Perspectives in Medicine, 7*, a024117. https://doi.org/10.1101/cshperspect.a024117.

Granatiero, V., & Manfredi, G. (2019). Mitochondrial transport and turnover in the pathogenesis of amyotrophic lateral sclerosis. *Biology (Basel), 8*, 36. https://doi.org/10.3390/biology8020036.

Hamblin, M. R. (2016). Shining light on the head: Photobiomodulation for brain disorders. *BBA Clinical, 6*, 113–124. https://doi.org/10.1016/j.bbacli.2016.09.002.

Hamblin, M. R, (2019). Photobiomodulation for Alzheimer's disease: Has the light dawned? *Photonics, 6*, 77. https://doi.org/10.3390/photonics6030077.

Hamilton, C., El Khoury, H., Hamilton, D., Nicklason, F., & Mitrofanis, J. (2019). The "buckets": The use of red and infrared light helmets in Parkinson's disease patients. *Photobiomodulation, Photomedicine, and Laser Surgery, 37*, 615–622.

Hardiman, O., Al-Chalabi, A., Chio, A., Corr, E. M., Logroscino, G., Robberecht, W., ... Van Den Berg, L. H. (2017). Amyotrophic lateral sclerosis. *Nature Reviews Disease Primers, 3*, 17071. https://doi.org/10.1038/nrdp.2017.71.

Hariz, M., & Blomstedt, P. (2022). Deep brain stimulation for Parkinson's disease. *Journal of Internal Medicine, 292*, 764–778. https://doi.org/10.1111/joim.13541.

Hirokawa, N., Niwa, S., & Tanaka, Y. (2010). Molecular motors in neurons: Transport mechanisms and roles in brain function, development, and disease. *Neuron, 68*, 610–638. https://doi.org/10.1016/j.neuron.2010.09.039.

Hong, Y., Song, Y., Wang, W., Shi, J., & Chen, X. (2025). Mitochondrial DNA editing: Key to the treatment of neurodegenerative diseases. *Genes & Diseases, 12*, 101437. https://doi.org/10.1016/j.gendis.2024.101437.

Hooten, K. G., Beers, D. R., Zhao, W., & Appel, S. H. (2015). Protective and toxic neuroinflammation in amyotrophic lateral sclerosis. *Neurotherapeutics: the Journal of the American Society for Experimental NeuroTherapeutics, 12*, 364–375. https://doi.org/10.1007/s13311-014-0329-3.

Hulisz, D. (2018). Amyotrophic lateral sclerosis: Disease state overview. *The American Journal of Managed Care, 24*.

Jiang, Q., Lin, J., Wei, Q., Li, C., Hou, Y., Zhang, L., ... Shang, H. (2023). Genetic and clinical characteristics of ALS patients with NEK1 gene variants. *Neurobiology of Aging, 123*, 191–199. https://doi.org/10.1016/j.neurobiolaging.2022.11.001.

Jo, M., Lee, S., Jeon, Y.-M., Kim, S., Kwon, Y., & Kim, H.-J. (2020). The role of TDP-43 propagation in neurodegenerative diseases: Integrating insights from clinical and experimental studies. *Experimental & Molecular Medicine, 52*, 1652–1662. https://doi.org/10.1038/s12276-020-00513-7.

Johnstone, D. M., Moro, C., Stone, J., Benabid, A.-L., & Johnstone, J. (2016). Turning on lights to stop neurodegeneration: The potential of near infrared light therapy in Alzheimer's and Parkinson's disease. *Frontiers in Neuroscience, 9*, 500. https://doi.org/10.3389/fnins.2015.00500.

Karle, K. N., Möckel, D., Reid, E., & Schöls, L. (2012). Axonal transport deficit in a KIF5A–/–mouse model. *Neurogenetics, 13*, 169–179. https://doi.org/10.1007/s10048-012-0324-y.

Kenna, K. P., van Doormaal, P. T. C., Dekker, A. M., Ticozzi, N., Kenna, B. J., Diekstra, F. P., ... Landers, J. E. (2016). NEK1 variants confer susceptibility to amyotrophic lateral sclerosis. *Nature Genetics, 48*, 1037–1042. https://doi.org/10.1038/ng.3626.

Kiernan, M. C., Vucic, S., Cheah, B. C., Turner, M. R., Eisen, A., Hardiman, O., ... Zoing, M. C. (2011). Amyotrophic lateral sclerosis. *The Lancet, 377*, 942–955. https://doi.org/10.1016/S0140-6736(10)61156-7.

Kotni, M. K., Zhao, M., & Wei, D.-Q. (2016). Gene expression profiles and protein-protein interaction networks in amyotrophic lateral sclerosis patients with C9orf72 mutation. *Orphanet Journal of Rare Diseases, 11*, 148. https://doi.org/10.1186/s13023-016-0531-y.

Lamptey, R. N. L., Chaulagain, B., Trivedi, R., Gothwal, A., Layek, B., & Singh, J. (2022). A review of the common neurodegenerative disorders: Current therapeutic approaches and the potential role of nanotherapeutics. *International Journal of Molecular Sciences, 23*, 1851. https://doi.org/10.3390/ijms23031851.

Larrea, D., Tamucci, K. A., Kabra, K., Velasco, K. R., Yun, T. D., Pera, M., ... Area-Gomez, E. (2025). Altered mitochondria-associated ER membrane (MAM) function shifts mitochondrial metabolism in amyotrophic lateral sclerosis (ALS). *Nature Communications, 16*, 379. https://doi.org/10.1038/s41467-024-51578-1.

Leigh, P. N., Whitwell, H., Garofalo, O., Buller, J., Swash, M., Martin, J. E., ... Anderton, B. H. (1991). Ubiquitin-immunoreactive intraneuronal inclusions in amyotrophic lateral sclerosis. Morphology, distribution, and specificity. *Brain, 114*(Pt 2), 775–788. https://doi.org/10.1093/brain/114.2.775.

Liao, B., Zhao, W., Beers, D. R., Henkel, J. S., & Appel, S. H. (2012). Transformation from a neuroprotective to a neurotoxic microglial phenotype in a mouse model of ALS. *Experimental Neurology, 237*, 147–152. https://doi.org/10.1016/j.expneurol.2012.06.011.

Liebert, A., Bicknell, B., Laakso, E.-L., Heller, G., Jalilitabaei, P., Tilley, S., ... Kiat, H. (2021). Improvements in clinical signs of Parkinson's disease using photobiomodulation: a prospective proof-of-concept study. *BMC Neurology, 21*, 256. https://doi.org/10.1186/s12883-021-02248-y.

Lin, B. C., Higgins, N. R., Phung, T. H., & Monteiro, M. J. (2022). UBQLN proteins in health and disease with a focus on UBQLN2 in ALS/FTD. *The FEBS Journal, 289*, 6132–6153. https://doi.org/10.1111/febs.16129.

Lindamood, H. L., Liu, T. M., Read, T.-A., & Vitriol, E. A. (2025). Using ALS to understand profilin 1's diverse roles in cellular physiology. *Cytoskeleton, 82*, 111–129. https://doi.org/10.1002/cm.21896.

Liu, J., & Wang, F. (2017). Role of neuroinflammation in amyotrophic lateral sclerosis: Cellular mechanisms and therapeutic implications. *Frontiers in Immunology, 8*, 1005. https://doi.org/10.3389/fimmu.2017.01005.

Loeffler, J.-P., Picchiarelli, G., Dupuis, L., & Gonzalez De Aguilar, J.-L. (2016). The role of skeletal muscle in amyotrophic lateral sclerosis. *Brain Pathology (Zurich, Switzerland), 26*, 227–236. https://doi.org/10.1111/bpa.12350.

Longo, L., Postiglione, M., Gabellini, M., & Longo, D. (2009). Amyotrophic lateral sclerosis (ALS) treated with low level LASER therapy (LLLT): A case report. *AIP Conference Proceedings, 1142*, 96–98. https://doi.org/10.1063/1.3175636.

Madji Hounoum, B., Blasco, H., Coque, E., Vourc'h, P., Emond, P., Corcia, P., ... Mavel, S. (2018). The metabolic disturbances of motoneurons exposed to glutamate. *Molecular Neurobiology, 55*, 7669–7676. https://doi.org/10.1007/s12035-018-0945-8.

Madji Hounoum, B., Mavel, S., Coque, E., Patin, F., Vourc'h, P., Marouillat, S., ... Blasco, H. (2017). Wildtype motoneurons, ALS-Linked SOD1 mutation and glutamate profoundly modify astrocyte metabolism and lactate shuttling. *Glia, 65*, 592–605. https://doi.org/10.1002/glia.23114.

Magaña, J. J., Velázquez-Pérez, L., & Cisneros, B. (2013). Spinocerebellar ataxia type 2: Clinical presentation, molecular mechanisms, and therapeutic perspectives. *Molecular Neurobiology, 47*, 90–104. https://doi.org/10.1007/s12035-012-8348-8.

Marogianni, C., Rikos, D., Provatas, A., Dadouli, K., Ntellas, P., Tsitsi, P., ... Xiromerisiou, G. (2019). The role of C9orf72 in neurodegenerative disorders: A systematic review, an updated meta-analysis, and the creation of an online database. *Neurobiology of Aging, 84*, 238.e25–238.e34. https://doi.org/10.1016/j.neurobiolaging.2019.04.012.

Miller, C. C. J., & Gomez-Suaga, P. (2023). Poor communication between ER and mitochondria: A signature of ALS/FTD? *Aging (Albany NY), 15*, 10814–10816. https://doi.org/10.18632/aging.205206.

Miller, L. A., Torraca, D. (Gross), & De Taboada, L. (2020). Retrospective observational study and analysis of two different photobiomodulation therapy protocols combined with rehabilitation therapy as therapeutic interventions for canine degenerative myelopathy. *Photobiomodul Photomed Laser Surg, 38*, 195–205. https://doi.org/10.1089/photob.2019.4723.

Mitrofanis, J. (2019). Run in the light: Exploring exercise and photobiomodulation in Parkinson's disease. Morgan & Claypool Publishers, https://doi.org/10.1088/2053-2571/ab2f70.

Mitrofanis, J., Stone, J., Hamblin, M. R., Magistretti, P., Hamblin, A.-L., & Jeffery, G. (2024). A spotlight on dosage and subject selection for effective neuroprotection: Exploring the central role of mitochondria. *Neural Regeneration Research, 20*, 1081–1082. https://doi.org/10.4103/NRR.NRR-D-24-00222.

Moens, T. G., Da Cruz, S., Neumann, M., Shelkovnikova, T. A., Shneider, N. A., & Van Den Bosch, L. (2025). Amyotrophic lateral sclerosis caused by FUS mutations: Advances with broad implications. *The Lancet Neurology, 24*, 166–178. https://doi.org/10.1016/S1474-4422(24)00517-9.

Moges, H., Vasconcelos, O. M., Campbell, W. W., Borke, R. C., McCoy, J. A., Kaczmarczyk, L., ... Anders, J. J. (2009). Light therapy and supplementary Riboflavin in the SOD1 transgenic mouse model of familial amyotrophic lateral sclerosis (FALS). *Lasers in Surgery and Medicine, 41*, 52–59. https://doi.org/10.1002/lsm.20732.

Monteiro Neto, J. R., de Souza, G. F., dos Santos, V. M., de Holanda Paranhos, L., Ribeiro, G. D., Magalhães, R. S. S., ... Eleutherio, E. C. A. (2025). SOD1, A crucial protein for neural biochemistry: Dysfunction and risk of amyotrophic lateral sclerosis. *Molecular Neurobiology*. https://doi.org/10.1007/s12035-025-05067-1.

Moskvin, S. V. (2024). A brief literature review of low-level laser therapy for treating amyotrophic lateral sclerosis and confirmation of its effectiveness. *Biomedicine (Taipei), 14*. https://doi.org/10.37796/2211-8039.1430.

Naeser, M. A., Martin, P. I., Ho, M. D., Krengel, M. H., Bogdanova, Y., Knight, J. A., ... Koo, B. (2023). Transcranial photobiomodulation treatment: Significant improvements in four ex-football players with possible chronic traumatic encephalopathy. *ADR, 7*, 77–105. https://doi.org/10.3233/ADR-220022.

Nicholas, R., Magliozzi, R., Marastoni, D., Howell, O., Roncaroli, F., Muraro, P., ... Friede, T. (2024). High levels of perivascular inflammation and active demyelinating lesions at time of death associated with rapidly progressive multiple sclerosis disease course: A retrospective postmortem cohort study. *Annals of Neurology, 95*, 706–719. https://doi.org/10.1002/ana.26870.

Nicolas, A., Kenna, K. P., Renton, A. E., Ticozzi, N., Faghri, F., Chia, R., Dominov, ... Landers, J. E. (2018). Genome-wide analyses identify KIF5A as a novel ALS gene. *Neuron, 97*, 1267–1288. https://doi.org/10.1016/j.neuron.2018.02.027.

Nogueira-Machado, J. A., Rocha-Silva, F., & Gomes, N. A. (2025). The role of mTOR in amyotrophic lateral sclerosis. *Biomedicines, 13*, 952. https://doi.org/10.3390/biomedicines13040952.

Obrador, E., Salvador-Palmer, R., López-Blanch, R., Jihad-Jebbar, A., Vallés, S. L., & Estrela, J. M. (2021). The link between oxidative stress, redox status, bioenergetics and mitochondria in the pathophysiology of ALS. *International Journal of Molecular Sciences, 22*, 6352. https://doi.org/10.3390/ijms22126352.

Okamoto, K., Mizuno, Y., & Fujita, Y. (2008). Bunina bodies in amyotrophic lateral sclerosis. *Neuropathology: Official Journal of the Japanese Society of Neuropathology, 28*, 109–115. https://doi.org/10.1111/j.1440-1789.2007.00873.x.

Onesto, E., Colombrita, C., Gumina, V., Borghi, M. O., Dusi, S., Doretti, A., ... Ratti, A. (2016). Gene-specific mitochondria dysfunctions in human TARDBP and C9ORF72 fibroblasts. *Acta Neuropathologica Communications, 4*, 47. https://doi.org/10.1186/s40478-016-0316-5.

O'Rourke, J. G., Bogdanik, L., Yáñez, A., Lall, D., Wolf, A. J., Muhammad, A. K. M. G., ... Baloh, R. H. (2016). C9orf72 is required for proper macrophage and microglial function in mice. *Science (New York, N. Y.), 351*, 1324–1329. https://doi.org/10.1126/science.aaf1064.

Parisi, C., Arisi, I., D'Ambrosi, N., Storti, A. E., Brandi, R., D'Onofrio, M., & Volonté, C. (2013). Dysregulated microRNAs in amyotrophic lateral sclerosis microglia modulate genes linked to neuroinflammation. *Cell Death & Disease, 4*, e959. https://doi.org/10.1038/cddis.2013.491.

Pellerin, L., & Magistretti, P. J. (1994). Glutamate uptake into astrocytes stimulates aerobic glycolysis: A mechanism coupling neuronal activity to glucose utilization. *Proceedings of the National Academy of Sciences of the United States of America, 91*, 10625–10629. https://doi.org/10.1073/pnas.91.22.10625.

Porciatti, V., & Chou, T.-H. (2022). Using noninvasive electrophysiology to determine time windows of neuroprotection in optic neuropathies. *IJMS, 23*, 5751. https://doi.org/10.3390/ijms23105751.

Provenzano, F., Torazza, C., Bonifacino, T., Bonanno, G., & Milanese, M. (2023). The key role of astrocytes in amyotrophic lateral sclerosis and their commitment to glutamate excitotoxicity. *International Journal of Molecular Sciences, 24*, 15430. https://doi.org/10.3390/ijms242015430.

Rajabi, D., Khanmohammadi, S., & Rezaei, N. (2024). The role of long noncoding RNAs in amyotrophic lateral sclerosis. *Reviews in the Neurosciences, 35*, 533–547. https://doi.org/10.1515/revneuro-2023-0155.

Renaud, L., Picher-Martel, V., Codron, P., & Julien, J.-P. (2019). Key role of UBQLN2 in pathogenesis of amyotrophic lateral sclerosis and frontotemporal dementia. *Acta Neuropathologica Communications,* 7, 103. https://doi.org/10.1186/s40478-019-0758-7.

Rizea, R. E., Corlatescu, A.-D., Costin, H. P., Dumitru, A., & Ciurea, A. V. (2024). Understanding amyotrophic lateral sclerosis: Pathophysiology, diagnosis, and therapeutic advances. *IJMS, 25*, 9966. https://doi.org/10.3390/ijms25189966.

Saberi, S., Stauffer, J. E., Schulte, D. J., & Ravits, J. (2015). Neuropathology of amyotrophic lateral sclerosis and its variants. *Neurologic Clinics, 33*, 855–876. https://doi.org/10.1016/j.ncl.2015.07.012.

Salehpour, F., Sadigh-Eteghad, S., Mahmoudi, J., Kamari, F., Cassano, P., & Hamblin, M. R. (2023). Photobiomodulation therapy for stroke. In F. Salehpour, S. Sadigh-Eteghad, J. Mahmoudi, F. Kamari, P. Cassano, & M. R. Hamblin (Eds.). *Photobiomodulation for the brain: Photobiomodulation therapy in neurology and neuropsychiatry, synthesis lectures on biomedical engineering* (pp. 221–258) Cham: Springer Nature Switzerland. https://doi.org/10.1007/978-3-031-36231-6_11.

Saltmarche, A. E., Naeser, M. A., Ho, K. F., Hamblin, M. R., & Lim, L. (2017). Significant improvement in cognition in mild to moderately severe dementia cases treated with transcranial plus intranasal photobiomodulation: Case series report. *Photomedicine and Laser Surgery, 35*, 432–441. https://doi.org/10.1089/pho.2016.4227.

San Miguel, M., Martin, K. L., Stone, J., & Johnstone, D. M. (2019). Photobiomodulation mitigates cerebrovascular leakage induced by the parkinsonian neurotoxin MPTP. *Biomolecules, 9*. https://doi.org/10.3390/biom9100564.

Shang, Y., & Huang, E. J. (2016). Mechanisms of FUS mutations in familial amyotrophic lateral sclerosis. *Brain Research, 1647*, 65–78. https://doi.org/10.1016/j.brainres.2016.03.036.

Shen, Q., Guo, H., & Yan, Y. (2024). Photobiomodulation for neurodegenerative diseases: A scoping review. *International Journal of Molecular Sciences, 25*, 1625. https://doi.org/10.3390/ijms25031625.

Smeyers, J., Banchi, E.-G., & Latouche, M. (2021). C9ORF72: What it is, what it does, and why it matters. *Frontiers in Cellular Neuroscience, 15*, 661447. https://doi.org/10.3389/fncel.2021.661447.

Smith, E. F., Shaw, P. J., & De Vos, K. J. (2019). The role of mitochondria in amyotrophic lateral sclerosis. *Neuroscience Letters, Mitochondria in degenerative and inflammatory neurological disorders, 710*, 132933. https://doi.org/10.1016/j.neulet.2017.06.052.

Sorrentino, S., Polini, A., Arima, V., Romano, A., Quattrini, A., Gigli, G., ... Moroni, L. (2022). Neurovascular signals in amyotrophic lateral sclerosis. *Current Opinion in Biotechnology, 74*, 75–83. https://doi.org/10.1016/j.copbio.2021.10.021.

Sproviero, W., La Bella, V., Mazzei, R., Valentino, P., Rodolico, C., Simone, I. L., ... Conforti, F. L. (2012). FUS mutations in sporadic amyotrophic lateral sclerosis: clinical and genetic analysis. *Neurobiology of Aging, 33*, 837.e1–837.e5. https://doi.org/10.1016/j.neurobiolaging.2011.10.005.

Stone, J. (2008). What initiates the formation of senile plaques? The origin of Alzheimer-like dementias in capillary haemorrhages. *Medical Hypotheses, 71*, 347–359. https://doi.org/10.1016/j.mehy.2008.04.007.

Stone, J., Mitrofanis, J., Johnstone, D. M., & Robinson, S. R. (2024). The catastrophe of intracerebral hemorrhage drives the capillary-hemorrhage dementias, including Alzheimer's disease. *JAD, 97*, 1069–1081. https://doi.org/10.3233/JAD-231202.

Sudhakar, V., & Richardson, R. M. (2019). Gene therapy for neurodegenerative diseases. *Neurotherapeutics: the Journal of the American Society for Experimental NeuroTherapeutics, 16*, 166–175. https://doi.org/10.1007/s13311-018-00694-0.

Tak, Y. J., Park, J.-H., Rhim, H., & Kang, S. (2020). ALS-related mutant SOD1 aggregates interfere with mitophagy by sequestering the autophagy receptor optineurin. *International Journal of Molecular Sciences, 21*, 7525. https://doi.org/10.3390/ijms21207525.

Tang, X., Toro, A., T.G, S., Gao, J., Chalk, J., Oskarsson, B. E., & Zhang, K. (2020). Divergence, convergence, and therapeutic implications: A cell biology perspective of C9ORF72-ALS/FTD. *Molecular Neurodegeneration, 15*, 34. https://doi.org/10.1186/s13024-020-00383-7.

Tefera, T. W., Steyn, F. J., Ngo, S. T., & Borges, K. (2021). CNS glucose metabolism in amyotrophic lateral sclerosis: A therapeutic target? *Cell & Bioscience, 11*, 14. https://doi.org/10.1186/s13578-020-00511-2.

Tolochko, C., Shiryaeva, O., Alekseeva, T., & Dyachuk, V. (2025). Amyotrophic lateral sclerosis: Pathophysiological mechanisms and treatment strategies (Part 2). *International Journal of Molecular Sciences, 26*, 5240. https://doi.org/10.3390/ijms26115240.

Tsekrekou, M., Giannakou, M., Papanikolopoulou, K., & Skretas, G. (2024). Protein aggregation and therapeutic strategies in SOD1- and TDP-43- linked ALS. *Frontiers in Molecular Biosciences, 11*, 1383453. https://doi.org/10.3389/fmolb.2024.1383453.

van Damme, P., van den Berg, L.H., & Robberecht, W. (2011). Expanded ATXN2 CAG repeat size in ALS identifies genetic overlap between ALS and SCA2 | Neurology [WWW Document]. URL https://www.neurology.org/doi/10.1212/WNL.0b013e31821f445b.

Wiedemann, R., Flint Beal, M., & A.Schon, E. (2002). Mitochondrial DNA and respiratory chain function in spinal cords of ALS patients - Wiedemann - 2002. *Journal of Neurochemistry*. https://onlinelibrary.wiley.com/doi/full/10.1046/j.0022-3042.2001.00731.x?sid=nlm%3Apubmed.

Wilson, D. M., Cookson, M. R., Van Den Bosch, L., Zetterberg, H., Holtzman, D. M., & Dewachter, I. (2023). Hallmarks of neurodegenerative diseases. *Cell, 186*, 693–714. https://doi.org/10.1016/j.cell.2022.12.032.

Witke, W. (2004). The role of profilin complexes in cell motility and other cellular processes. *Trends in Cell Biology, 14*, 461–469. https://doi.org/10.1016/j.tcb.2004.07.003.

Wu, C.-H., Fallini, C., Ticozzi, N., Keagle, P. J., Sapp, P. C., Piotrowska, K., ... Landers, J. E. (2012). Mutations in the profilin 1 gene cause familial amyotrophic lateral sclerosis. *Nature, 488*, 499–503. https://doi.org/10.1038/nature11280.

Yamanaka, K., Chun, S. J., Boillee, S., Fujimori-Tonou, N., Yamashita, H., Gutmann, D. H., ... Cleveland, D. W. (2008). Astrocytes as determinants of disease progression in inherited ALS. *Nature Neuroscience, 11*, 251–253. https://doi.org/10.1038/nn2047.

Yerbury, J. J., & Cashman, N. R. (2020). Selenium-based compounds: Emerging players in the ever-unfolding story of SOD1 in amyotrophic lateral sclerosis. *EBioMedicine, 59*, 102997. https://doi.org/10.1016/j.ebiom.2020.102997.

Younger, D., & Brown, R. (2023). Motor system disorders, part II: Spinal cord, neurodegenerative, and cerebral disorders and treatment. *Handbook of clinical neurology*. Elsevier, https://doi.org/10.1016/C2021-0-01304-3.

Zaccai, S., Nemirovsky, A., Lerner, L., Alfahel, L., Eremenko, E., Israelson, A., & Monsonego, A. (2024). CD4 T-cell aging exacerbates neuroinflammation in a late-onset mouse model of amyotrophic lateral sclerosis. *Journal of Neuroinflammation, 21*, 17. https://doi.org/10.1186/s12974-023-03007-1.

Zhao, J., Wang, X., Huo, Z., Chen, Y., Liu, J., Zhao, Z., ... Zhou, S. (2022). The impact of mitochondrial dysfunction in amyotrophic lateral sclerosis. *Cells, 11*, 2049. https://doi.org/10.3390/cells11132049.

Zhou, T., Ohulchanskyy, T. Y., & Qu, J. (2021). Effect of NIR light on the permeability of the blood-brain barriers in in vitro models. *Biomedical Optics Express, 12*, 7544–7555. https://doi.org/10.1364/BOE.438445.

CHAPTER EIGHT

# Impact of light on inner workings of neuronal functions

**Jose Martinez-Hernandez[a,*] and Alain Buisson[b,*]**
[a]Fonds de Dotation Clinatec, University Grenoble Alpes, Grenoble, France
[b]Grenoble Institut Neurosciences, University Grenoble Alpes, Inserm, U1216, Grenoble, France
*Corresponding authors. e-mail address: jose.martinezhernandez@cea.fr; alain.buisson@univ-grenoble-alpes.fr

## Contents



## Abstract

The aim of this chapter is to review the molecular mechanisms that underlie the effect of red and near-infrared light –collectively referred to as photobiomodulation (PBM)– on brain tissue. We discuss the therapeutic potential of PBM in the brain, considering both systemic actions and cell-type-specific responses. The prevailing hypothesis is that PBM enhances mitochondrial activity, leading to increased cellular energy production. While this mechanism may benefit many cell types, the central nervous system (CNS) appears particularly responsive to PBM due to either cell- specific signalling pathways that are highly energy-dependent or the existence of several mechanisms of action for PBM, some of them could be specific to CNS cells. In this chapter, we highlighted studies demonstrating PBM's functional impact on different CNS cell populations in order to identify the molecular and cellular mediators of its effects.

## 1. Introduction

Light is able to influence the inner working of all kind of eukaryotic cells, including plant and animal cells. As a source of energy, light can be captured and transformed to some form of chemical energy by specialised

*International Review of Neurobiology*, Volume 189
ISSN 0074-7742, https://doi.org/10.1016/bs.irn.2026.01.012

cells, but also light may be used by photoreceptor cells with specialised ion channels (opsins) to detect the environment and be the first-order neurones in a visual system necessary to navigate through our environment.

Does light change the inner working of other type of cells not necessarily specialised on detecting it and being part of the visual system? At high intensity, light provokes a thermal effect, and of course a temperature increase will affect the inner working of any kind of cell. However, if we stay on an intensity range that does not induce any major temperature increase, can we modify functions in cells that are not specialised in detecting light?

This is the concept of photobiomodulation (PBM), a light therapy approach that uses low-intensity, non-thermal light in the red or near-infrared wavelengths to stimulate biological processes in cells and tissues. It is well-known that ultraviolet light has deleterious effect on living organisms, usually by provoking DNA damage but PBM is based on the application of red-infrared light (typically in the 600–1400 nm range) on biological tissues. PBM has been shown to induce many beneficial effects on cell functions and improve cell survival leading to the idea that PBM may represent an alternative therapeutic approach to various diseases (for review Arany, 2025). Despite the growing body of evidence supporting PBM's therapeutic potential, the precise mechanisms by which red light produces these outcomes are still poorly understood. The most consensual hypothesis to explain these beneficial effects is the absorption of photons by endogenous chromophores, most notably cytochrome C oxidase (CCO) in the mitochondrial respiratory chain (Salehpour et al., 2018; Moro et al., 2022). This absorption could trigger an increased production of energy, reduce oxidative stress and on the long term, modulate gene expression and promote protein adaptations (Silva et al., 2025).

These effects can be achieved in most of the cells that compose the body; but it could be of major importance if we consider the nervous system, and more specifically neurones. Neurones have high energetic demands to maintain their activity and they contain a high number of mitochondria compared to other cell types. So, it can be assumed that in neurodegenerative diseases, but also after acute brain injuries, a boost of energy and the reduction of the oxidative stress will help the neurones to better recover from an alteration of their integrity (Huang et al., 2013; Xuan et al., 2014; Bathini et al., 2022; Stevens et al., 2024).

This is why PBM has gained considerable attention for its capacity to modulate cellular pathways, to enhance tissue repair, to provide neuroregenerative and anti-inflammatory responses with minimal side effects. Its potential therapeutic applications span a wide range of conditions, including neurological and neurodegenerative disorders, wound healing, and peripheral nerve regeneration (Rosso et al., 2018; Nairuz et al., 2024; Arany, 2025).

However, this simple statement becomes a complex issue when we try to translate into a suitable clinical treatment. What is the best wavelength of light to treat different diseases or for optimal tissue penetration? By what means, namely transcranial, intranasal illumination or by implanting an optical probe? What about the light scatter once within the body tissue? What are the energy levels that are necessary to induce beneficial effects? Is one single dose sufficient or should the treatment be maintained over time? Are all the cell types equally sensitive to PBM? There are still many open questions to reach a consensus on how to use PBM in neurones. Because even if we know the potential beneficial effects of applying red-infrared light in living tissues, we do not know how to translate PBM into a suitable clinical treatment. By better understanding the beneficial effects on neuronal functions, we will be able in a better position to optimise the PBM influence on the brain and obtain the most efficient clinical improvements in specific brain injuries.

One of the mechanisms underlying PBM's effects, involve the absorption of photons by mitochondrial cytochrome C oxidases. This absorption causes photodissociation of nitric oxide (NO) from its CCO binding sites, increasing the availability of electrons to reduce oxygen in the catalytic centre of CCO (Wu et al., 2014). As a consequence, the mitochondrial membrane potential (MMP) is increased leading to augmented adenosine-triphosphate (ATP) synthesis and to a reduction of reactive oxygen species (ROS) levels (Hamblin & Liebert 2022; Mitrofanis, 2019; Hoh Kam & Mitrofanis, 2023). In a longer-time scale, PBM has been shown to induce modulations of gene expression changing cellular protein content and metabolic pathways (Bathini et al., 2022).

An important question is whether these molecular responses to PBM are the only mechanism of action responsible for its beneficial effect, or if the red-infrared light affects other processes, such as the modification of water particles viscosity (Sommer, 2019), or affect some specific neuronal process, like a direct aperture of selective ion channels (Zhang et al.,2024); both of them potential mechanisms that would rapidly modify neuronal activity.

Interestingly, several non-mitochondrial mediated effects have been described to explain PBM-induced cellular benefits. For instance, cell proliferation induced by 660 nm wavelength has been shown to occur in cell lines that do not express assembled CCO, suggesting alternative mitochondrial or non-mitochondrial chromophores may also mediate PBM effects (Lima et al., 2019). These cell lines lacking CCO also showed an increase of ATP production and citrate synthase levels after PBM. This diversity in mechanistic pathways underscores both the complexity and therapeutic promise of PBM across a large range of cell types and conditions.

Nevertheless, it is important to know if these changes in cellular mechanisms provoked by PBM can help the brain to cope with neurodegenerative diseases in a direct or indirect pathway to optimise PBM protocol. An indirect pathway would involve the response of peripheral tissue to PBM that promotes a global response such as a reduction of inflammation or an increase of energy production that would help the brain to cope with pathological situations. On the other hand, a direct impact of PBM would affect directly neurones and/or astrocytes by preserving their functions, giving a better cellular response to stressful situations (Mitrofanis, 2019).

The variability among the PBM parameters is considerable and there is still much uncertainty and debate over the optimal PBM protocols, limiting the efficacy of treatment and/or translation to clinical application. A biphasic low dose exposure has been frequently used to induce beneficial effect on neurones while increasing the power or the length of exposition to PBM treatment reduce its beneficial effects (Huang et al., 2011). So, for optimising neuronal PBM treatment parameters, it is necessary to validate on a specific targeted neuronal readout as a required step to settle parameters that would most efficiently improve neuronal functions since out of the optimal conditions no beneficial effects of PBM may be observed.

Finally, mitochondria are present in almost all the cell types and red and near infrared light illumination will stimulate them, independently of the cell type. We can imagine a scenario where there would be both systemic and direct, inherent changes on the illuminated cells. These changes can act synergistically or not, but we do not know if all the cell types respond equally to the different light parameters. Therefore, even if the treatment with red-infrared light seems a rather, simple and innocuous treatment, there are still many open questions, that make PBM study complex. For this reason, we are going to go through the literature of PBM effect at different levels, from a systemic point of view to focus on what kind of effect and on different kind of cells, and specially on the neuronal activity and plasticity.

## 2. Complex systems, systemic effects

The CNS is a highly sophisticated system with specialised cell types exclusively present in the brain and in the spinal cord. In addition to neurones, the cells responsible to transmit and process the information, there are glial cells, supporting neurones in many different ways. The immune system has a specialised version involving microglia as key actor and even the blood vessels going through the CNS display a specialised features to generate the blood brain barrier, with the function to limit any harmful element from penetrating into the CNS and to maintain the appropriate composition in ions and nutrients of the brain environment (Daneman & Prat, 2015).

These cell types may respond differently to light exposure, but in addition, this distinct CNS environment is not disconnected to the rest of the body. Any effect on the vascular system will have an indirect effect on the exposure of light of the brain with a potential severe impact on the neuronal activity (Mitrofanis, 2019).

It has been reported that PBM induces NO release in primary neurones (Sharma et al., 2011) and in human endothelial cells (Yokomizo et al., 2022) cultures in a dose-response dependent way. PBM provokes photo-dissociation of NO from the mitochondrial electron transport chain (Boelens et al., 1983), allowing an increased generation of ATP in the illuminated cell. NO has also been characterised as a neuronal signalling molecule that can diffuse freely through cell membrane, with among other functions, the ability to couple neuronal activity and blood vessels vaso-dilation, improving cerebral blood flow (CBF) in a brain active zone (Uozumi et al., 2010; Wang et al., 2017). This improved cerebral blood perfusion may further support neuronal health and function. Uozumi and colleagues showed that PBM treatment (1.6 W/cm$^2$, 15–45 min) increased local CBF by 30 % compared to control mice. This increase on the CBF was concomitant with an increase in cerebral NO concentration. Similar results were obtained in humans, by using near-infrared spectroscopy, Wang et al. (2019) recorded changes in concentrations of oxidised cytochrome c oxidase, as well as oxyhemoglobin and deoxyhemoglobin after applying PBM on the forehead of healthy human subjects.

In various animal models of disease, it has been shown that PBM can increase neuronal NO production, expand blood vessel diameter leading to an enhanced CBF subsequently reducing ischaemic brain injury (Lee et al., 2017). The PBM-induced increased CBF has also been described in humans, even if the causal link between improved CBF and NO was not established. In prolonged vegetative state patients, repeated exposure to PBM sessions in the forehead provoked an enhanced local blood perfusion in the frontal areas of cortex (Nawashiro et al., 2012). Similarly, Salgado et al. (2015) showed in healthy elderly women that there is an improvement in the blood flow (in terms of speed, resistance and pulsatility) on the cerebral arteries analysed after a long-term PBM treatment (4 weeks/twice a week/4 head points).

In addition to vasodilation and an increased CBF, PBM treatment showed another beneficial effect through the vascular system by the modulation of neuroinflammation. PBM has been shown to decrease pro-inflammatory cytokines and to increase anti-inflammatory cytokines productions (Nairuz et al., 2024). This intriguing effect of PBM may occur through direct effects on glial cells, as a reduced level of reactive glia a key player in the production of neuroinflammatory response in the CNS, was observed. Dos Santos Cardoso et al. (2022) illuminated aged rats transcranially at 660 nm and 100 mW power for 10 consecutive days. The treatment promoted an increased level of Interleukin (IL)−1 alpha, a ubiquitous pro-inflammatory cytokine, and a decreased level of IL-5, also a pro-inflammatory cytokine, in the cerebral cortex. PBM treatment also increased similarly in the hippocampus the level of IL-1 alpha and decreased levels of IL-5, IL-18 (a pro-inflammatory cytokine) and fractalkine, a chemokine participating in neuroinflammation.

This reduction in the inflammatory status by PBM is not restricted to the CNS. PBM has been shown to change inflammatory cytokines expression in other cell types outside the central nervous system. In human synovial cell MH7A cultures, PBM decreases inflammatory factors (IL-6, IL-1β, IL-8, and Matrix MetalloProteinase-3) and increased anti-inflammatory factor IL-10 (Meng et al. 2020). In human gingival fibroblast cells, Lim et al. (2013) found that 635 nm light irradiation lead the inhibition of several pro-inflammatory cytokines, including I-κB kinase (IKK) and nuclear factor-kappa B (NF-κB), part of a signalling pathways regulating inflammatory responses. This anti-inflammatory effect observed in peripheral tissues may decrease the impact of circulating cytokines on the brain and promotes beneficial effects of PBM in the brain even if the targeted zones by PBM were not including the brain itself (Hennessy & Hamblin, 2017). This abscopal neuroprotective effect has been shown to induce neuroprotection in a mouse model of parkinsonism (Johnstone et al., 2014). Mice that received MPTP injection, a neurotoxin that selectively destroys dopaminergic neurones, were treated by PBM on body area except the brain, had a better survival of dopaminergic neurones than control, even if it was not as effective as treating directly the lesioned brain area. Similarly, Tobelem et al. (2023) treated with PBM at caudal vein/ artery rats with spinal cord injury showed a better functional and histological recovery of muscle and nerve tissue. So, even if the PBM was targeting body's area different than lesioned brain regions, the circulating beneficial cytokines or the reduction of deleterious cytokines participate to the improved recovery of the remote brain area.

## 3. Enhancing metabolism: Direct effects

The main hypothesis of PBM relies on the interaction of red or near-infrared light with cellular chromophores, primarily CCO in the mitochondrial respiratory chain which promotes an enhanced mitochondrial membrane potential, and consequently elevated ATP production. This boost in ATP fuels cellular metabolism and repair processes. We have to take in consideration that different cell types display considerable variation in both their mitochondrial number and activity, meaning that alterations of the mitochondrial respiratory chain can have cell-type specific consequences. For example, a neurone with high oxidative metabolism may be more responsive to PBM than a fibroblast which displays lower respiratory demands. Most of the cells can use as energy source glucose, fatty acids, and amino acids, but in the case of neurones they mostly depend on glucose for energy metabolism (Mergenthaler et al., 2013). So, it is possible that enhancing mitochondrial oxidative phosphorylation by PBM will have a major impact on neuronal metabolism.

In neuronal cultures, it has been shown that PBM at specific wavelengths and fluences results in a dose-dependent up-regulation of intracellular calcium concentration, ATP production, MMP and lowering ROS. Overall PBM promotes beneficial effects but only at lower doses with potential detrimental consequences at higher exposures (Sharma et al., 2011). Similar results indicate that when MMP is compromised due to oxidative stress or electron transport inhibition, PBM can restore MMP toward physiological levels and thereby reduce ROS production, supporting its therapeutic potential in managing oxidative pathologies (Huang et al., 2013).

Further supporting these mechanisms, research has shown that 670 nm light treatment elevates ATP synthase activity in the human brain of old adults, as revealed by in vivo magnetic resonance spectroscopy studies. These findings suggest that PBM may enhance mitochondrial function and brain energy metabolism, promoting beneficial effects against aging and for neuroprotection (Fear et al., 2023).

This increase production of ATP has an impact on the general metabolism, but also directly on the neurone's activity through the opening and regulation of several ion channels, including ATP-sensitive potassium channels, that relies on intracellular ATP. This effect was demonstrated in studies using HEK cells expressing Kir6.2 channels, where modulation of ATP by PBM contributed to altered channel activity (Zhong et al., 2014). Increased extracellular ATP levels following PBM also regulate calcium influx via ATP-gated P2X receptors and modulate cyclic AMP signalling, illustrating multiple avenues by which PBM-driven mitochondrial activity impacts cellular physiology (Jiang et al., 2013).

## 4. Non-neuronal cells activity

The CNS is composed by different kind of cells, mainly neurones, astrocytes, and microglia. Efficient neuronal activity very much depends on the supporting environment that the astrocytes and microglia provide; in addition to the neurones, both these glia cell types are potential targets of PBM treatment.

El Massri et al. (2018) studied the number of both astrocytes and microglia in the striatum after PBM in aged mice. 5-month-old mice were treated with 670 nm light for 20 min per day until they were 12-month-old. Using immunohistochemical markers, they observed a clear reduction in glial cell number, both astrocytes and microglia, compared to 12-month-old control animals without affecting the number of neurones or axonal terminations in this nucleus. Long-term treatment with PBM was enough to reduce the number of supporting glia on aged mice.

On the other hand, using a 660 nm light treatment over an optimised astrocyte culture Yoon et al. (2021) observed an astrocyte-specific proliferation by the increased co-expression of glial fibrillary acidic protein (GFAP) and bromodeoxyuridine (BrdU), without affecting other cell types in the culture. On a follow-up paper, the same group studied the effect of PBM on astrocyte migration (Yoon et al., 2023) showing significant difference in astrocyte migration on two tests depending on PBM fluence. With these specific fluence conditions, they also observed differences cell morphology and in drebrin expression, an actin-binding protein, essential for scar formation. Hence, on a single dose or short treatment, PBM is able to increase astrocytic numbers and help to their differentiation and migration.

Several other studies claim that PBM can have a pivotal role in the action or role of the astrocytes, specially after traumatic situations. For example, after mild traumatic brain injury (mTBI) on rats, the animals treated with 660, 810 nm or both wavelengths simultaneously, performed better on the behavioural tests. Further, brain sections showed significantly reduced activation of both CD11b+ microglia and GFAP+ astrocytes at 3 days post-injury (Stevens et al., 2024). This reduction of the astrocyte/microglial activation would provoke that there is less apoptosis on the area affected. PBM could inhibit neurotoxic astrocytic polarisation. Similarly, in a stroke animal model, Feng et al. (2023) showed that after 2-min 808 nm light treatment for 7 days significantly increased synaptic proteins but also increased neuroprotective astrocytic marker S100 Calcium Binding Protein A10 (S100A10) and inhibited neurotoxic astrocytic marker C3d in the peri-infarct region after ischaemic stroke. On cell culture studies, one single dose was enough to reduce C3d on astrocytes exposed to oxygen-glucose deprivation (Feng et al., 2023).

It has been described that after spinal cord injury (SCI), reactive astrocytes can be classified into two phenotypes depending on their different functions: neurotoxic (A1) astrocytes and neuroprotective (A2) astrocytes (Liddelow et al., 2017). PBM promotes functional recovery, motor function recovery and improve tissue repair after SCI through polarisation of macrophage/microglia (Song et al., 2017). Wang et al. (2021) showed that PBM was upregulating the expression of basic fibroblast growth factor (bFGF) and transforming growth factor-β (TGF-β) and both factors contributed to the transformation of A1/A2 astrocytes in a dose-dependent manner. (Wang et al., 2021). It seems that this expression could be regulated by enzymatic cascades, since they publish the PBM was able to inhibit Lcn2/JAK2-STAT3 crosstalk after spinal cord injury in male rats and inhibiting the activation of neurotoxic microglia and astrocytes by (Wang et al., 2021b).

## 5. Ion channels

Given the central role of ion channels in neuronal physiology and disease, a deeper understanding of PBM's effects at the ion channel level is essential to optimise its clinical applications and developing targeted therapies. The ability of PBM to influence ion channels offers a non-invasive tool for neurostimulation that might be crucial for neurological disease treatments (Ghaderi et al., 2021).

PBM can indirectly regulate channel function by increasing mitochondrial ATP generation, but also it is hypothesised that it could directly alter ion channel protein structure via photon-induced physical interactions or through some other molecular targets (Hamblin & Liebert 2022). Regulating the flow of ions across the cell membrane has a direct impact on cellular excitability, signalling, and homeostasis. All of these are vital processes that could be modulated by light through the ion channels (Zhang et al., 2024). Recent research indicates that in various ion channels, PBM directly affects their activity including ion channels for potassium, sodium, calcium, as well as neurotransmitter receptors such as glutamate and acetylcholine receptors (Zhang et al., 2024). These changes in ion channels properties are thought to contribute to PBM's therapeutic effects in the CNS, where ion channel activity underlies synaptic transmission, neuroplasticity, and neuroprotection (Hamblin & Liebert, 2022).

Neurones are electrical units that have to maintain a negative membrane potential on their membranes. This is because they need to be able to transmit signals rapidly between their different constituent, in order to create action potentials, as well as the final synaptic release of neurotransmitters. In this entire process, the control and properties of ion channels are fundamental. The membrane potential and its properties depend on the ion channels present on the plasma membrane and their dynamics. Astrocytes, even if they are non-excitable cells, have a hyperpolarised membrane (Du et al., 2016) and have many ion channels that play supportive roles to the neuronal activity. Astrocytes regulate neuronal activity through different mechanisms, including the homeostatic control of extracellular levels of ions and neurotransmitters and the secretion of signalling molecules that regulate synaptic formation, structure, and function (Chung et al., 2015). The diversity of mechanisms involved in the regulation of membrane potential and the different types of treatments applied hampers a comprehensive meta-analysis for evaluating light treatment strategies effectively.

Of all the neural receptors impacted by PBM treatment, the ones associated with glutamate are the most comprehensively studied, given their centrality in excitatory neurotransmission and plasticity. Golovynska et al. (2019), showed in HeLa cells that 808 nm laser irradiation induced a substantial $Ca2+$ influx depending on the subtype of ionotropic glutamatergic receptors important for synaptic plasticity, the N-methyl-D-aspartate receptors (NMDAR), since this effect was abolished by using memantine, a non-competitive NMDAR antagonist. In a more recent study, (Golovynska et al., 2021) they characterised the impact of 650 and 808 nm laser irradiation on cultured neurones (among other cells types), and reported that these wavelengths of light cause cell membrane depolarisation, that promotes NMDARs activation.

NMDAR's rapid response and low depolarisation potential highlights their pivotal role in the initial increase in $Ca2+$ influx that triggers several neuronal signalling cascades. Indeed, PBM's influence extends to intracellular signalling pathways activation such as MAPK/ERK and AKT/PI3K, both essential for neuronal differentiation and plasticity (Bathini et al., 2022). Similarly, it has been shown that there is a PBM-induced upregulation of vesicular monoamine transporter 2 (VMAT2), via activation of ERK/CREB/VMAT2 pathway. This supports the notion that an increase in VMAT2 expression would change the dopamine synaptic vesicle content and therefore would reduce neurotoxicity in $MPP^{+}$-induced neuronal injury in a cellular model of Parkinson's disease (Gu et al., 2017).

It has also been reported that a direct effect of PBM of channel proteins changes the kinetics, action potential dynamics and the spiking rate on single neurones, even if an effect of temperature mediated changes cannot be totally excluded (Garrido-Peña et al., 2024). The authors of this study identified two different population of neurones of the water snail Lymnaea stagnalis using sharp electrodes to obtain intracellular electrophysiological recordings. They showed that short PBM exposure induced a rapid modification of neuronal activity and a fast recovery of the neuronal dynamics at the end of the stimulation. They claimed that such a rapid effect of PBM on neuronal activity suggested a direct effect on the ionic channels not mediated by an increase of ATP after stimulating CCO but with a key role of temperature on specific channel properties.

A recent study by Armăşescu et al. (2025) using whole-cell patch-clamp recordings on primary cultured rat neurones reported an increase in the resting membrane potential following a 3-minute exposure to

808 nm laser light. This effect persisted for approximately 10 min before returning to baseline values. Such an increase in membrane potential could enhance neuronal excitability. To further investigate the excitability and functional properties of voltage-gated $Na^+$ channels that contribute to the resting membrane potential, the authors conducted voltage-clamp experiments on human embryonic kidney (HEK293) cells stably expressing human voltage-dependent $Na^+$ channels (Nav1.5). They observed a slight increase in the amplitude of the $Na^+$ current peak following laser application. Although the underlying mechanism remains unclear, the authors proposed that enhanced ATP production induced by PBM could support more efficient phosphorylation of $Na^+$ channels. Alternatively, the intense light might directly influence macromolecular polarisation, including ion channels, thereby modulating their gating properties.

In other non-neural cells (as adipocytes or mast cells, but also Xenopus oocytes expressing TRPV1) it has been proposed that PBM affects directly TRPV channels, activating different subtypes of TRPV channels at −60 mV (Wang et al., 2014). The authors suggest that TRPV channels by absorbing directly light through various chromophores in their structures may act as switches to influence the function of membrane proteins, a molecular mechanism that could be also effective in the TRPV channels present in the nervous system.

In summary, notwithstanding major advances in recent years, the precise molecular underpinnings by which molecular mechanism PBM affects ion channel function are not entirely clear. This highlights the importance of future studies using integrative approaches such as patch-clamp electrophysiology.

## 6. Neuronal activity

There are several ways to monitor neuronal activity, from a single neurone (patch-clamp) to complex network activities (field recordings, early genes biomarkers) and to whole brain activity recording (electroencephalography, functional magnetic resonance imaging).

Coupling PBM with neuronal patch-clamp approaches offer a powerful tool to dissect the mechanisms by which light may modulate neuronal activity, enabling researchers to observe direct impact of specific wavelengths, intensities, and durations of light exposure on ion channel behaviour in health and disease. Further, the use of selective pharmacological compounds that block ion channels helps to decipher the dynamic modifications and spiking characteristics induced by PBM (Garrido-Peña et al., 2024).

Patch-clamp has been used to measure the effect of PBM on the firing ratio of single neurones of the dorsal horn of the spinal cord. Uta et al. (2023) found that PBM inhibits neuronal firing of neurones located in lamina II of the dorsal horn, suggesting that laser irradiation can selectively influence Aδ and/or C fibres that conduct nociceptive stimuli. By using extracellular recording techniques, Ishibashi et al. (2024) found that exposing light directly on the sciatic nerve, transiently increases the frequency of spontaneous firing in the superficial layer of neurones in the dorsal horn without affecting the deep layers of the dorsal horn. This PBM-induced effect relates to a stimulus that does not generate pain and highlights the potential therapeutical interests of PBM in the management of pain (Chow & Armati, 2016).

Yoo et al. (2013) has studied the effect of near infrared light on deep brain structures in vivo, by using extracellular recordings. They stereotaxically placed optrodes, an extracellular recording device, in the globus pallidus or in the subthalamic nucleus (STN) of the basal ganglia. They showed that, in most of the animals treated by 880 nm laser, there was an increase in both excitatory and inhibitory neurotransmission. These authors also reported that, only after radiations above 20 mW, there was an accompanying temperature increase of more than 2 °C, and that this elicited a statistically significant neural response. This study demonstrated the feasibility of near infrared laser stimulation in in-vivo deep brain structures, suggesting the possibility of this technique as a promising new tool to potentially supplement or replace the current method of electrical deep brain stimulation in humans.

Transcranial PBM in a parkinsonian model of mice treated with MPTP, in addition to have a neuroprotective effect, also has also an effect on the global pattern of brain activity. After MPTP insult, the authors observed a PBM-induced reduction of 30–45 % of the number in Fos(+) cell in the STN and in the zona incerta. This effect was considered as a "repairing" mechanism that partially reduced abnormal neuronal activity generated by the toxin injection (Shaw et al., 2012). Similarly, by transcranial application of PBM in mice, Pan et al. (2025) observed an increase in c-Fos expression and calcium signalling intensity in hippocampal neurones of the dentate gyrus. This activation promotes an improved cognitive function and learning behaviour of the animals treated by PBM; they each displayed a better score in spatial memory tests. This effect occurred in naïve animals, but also helps to attenuate cognitive dysfunction observed in animals recovering from anaesthesia.

Finally in humans, transcranial PBM has been shown to increase neuronal activity, as evidenced by changes in EEG power spectra (Vargas et al.,2017; Zomorrodi et al., 2019; Wang et al., 2019) and improved cortical function monitored by fMRI (Vargas et al.,2017; Dole et al., 2023). Vargas et al. (2017) in an exploratory experiment with 12 healthy old adults treated over 5 weeks, showed an improved cognitive score after PBM treatment in all participants, but also found an increased resting-state EEG alpha, beta, and gamma power and a more efficient prefrontal blood-oxygen-level dependent (BOLD)-fMRI response. In another pilot study with 20 healthy subjects (Wang et al., 2019) compared the effect of a single dose of PBM to the right forehead on the brain rhythms before, during and after the treatment. They observed an increase of the oscillations on the alpha and beta bands on a time scale of minutes. Similarly, Zomorrodi et al. (2019) measured the resting state of healthy subjects in response to a single dose of transcranial PBM. They observed significant effects on neural oscillations with an increased power of alpha, beta and gamma (the higher oscillatory frequencies), but a reduced power in delta and theta (the slower frequencies).

Overall, these studies bring an interesting perspective: that PBM exerts a stimulating effect on neuronal activity observable at a single neurone level or by recording network or global activity of the CNS. The molecular mechanism is not well defined, but this stimulating effect presents the advantage to stimulate deep brain structures without the implantation of optic probes and to be efficient in humans. The challenge to use PBM for treating specific neuronal activity alterations induced by neurological or neurodegenerative diseases will depend on our understanding of which brain's structures to target to generate the most suitable effects for the patients.

## 7. Neuronal plasticity

Neuroplasticity can be defined as the brain's ability to reorganise its neural circuitry in response to experiences or injury. PBM has been shown to facilitate considerable neuroplasticity . PBM acts on multiple levels: from fuelling energy-demanding restorative processes in neurones and glia to promote changes in electrical properties (Garrido-Peña et al., 2024), synaptic neurotransmitter release and reduced neurotoxicity (Gu et al., 2017), neuronal morphology (Meng et al., 2013) or neurogenesis (Oron et al., 2006).

A common method to study neuronal plasticity is by applying an electrophysiological stimulus to generate long-term potentiation (LTP), usually in the hippocampus. This stimulus leads to a persistent strengthening of synapses that can be monitored by extracellular recording. Previous studies have shown that PBM treatment increases the magnitude of LTP in wild type animals. More strikingly, it has been reported that, in a chronic (9 months) mouse model of Alzheimer's disease, treatment with PBM at a dose of 100 J over five consecutive days generates a rescue LTP, together with a partial long-term depression (LTD), another type neuronal plasticity (Buendía et al., 2022).

Other key findings involve the effect of PBM on neuronal morphology, since this modification may sustain the rewiring necessary for long term synaptic plasticity and improved cognitive function. Cardoso et al. (2022) showed that PBM treatment increases the number of both dendritic nodes and ends, while reducing the total dendritic length of neurones in the cerebral cortex of rats. Others have shown that dendrite growth was increased after PBM, as revealed by the upregulation of Rac1 activity and the increased expression of PSD-95, a scaffolding synaptic protein. In essence, PBM treatment has been shown to promote an increase in length, branching, and spine density in hippocampal neurones (Meng et al., 2013).

These modifications of neuronal morphology are determined by the organisation of the cytoskeleton. So, it is not unreasonable to suggest that PBM impacts on the cytoskeleton properties of neurones. It has been shown that PBM can modify the dynamics of microtubules and actin filaments. Both cytoskeletal components play crucial roles in maintaining neuronal structure, intracellular transport, plasticity, and synaptic function. Experimental evidence indicates that PBM induces reversible changes in neuronal microtubule assembly and disassembly, forming small varicosities (swellings) along axons that disappear after the treatment, suggesting a regulatory but non-damaging effect on cytoskeletal structures (Liebert et al., 2016). One possible explanation for this effect is the increase of ATP production, that can influence cytoskeletal remodelling by supplying energy to actomyosin motors and promoting the polymerisation and the stability of cytoskeletal components. Other studies have also suggested that

PBM may modulate tubulin polymerisation directly and/or through signalling pathways that control cytoskeleton organisation (Staelens et al., 2022). These studies reported on the effect of PBM on living cells and also isolated microtubules. They observed a variety of results, depending on the cell type studied, but also depending on the tubulin concentration. PBM increased depolymerisation or polymerisation, suggesting that it has a complex control with direct and indirect mechanisms that could be exerting opposing effects on the microtubule cytoskeleton.

Another explanation for these morphological changes is that they occur through an activation of transcription factors that upregulate the expression of protective and functional genes, including those encoding for neurotrophic factors such as brain-derived neurotrophic factor (BDNF; Meng et al., 2013), or glial cell-derived neurotrophic factor (GDNF; El Massri et al., 2017). These gene activation cascades work via vital signalling pathways, such as MAPK/ERK and AKT/PI3K, and are critical for neuronal growth, survival, and synaptic remodelling (Yan et al., 2017; Rhee et al., 2019). Taken together, these PBM effects may contribute substantially to the repair and regeneration of neuronal cells. The PBM effects described here may facilitate the restoration of neural function by promoting cellular resilience and enhancing the intrinsic regenerative mechanisms of the neurones.

In addition to the morphological changes, experimental evidence in animal models shows that PBM can stimulate neurogenesis in regions like the hippocampus and subventricular zone (SVZ). This is particularly the case in pathological states, such as traumatic brain injury, stroke, or in neurodegenerative disease, potentiating endogenous repair processes or enhancing the probability of success of cell-based therapies (Oron et al., 2006; Xuan et al., 2014, 2015; Hong et al. 2022). Oron et al. (2006) provided the first evidence of PBM stimulated neurogenesis and neuronal progenitor cell migration in rats. Their study demonstrated that PBM at 808 nm significantly increased proliferating cell numbers (by counting cells incorporating bromodeoxyuridine) in the subventricular zone (SVZ) of the hemisphere affected by stroke. Hong et al. (2022) showed in a mouse model of status epilepticus that PBM was able to restore normal hippocampal neurogenesis by enhancing the migration of maturing granular cells avoiding the characteristic neuronal loss observed in this epilepsy model. In a more traumatic context, Xuan et al. (2014) reported that transcranial PBM could stimulate the process of neurogenesis in the hippocampus and SVZ of mice subjected to traumatic brain injury (TBI). These newly produced neuroprogenitors could travel to the injured area of the cortex

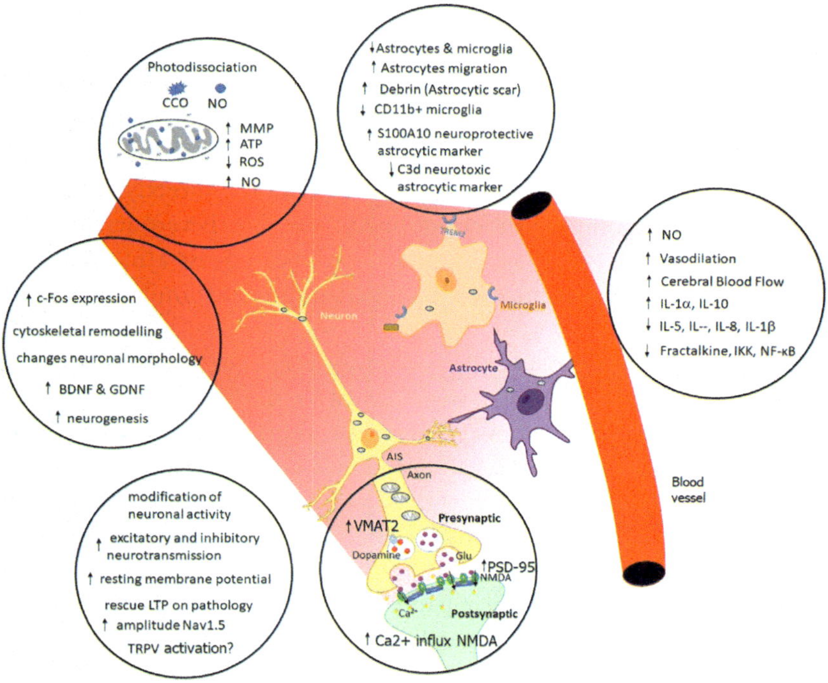


**Fig. 1 Schematic illustration of red and near-infrared light triggering molecular mechanisms in the CNS.** This figure depicts different effects on neurons, glial cells, and blood vessels following photobiomodulation, leading to increased mitochondrial activity, nitric oxide production, and synaptic modulation. It highlights enhanced neurotrophic support, neurogenesis, astrocyte and microglial changes, and restored synaptic plasticity. Vascular effects include nitric oxide–mediated vasodilation, increased cerebral blood flow, and modulation of inflammatory cytokines.

and improved the restoration processes of the damaged region. In another study (Xuan et al., 2015), the same group showed that PBM treatment promotes BDNF increase in the hippocampus and in the SVZ at one-week post TBI. At 4 weeks post TBI, they showed an increase in synaptogenesis in the cortex of PBM treated animals showing that new connections between existing brain cells could be stimulated by light.

## 8. Conclusions

The global aim of this review was to consider the literature associated with the cellular mechanisms that underpin the benefits of PBM on the central nervous system. We have highlighted studies that have reported the impact of PBM on the various cell types present in the nervous system, whether neurones or glial cells. In terms of the molecular mechanisms responsible for the effects of PBM, its main influence is on cellular metabolism by increasing ATP production, but it may also have direct effects on neuronal activity by directly or indirectly influencing membrane permeability to different ion species. The heterogeneity of the effects observed and of the cell types impacted suggest a mechanism of action on multiple molecular targets with an overall stimulating effect of PBM (Fig. 1).

The ever-increasing data collected across a number of labs world-wide, suggest that PBM represents a powerful tool for facilitating neuroplasticity, neuroprotection, and functional restoration, with a strong safety profile highlighted in current preclinical and clinical studies. However, the significant heterogeneity in wavelength, dose, pulse structure, and exposure protocols across studies has so far prevented the standardisation of PBM for widespread clinical adoption.

Depending on the particular pathology to be treated and the kind of effect expected, one would expect to have different protocols. As highlighted in many systematic reviews (Salehpour et al., 2018; Bathini et al., 2022; Zhang et al., 2024), the effect on CCO is well-demonstrated, but possibly not the only mechanism. A better understanding of the relationship between protocol parameters, mechanisms of action, ion channel subtypes and cellular context is essential to unlock PBM's full therapeutic potential.

## References

Arany, P. R. (2025). Photobiomodulation therapy. *JADA Foundational Science, 4*, 100045. https://doi.org/10.1016/j.jfscie.2025.100045.

Armăşescu, F., Amuzescu, B., Gheorghe, R. O., Ghenghea, M., Ristoiu, V., Ciurea, J., & Gruia, I. (2025). Fiber-optic-guided near-infrared laser exposure induces depolarization of cultured primary sensory neurons and modifies biophysical properties of human Nav1.5 channels. *Journal of Photochemistry and Photobiology. B, Biology, 269*, 113191. https://doi.org/10.1016/j.jphotobiol.2025.113191.

Bathini, M., Raghushaker, C. R., & Mahato, K. K. (2022). The molecular mechanisms of action of photobiomodulation against neurodegenerative diseases: A systematic review. *Cellular and Molecular Neurobiology, 42*(4), 955–971. https://doi.org/10.1007/s10571-020-01016-9.

Boelens, R., Wever, R., Van Gelder, B. F., & Rademaker, H. (1983). An EPR study of the photodissociation reactions of oxidised cytochrome c oxidase-nitric oxide complexes. *Biochimica et Biophysica Acta, 724*(2), 176–183. https://doi.org/10.1016/0005-2728(83)90136-6.

Buendía, D., Guncay, T., Oyanedel, M., Lemus, M., Weinstein, A., Ardiles, Á. O., ... Muñoz, P. (2022). The transcranial light therapy improves synaptic plasticity in the Alzheimer's disease mouse model. *Brain Sciences, 12*(10), 1272. https://doi.org/10.3390/brainsci12101272.

Cardoso, F. D. S., Serra, F. T., Coimbra, N. C., Gonzalez-Lima, F., & Gomes da Silva, S. (2022). Transcranial photobiomodulation changes neuronal morphology in the cerebral cortex of rats. *Neuroscience Letters, 781*, 136681. https://doi.org/10.1016/j.neulet.2022.136681.

Chow, R. T., & Armati, P. J. (2016). Photobiomodulation: Implications for anesthesia and pain relief. *Photomedicine and Laser Surgery, 34*(12), 599–609. https://doi.org/10.1089/pho.2015.4048.

Chung, W. S., Allen, N. J., & Eroglu, C. (2015). Astrocytes control synapse formation, function, and elimination. *Cold Spring Harbor Perspectives in Biology,* 7(9), a020370. https://doi.org/10.1101/cshperspect.a020370.

Daneman, R., & Prat, A. (2015). The blood-brain barrier. *Cold Spring Harbor Perspectives in Biology,* 7(1), a020412. https://doi.org/10.1101/cshperspect.a020412.

Dole, M., Auboiroux, V., Langar, L., & Mitrofanis, J. (2023). A systematic review of the effects of transcranial photobiomodulation on brain activity in humans. *Reviews in the Neurosciences, 34*(6), 671–693. https://doi.org/10.1515/revneuro-2023-0003.

Dos Santos Cardoso, F., Mansur, F. C. B., Araújo, B. H. S., Gonzalez-Lima, F., & Gomes da Silva, S. (2022). Photobiomodulation improves the inflammatory response and intracellular signaling proteins linked to vascular function and cell survival in the brain of aged rats. *Molecular Neurobiology, 59*(1), 420–428. https://doi.org/10.1007/s12035-021-02606-4.

Du, Y., Kiyoshi, C. M., Wang, Q., Wang, W., Ma, B., Alford, C. C., et al. (2016). Genetic deletion of TREK-1 or TWIK-1/TREK-1 potassium channels does not alter the basic electrophysiological properties of mature hippocampal astrocytes in situ. *Frontiers in Cellular Neuroscience, 10*, 13.

El Massri, N., Lemgruber, A. P., Rowe, I. J., Moro, C., Torres, N., Reinhart, F., ... Mitrofanis, J. (2017). Photobiomodulation-induced changes in a monkey model of Parkinson's disease: Changes in tyrosine hydroxylase cells and GDNF expression in the striatum. *Experimental Brain Research, 235*(6), 1861–1874. https://doi.org/10.1007/s00221-017-4937-0.

El Massri, N., Weinrich, T. W., Kam, J. H., Jeffery, G., & Mitrofanis, J. (2018). Photobiomodulation reduces gliosis in the basal ganglia of aged mice. *Neurobiology of Aging, 66*, 131–137. https://doi.org/10.1016/j.neurobiolaging.2018.02.019.

Fear, E. J., Torkelsen, F. H., Zamboni, E., Chen, K. J., Scott, M., Jeffery, G., ... Kennerley, A. J. (2023). Use of $^{31}$ P magnetisation transfer magnetic resonance spectroscopy to measure ATP changes after 670 nm transcranial photobiomodulation in older adults. *Aging Cell, 22*(11), e14005. https://doi.org/10.1111/acel.14005.

Feng, Y., Yang, L., Ma, X., Huang, Z., Zong, X., Citadin, C. T., ... Zhang, Q. (2023). Photobiomodulation treatment inhibits neurotoxic astrocytic polarization and protects neurons in in vitro and in vivo stroke models. *Neurochemistry International, 162*, 105464. https://doi.org/10.1016/j.neuint.2022.105464.

Garrido-Peña, A., Sanchez-Martin, P., Reyes-Sanchez, M., Levi, R., Rodriguez, F. B., Castilla, J., ... Varona, P. (2024). Modulation of neuronal dynamics by sustained and activity-dependent continuous-wave near-infrared laser stimulation. *Neurophotonics, 11*(2), 024308. https://doi.org/10.1117/1.NPh.11.2.024308.

Ghaderi, A. H., Jahan, A., Akrami, F., & Moghadam Salimi, M. (2021). Transcranial photobiomodulation changes topology, synchronizability, and complexity of resting state brain networks. *Journal of Neural Engineering, 18*(4), https://doi.org/10.1088/1741-2552/abf97c.

Golovynska, I., Golovynskyi, S., Stepanov, Y. V., Garmanchuk, L. V., Stepanova, L. I., Qu, J., & Ohulchanskyy, T. Y. (2019). Red and near-infrared light induces intracellular Ca2+ flux via the activation of glutamate N-methyl-D-aspartate receptors. *Journal of Cellular Physiology, 234*(9), 15989–16002. https://doi.org/10.1002/jcp.28257.

Golovynska, I., Golovynskyi, S., Stepanov, Y. V., Stepanova, L. I., Qu, J., & Ohulchanskyy, T. Y. (2021). Red and near-infrared light evokes Ca2+ influx, endoplasmic reticulum release and membrane depolarization in neurons and cancer cells. *Journal of Photochemistry and Photobiology. B, Biology, 214*, 112088. https://doi.org/10.1016/j.jphotobiol.2020.112088.

Gu, X., Liu, L., Shen, Q., & Xing, D. (2017). Photoactivation of ERK/CREB/VMAT2 pathway attenuates MPP+-induced neuronal injury in a cellular model of Parkinson's disease. *Cellular Signalling, 37*, 103–114. https://doi.org/10.1016/j.cellsig.2017.06.007.

Hamblin, M. R., & Liebert, A. (2022). Photobiomodulation therapy mechanisms beyond cytochrome c oxidase. *Photobiomodulation, Photomedicine, and Laser Surgery, 40*(2), 75–77. https://doi.org/10.1089/photob.2021.0119.

Hennessy, M., & Hamblin, M. R. (2017). Photobiomodulation and the brain: a new paradigm (2010) *Journal of optics, 19*(1), 013003. https://doi.org/10.1088/2040-8986/19/1/013003.

Hoh Kam, J., & Mitrofanis, J. (2023). Glucose improves the efficacy of photobiomodulation in chlevels in mouse fibroblast cell culanging ATP and ROS levels in mouse fibroblast cell cultures. *Cells, 12*(21), 2533. https://doi.org/10.3390/cells12212533.

Hong, N., Kang, G. W., Park, J. O., Chung, P. S., Lee, M. Y., & Ahn, J. C. (2022). Photobiomodulation regulates adult neurogenesis in the hippocampus in a status epilepticus animal model. *Scientific Reports, 12*(1), 15246. https://doi.org/10.1038/s41598-022-19607-5.

Huang, Y. Y., Nagata, K., Tedford, C. E., McCarthy, T., & Hamblin, M. R. (2013). Low-level laser therapy (LLLT) reduces oxidative stress in primary cortical neurons in vitro. *Journal of Biophotonics, 6*(10), 829–838. https://doi.org/10.1002/jbio.201200157.

Huang, Y. Y., Sharma, S. K., Carroll, J., & Hamblin, M. R. (2011). Biphasic dose response in low level light therapy - An update. *Dose-Response: a publication of International Hormesis Society, 9*(4), 602–618. https://doi.org/10.2203/dose-response.11-009.Hamblin.

Ishibashi, N., Uta, D., Sawahata, M., & Kume, T. (2024). Photobiomodulation transiently increases the spontaneous firing in the superficial layer of the rat spinal dorsal horn. *Biochemical and Biophysical Research Communications, 729*, 150362. https://doi.org/10.1016/j.bbrc.2024.150362.

Jiang, R., Taly, A., & Grutter, T. (2013). Moving through the gate in ATP-activated P2X receptors. *Trends in Biochemical Sciences, 38*(1), 20–29. https://doi.org/10.1016/j.tibs.2012.10.006.

Johnstone, D. M., elMassri, N., Moro, C., Spana, S., Wang, X. S., Torres, N., ... Mitrofanis, J. (2014). Indirect application of near infrared light induces neuroprotection in a mouse model of parkinsonism - An abscopal neuroprotective effect. *Neuroscience, 274*, 93–101. https://doi.org/10.1016/j.neuroscience.2014.05.023.

Lee, H. I., Lee, S. W., Kim, S. Y., Kim, N. G., Park, K. J., Choi, B. T., & Shin, H. K. (2017). Pretreatment with light-emitting diode therapy reduces ischemic brain injury in mice through endothelial nitric oxide synthase-dependent mechanisms. *Biochemical and Biophysical Research Communications, 486*(4), 945–950. https://doi.org/10.1016/j.bbrc.2017.03.131.

Liddelow, S. A., Guttenplan, K. A., Clarke, L. E., Bennett, F. C., Bohlen, C. J., Schirmer, L., ... Barres, B. A. (2017). Neurotoxic reactive astrocytes are induced by activated microglia. *Nature, 541*(7638), 481–487. https://doi.org/10.1038/nature21029.

Liebert, A. D., Chow, R. T., Bicknell, B. T., & Varigos, E. (2016). Neuroprotective Effects Against POCD by Photobiomodulation: Evidence from Assembly/Disassembly of the Cytoskeleton. *Journal of experimental neuroscience, 10*, 1–19. https://doi.org/10.4137/JEN.S33444.

Lim, W., Kim, J., Kim, S., Karna, S., Won, J., Jeon, S. M., ... Kim, O. (2013). Modulation of lipopolysaccharide-induced NF-κB signaling pathway by 635 nm irradiation via heat shock protein 27 in human gingival fibroblast cells. *Photochemistry and Photobiology, 89*(1), 199–207. https://doi.org/10.1111/j.1751-1097.2012.01225.x.

Lima, P. L. V., Pereira, C. V., Nissanka, N., Arguello, T., Gavini, G., Maranduba, C. M. D. C., ... Moraes, C. T. (2019). Photobiomodulation enhancement of cell proliferation at 660 nm does not require cytochrome c oxidase. *Journal of Photochemistry and Photobiology. B, Biology, 194*, 71–75. https://doi.org/10.1016/j.jphotobiol.2019.03.015.

Meng, C., He, Z., & Xing, D. (2013). Low-level laser therapy rescues dendrite atrophy via upregulating BDNF expression: implications for Alzheimer's disease. *The Journal of neuroscience: the official journal of the Society for Neuroscience, 33*(33), 13505–13517. https://doi.org/10.1523/JNEUROSCI.0918-13.2013.

Meng, C., Xia, Q., Wu, H., Huang, H., Liu, H., Li, Y., ... Song, W. (2020). Photobiomodulation with 630-nm LED radiation inhibits the proliferation of human synoviocyte MH7A cells possibly via TRPV4/PI3K/AKT/mTOR signaling pathway. *Lasers in Medical Science, 35*(9), 1927–1936. https://doi.org/10.1007/s10103-020-02977-5.

Mergenthaler, P., Lindauer, U., Dienel, G. A., & Meisel, A. (2013). Sugar for the brain: The role of glucose in physiological and pathological brain function. *Trends in Neurosciences, 36*(10), 587–597. https://doi.org/10.1016/j.tins.2013.07.001.

Mitrofanis, J. (2019). *Run in the Light.* Morgan & Claypool Publishers, https://doi.org/10.1088/2053-2571/ab2f70.

Moro, C., Valverde, A., Dole, M., Hoh Kam, J., Hamilton, C., Liebert, A., ... Mitrofanis, J. (2022). The effect of photobiomodulation on the brain during wakefulness and sleep. *Frontiers in Neuroscience, 16*, 942536. https://doi.org/10.3389/fnins.2022.942536.

Nairuz, T., Sangwoo-Cho, & Lee, J. H. (2024). Photobiomodulation therapy on brain: Pioneering an innovative approach to revolutionize cognitive dynamics. *Cells, 13*(11), 966. https://doi.org/10.3390/cells13110966.

Nawashiro, H., Wada, K., Nakai, K., & Sato, S. (2012). Focal increase in cerebral blood flow after treatment with near-infrared light to the forehead in a patient in a persistent vegetative state. *Photomedicine and Laser Surgery, 30*(4), 231–233. https://doi.org/10.1089/pho.2011.3044.

Oron, A., Oron, U., Chen, J., Eilam, A., Zhang, C., Sadeh, M., ... Chopp, M. (2006). Low-level laser therapy applied transcranially to rats after induction of stroke significantly reduces long-term neurological deficits. *Stroke; a Journal of Cerebral Circulation, 37*(10), 2620–2624. https://doi.org/10.1161/01.STR.0000242775.14642.b8.

Pan, W. T., Gu, H. W., Zhou, Y. B., Ma, S., Ma, D., Liu, P. M., & Yang, J. J. (2025). Photo-biomodulation of the hippocampus using near-infrared laser to enhance cognitive function in mice. *Advanced Science (Weinheim, Baden-Wurttemberg, Germany), 12*(22), e2417380. https://doi.org/10.1002/advs.202417380.

Rhee, Y. H., Moon, J. H., Jung, J. Y., Oh, C., Ahn, J. C., & Chung, P. S. (2019). Effect of photobiomodulation therapy on neuronal injuries by ouabain: The regulation of Na, K-ATPase; Src; and mitogen-activated protein kinase signaling pathway. *BMC Neuroscience, 20*(1), 19. https://doi.org/10.1186/s12868-019-0499-3.

Rosso, M. P. O., Buchaim, D. V., Kawano, N., Furlanette, G., Pomini, K. T., & Buchaim, R. L. (2018). Photobiomodulation therapy (PBMT) in peripheral nerve regeneration: A systematic review. *Bioengineering (Basel, Switzerland), 5*(2), 44. https://doi.org/10.3390/bioengineering5020044.

Salehpour, F., Mahmoudi, J., Kamari, F., Sadigh-Eteghad, S., Rasta, S. H., & Hamblin, M. R. (2018). Brain photobiomodulation therapy: A narrative review. *Molecular Neurobiology, 55*(8), 6601–6636. https://doi.org/10.1007/s12035-017-0852-4.

Salgado, A. S., Zângaro, R. A., Parreira, R. B., & Kerppers, I. I. (2015). The effects of transcranial LED therapy (TCLT) on cerebral blood flow in the elderly women. *Lasers in Medical Science, 30*(1), 339–346. https://doi.org/10.1007/s10103-014-1669-2.

Sharma, S. K., Kharkwal, G. B., Sajo, M., Huang, Y. Y., De Taboada, L., McCarthy, T., & Hamblin, M. R. (2011). Dose response effects of 810 nm laser light on mouse primary cortical neurons. *Lasers in Surgery and Medicine, 43*(8), 851–859. https://doi.org/10.1002/lsm.21100.

Shaw, V. E., Peoples, C., Spana, S., Ashkan, K., Benabid, A. L., Stone, J., ... Mitrofanis, J. (2012). Patterns of cell activity in the subthalamic region associated with the neuroprotective action of near-infrared light treatment in MPTP-treated mice. *Parkinson's Disease, 2012*, 296875. https://doi.org/10.1155/2012/296875.

Silva, A. V. S., Sousa, F. D., Sousa, B. F., Santos, W. V., Oliveira, A. E. R., Lobo, M. D. P., ... de Sousa, J. S. (2025). Biomechanical insights into the proteomic profiling of cells in response to red light absorption. *Small (Weinheim an der Bergstrasse, Germany), 21*(9), e2410062. https://doi.org/10.1002/smll.202410062.

Sommer, A. P. (2019). Mitochondrial cytochrome c oxidase is not the primary acceptor for near infrared light-it is mitochondrial bound water: The principles of low-level light therapy. *Annals of Translational Medicine,* 7(1), S13. https://doi.org/10.21037/atm.2019.01.43.

Song, J. W., Li, K., Liang, Z. W., Dai, C., Shen, X. F., Gong, Y. Z., ... Wang, Z. (2017). Low-level laser facilitates alternatively activated macrophage/microglia polarization and promotes functional recovery after crush spinal cord injury in rats. *Scientific Reports,* 7(1), 620. https://doi.org/10.1038/s41598-017-00553-6.

Staelens, M., Di Gregorio, E., Kalra, A. P., Le, H. T., Hosseinkhah, N., Karimpoor, M., ... Tuszyński, J. A. (2022). Near-infrared photobiomodulation of living cells, tubulin, and microtubules in vitro. *Frontiers in Medical Technology, 4*, 871196. https://doi.org/10.3389/fmedt.2022.871196.

Stevens, A. R., Hadis, M., Thareja, A., Anderson, F. G., Milward, M. R., Di Pietro, V., ... Ahmed, Z. (2024). Photobiomodulation improves functional recovery after mild traumatic brain injury. *Bioengineering & Translational Medicine, 10*(2), e10727. https://doi.org/10.1002/btm2.10727.

Tobelem, D. D. C., Andreo, L., Silva, T., Malavazzi, T. C. S., Martinelli, A., Horliana, A. C. R. T., ... Mesquita-Ferrari, R. A. (2023). Systemic vascular photobiomodulation accelerates the recovery of motor activity in rats following spinal cord injury. *Lasers in Surgery and Medicine, 55*(6), 577–589. https://doi.org/10.1002/lsm.23665.

Uozumi, Y., Nawashiro, H., Sato, S., Kawauchi, S., Shima, K., & Kikuchi, M. (2010). Targeted increase in cerebral blood flow by transcranial near-infrared laser irradiation. *Lasers in Surgery and Medicine, 42*(6), 566–576. https://doi.org/10.1002/lsm.20938.

Uta, D., Ishibashi, N., Konno, T., Okada, Y., Kawase, Y., Tao, S., & Kume, T. (2023). Near-infrared photobiomodulation of the peripheral nerve inhibits the neuronal firing in a rat spinal dorsal horn evoked by mechanical stimulation. *International Journal of Molecular Sciences, 24*(3), 2352. https://doi.org/10.3390/ijms24032352.

Vargas, E., Barrett, D. W., Saucedo, C. L., Huang, L. D., Abraham, J. A., Tanaka, H., ... Gonzalez-Lima, F. (2017). Beneficial neurocognitive effects of transcranial laser in older adults. *Lasers in Medical Science, 32*(5), 1153–1162. https://doi.org/10.1007/s10103-017-2221-y.

Wang, X., Dmochowski, J. P., Zeng, L., Kallioniemi, E., Husain, M., Gonzalez-Lima, F., & Liu, H. (2019). Transcranial photobiomodulation with 1064-nm laser modulates brain electroencephalogram rhythms. *Neurophotonics, 6*(2), 025013. https://doi.org/10.1117/1.NPh.6.2.025013.

Wang, X., Li, X., Zuo, X., Liang, Z., Ding, T., Li, K., ... Wang, Z. (2021b). Photobiomodulation inhibits the activation of neurotoxic microglia and astrocytes by inhibiting Lcn2/JAK2-STAT3 crosstalk after spinal cord injury in male rats. *Journal of Neuroinflammation, 18*(1), 256. https://doi.org/10.1186/s12974-021-02312-x.

Wang, X., Tian, F., Reddy, D. D., Nalawade, S. S., Barrett, D. W., Gonzalez-Lima, F., & Liu, H. (2017). Up-regulation of cerebral cytochrome-c-oxidase and hemodynamics by transcranial infrared laser stimulation: A broadband near-infrared spectroscopy study. *Journal of Cerebral Blood Flow and Metabolism: official journal of the International Society of Cerebral Blood Flow and Metabolism, 37*(12), 3789–3802. https://doi.org/10.1177/0271678X17691783.

Wang, L., Zhang, D., & Schwarz, W. (2014). TRPV channels in mast cells as a target for low-level-laser therapy. *Cells, 3*(3), 662–673. https://doi.org/10.3390/cells3030662.

Wang, X., Zhang, Z., Zhu, Z., Liang, Z., Zuo, X., Ju, C., ... Wang, Z. (2021). Photobiomodulation promotes repair following spinal cord injury by regulating the transformation of A1/A2 reactive astrocytes. *Frontiers in Neuroscience, 15*, 768262. https://doi.org/10.3389/fnins.2021.768262.

Wu, S., Zhou, F., Wei, Y., Chen, W. R., Chen, Q., & Xing, D. (2014). Cancer phototherapy via selective photoinactivation of respiratory chain oxidase to trigger a fatal superoxide anion burst. *Antioxidants & Redox Signaling, 20*(5), 733–746. https://doi.org/10.1089/ars.2013.5229.

Xuan, W., Agrawal, T., Huang, L., Gupta, G. K., & Hamblin, M. R. (2015). Low-level laser therapy for traumatic brain injury in mice increases brain derived neurotrophic factor (BDNF) and synaptogenesis. *Journal of Biophotonics, 8*(6), 502–511. https://doi.org/10.1002/jbio.201400069.

Xuan, W., Vatansever, F., Huang, L., & Hamblin, M. R. (2014). Transcranial low-level laser therapy enhances learning, memory, and neuroprogenitor cells after traumatic brain injury in mice. *Journal of Biomedical Optics, 19*(10), 108003. https://doi.org/10.1117/1.JBO.19.10.108003.

Yan, X., Liu, J., Zhang, Z., Li, W., Sun, S., Zhao, J., ... Sun, H. (2017). Low-level laser irradiation modulates brain-derived neurotrophic factor mRNA transcription through calcium-dependent activation of the ERK/CREB pathway. *Lasers in Medical Science, 32*(1), 169–180. https://doi.org/10.1007/s10103-016-2099-0.

Yokomizo, S., Roessing, M., Morita, A., Kopp, T., Ogawa, E., Katagiri, W., ... Kashiwagi, S. (2022). Near-infrared II photobiomodulation augments nitric oxide bioavailability via phosphorylation of endothelial nitric oxide synthase. *FASEB Journal: official publication of the Federation of American Societies for Experimental Biology, 36*(9), e22490. https://doi.org/10.1096/fj.202101890R.

Yoo, M., Koo, H., Kim, M., Kim, H. I., & Kim, S. (2013). Near-infrared stimulation on globus pallidus and subthalamus. *Journal of Biomedical Optics, 18*(12), 128005. https://doi.org/10.1117/1.JBO.18.12.128005.

Yoon, S. R., Hong, N., Lee, M. Y., & Ahn, J. C. (2021). Photobiomodulation with a 660-nanometer light-emitting diode promotes cell proliferation in astrocyte culture. *Cells, 10*(7), 1664. https://doi.org/10.3390/cells10071664.

Zhang, Z., Zhang, Z., Liu, P., Xue, X., Zhang, C., Peng, L., ... Wang, F. (2024). The role of photobiomodulation to modulate ion channels in the nervous system: A systematic review. *Cellular and Molecular Neurobiology, 44*(1), 79. https://doi.org/10.1007/s10571-024-01513-1.

Zhong, F., Li, Y., & Mi, X. (2014). Photobiomodulation on KATP channels of Kir6.2-transfected HEK-293 Cells. *International Journal of Photoenergy, 2014*, 898752. https://doi.org/10.1155/2014/898752.

Zomorrodi, R., Loheswaran, G., Pushparaj, A., & Lim, L. (2019). Pulsed near infrared transcranial and intranasal photobiomodulation significantly modulates neural oscillations: A pilot exploratory study. *Scientific Reports, 9*(1), 6309. https://doi.org/10.1038/s41598-019-42693-x.

Printed and bound by CPI Group (UK) Ltd, Croydon, CR0 4YY
24/09/2026
02222378-0001